D0793561

PERGAMON INTERNATIONAL LIBRARY
of Science, Technology, Engineering and Social Studies
The 1000-volume original paperback library in aid of education, industrial training and the enjoyment of leisure
Publisher: Robert Maxwell, M.C.

INTERNATIONAL SERIES IN
PURE AND APPLIED MATHEMATICS
General Editor I. N. SNEDDON

VOLUME 106

SETS: NAÏVE, AXIOMATIC AND APPLIED

THE PERGAMON TEXTBOOK
INSPECTION COPY SERVICE

An inspection copy of any book published in the Pergamon International Library will gladly be sent to academic staff without obligation for their consideration for course adoption or recommendation. Copies may be retained for a period of 60 days from receipt and returned if not suitable. When a particular title is adopted or recommended for adoption for class use and the recommendation results in a sale of 12 or more copies, the inspection copy may be retained with our compliments. The Publishers will be pleased to receive suggestions for revised editions and new titles to be published in this important International Library.

SETS: NAÏVE, AXIOMATIC AND APPLIED

A Basic Compendium with Exercises for Use in Set Theory For Non Logicians, Working and Teaching Mathematicians and Students

by

D. VAN DALEN
UNIVERSITY OF UTRECHT, NETHERLANDS

H. C. DOETS
UNIVERSITY OF AMSTERDAM, NETHERLANDS

H. DE SWART
CATHOLIC UNIVERSITY, NIJMEGEN, NETHERLANDS

PERGAMON PRESS
OXFORD · NEW YORK · TORONTO · SYDNEY · PARIS · FRANKFURT

QA
248
D23413
1978

U.K.	Pergamon Press Ltd., Headington Hill Hall, Oxford OX3 0BW, England
U.S.A.	Pergamon Press Inc., Maxwell House, Fairview Park, Elmsford, New York 10523, U.S.A.
CANADA	Pergamon of Canada Ltd., 75 The East Mall, Toronto, Ontario, Canada
AUSTRALIA	Pergamon Press (Aust.) Pty. Ltd., 19a Boundary Street, Rushcutters Bay, N.S.W. 2011, Australia
FRANCE	Pergamon Press SARL, 24 rue des Ecoles, 75240 Paris, Cedex 05, France
FEDERAL REPUBLIC OF GERMANY	Pergamon Press GmbH, 6242 Kronberg-Taunus, Pferdstrasse 1, Federal Republic of Germany

Copyright © 1978 Pergamon Press Ltd

All Rights Reserved. No part of this publication may be reproduced, stored in a retrieval system or transmitted in any form or by any means: electronic, electrostatic, magnetic tape, mechanical, photocopying, recording or otherwise, without permission in writing from the publishers

First edition 1978

British Library Cataloguing in Publication Data

Van Dalen, D
Sets: naive, axiomatic, and applied.
(International series in pure and applied mathematics; v. 106)
Translation of Verzamelingen.
Bibliography: p.
1. Set theory. 2. Axiomatic set theory.
I. Doets, H. C., joint author. II. Swart,
H. C. M. de, joint author. III. Title.
QA248.D23413 511'.32 76-43992

ISBN 0-08-021166-6 (hardcover)
ISBN 0-08-023047-4 (flexicover)

In order to make this volume available as economically and as rapidly as possible the author's typescript has been reproduced in its original form. This method unfortunately has its typographical limitations but it is hoped that they in no way distract the reader.

Translated from the Dutch edition entitled "VERZAMELINGEN" published by OOSTHOEK, SCHELTEMA & HOLKEMA (Utrecht)

Printed in Great Britain by William Clowes & Sons Limited, London, Beccles and Colchester

CONTENTS

2-2-81 (3x7) HPCP
50360

50360

Chapter III. Applications

PREFACE

Set theory is a funny discipline. For ages and ages mathematics has managed without set theory, but nowadays one gets from the average textbook the impression that set theory is absolutely indispensable. Even texts for high schools (not to mention nursery schools) start with sets, unions, intersections, etc.

Among the professional mathematicians there are some who claim that "there are no things but sets " How do these extremists justify their opinion? We will try to unearth some of the motivation for a belief in the supremacy of set theory.

The original concept of a set was very liberal: a set could contain both frogs and functions. In mathematical language, why shouldn't a set contain points, numbers, and functions at the same time? A set was obtained by simply throwing together a number of things, or by giving some common characterization. Cantor presented the following definition of a set: A set is a collection of certain distinct objects of our intuition or of our thought into a whole [Cantor, 1895, 1966]. Clearly, therefore, one should first have objects to be able to form sets. Now then, mathematics provided a wealth of objects: numbers, points, functions, matrices, curves, etc.; so there seemed not to be any special reason to worry about the universe of set theory (apart from some disagreement concerning notions of infinity). The obvious role of set theory was one of utility and hygiene. Set theory enabled one to give clean and concise formulations. In short, the availability of the set theoretic apparatus proved to be a methodological blessing. In particular, set theory turned out to be indispensable in topology. At the same time the subject acquired, under the hands of Georg Cantor, an interest of its own. In his relatively short but extremely fruitful period of creativity Cantor introduced almost all the standard concepts of our present-day set theory: cardinal, ordinal, well-ordering, powerset, etc. (for an historical survey see Van Dalen-Monna, 1972). All the same, for the time being set theory depended on a number of expedients from various parts of mathematics such as "natural number" and "function". The concept of a function, in particular, was very prominent, if only because of its role in the concepts of "cardinal", "ordertype", etc. In the early stage, therefore, set theory was certainly not a suitable base for mathematics. Set theory itself was in need of extraneous elements.

The function concept already had a long history. In the nineteenth century it had evolved from "analytical expression" to Dirichlet's famous abstract version (cf. Monna, 1972), but there was no reduction to the concept of set. Terms like "correspondence" and "Zuordnung" were used without analyzing them. The definition of a function, as a set of ordered pairs became available only after the discovery that "ordered pair" could be defined intrinsically in terms of sets (Wiener, Hausdorff, Kuratowski).
Moreover, through practical experience mathematicians found that various well-known concepts could be defined in set theoretical terms (in particular the fundamental number systems). Gradually one got the impression that the whole of mathematics could be reduced to set theory, a kind of experimental hypothesis! In due course one could observe a singular phenomenon: genuine mathematical objects were replaced by their set theoretical codifications. The success of this program can be inferred from the fact that many mathematicians take the codification for the original object. The case of the function is paradigmatic: the starting point - viz. the law (or instruction) associating objects to objects - has a clear intuitive content. Through experience we learnt that in almost all cases the manipulations of functions were independent of the defining laws, that is to say, it was sufficient to know all pairs (input, output) ! From then on the course of things was reversed: the correspondence - concept was suppressed and a function was nothing but a set of ordered pairs (a graph). One can put a label on this phenomenon: replacement of *intension* by *extension*.

This process of extensionalizing has (correctly) been questioned since the rise of computerscience. From some view-points (especially the practical one) a law (program) is more basic than a graph. In pure mathematics the intensional aspects of the function concept have been traditionally exploited by recursion theory.

Let us try to summarize the arguments for the slogan "Everything is a set": all objects from mathematical practice turn out to be representable in terms of sets and all current mathematical reasoning can be formalized in set theory. The above statement, which is not a mathematical theorem but rather a hypothesis concerning mathematics, is rather vague. In order to make the statement (at least in principle) provable or disprovable, we would have to formalize "mathematics" and "set theory", so that subsequently an interpretation of the first in the latter could be given. There are however fundamental and technical obstacles. A formalization of the whole of mathematics is a chimera, since we cannot, once and for all, decide the extent of mathematics; and also because no formalization can produce all true statements (by Gödel's theorem).

Therefore we will not worry about the status of the above slogan. It suffices that the set theoretical apparatus is practical and elegant. We will take set theory to be "open ended", in the sense that, if necessary, new principles can be added.

In mathematics, several axiom systems for set theory are in use. We will stick to the set theoretical universe cosist of?

We have already argued that the familiar mathematical objects can be reduced to sets. This leaves the following question unanswered: which objects (i.e. sets) should we have in order to get everything we want? In other words: what does the set theoretical universe consist of?

The construction of the set theoretical universe happens to be an adventure fit for a baron Von Münchhausen. Even leaving the infinity axiom aside (i.e. without committing ourselves to the existence of infinite objects), we can, starting from the empty set, create by means of the usual set theoretical operations a surprisingly rich fauna of objects (the hereditarily finite sets, cf. II. 11.23). The cummulative hierarchy shows how far one gets with how little. ZF gives you a lot of mileage.

Recent developments in set theory, in particular those of Cohen and Gödel concerning the axiom of choice and the continuum hypothesis, made it clear that Cantor's Paradise is not the comfortable abode it was advertised to be. The independence of the axiom of choice and the continuum hypothesis of respectable systems, such as those of Zermelo-Fraenkel or Von Neumann - Bernays - Gödel, makes situations possible, which are familiar in older branches of mathematics. In geometry, for instance, one has to distinguish between Desarguian and non-Desarguian geometry. So maybe, we will eventually have to adopt the practice of specifying our set theoretical axioms in more detail. Think of doing group theory in a non-Cantorian set theory. These are not purely theoretical reflections as may be illustrated by the case of Lebesgue-measure. Assuming the axiom of choice, we can prove that not every set of reals is measurable, but assuming the axiom of determinateness we can show each set of reals to be measurable. So there are consequences for daily life.

In this book we will not venture to deal with the independence problems of the "higher" axioms of set theory. We will stick to those parts that do not require a refined metamathematical apparatus. Although the authors do not subscribe to the thesis that "everything is a set" (they are not sets themselves (after A. Mostowski)), they are convinced that the fruitfulness of set theory, as a mathematical discipline, is beyond dispute. So they welcome the reader, with a clear conscience, to Cantor's paradise.

ACKNOWLEDGEMENTS

The authors wish to thank Jeff Zucker for his very helpful criticism and advice, Mrs. E.L.M. Remers for her patient and excellent typing of the manuscript and Mr. F. Ribot and J.M. Geertsen for giving their kind permission to make use of the facilities of the Faculty of Mathematics and Science of the Catholic University at Nijmegen, the Netherlands.

INTRODUCTION

The addition of another text in set theory to those already existing is an act that needs justification. The authors of the present book are confident that the contents offer a sufficient choice of topics not covered by standard textbooks in set theory, to exclude the possibility of a mere repetition of well-known stories and tricks. It is hardly necessary to stress the fundamental role of set theory in the mathematics of these days. Ever since the days of Cantor, algebra, analysis, etc., have been absorbing the techniques and the flavour of set theory, so that a journey into modern mathematics without a firm background in set theory would be ill-advised.

This book consists of three chapters: The first contains the traditional material (e.g. Boolean operations, countability, Cantor-Bernstein, well-ordering), treated informally in such a way that everything can easily be adapted to an axiomatic treatment. At the end of the first chapter the axioms of Zermelo-Fraenkel are introduced and it is indicated how to rework the preceding sections.

The second chapter is strictly axiomatic, but rather along the lines of Hilbert's *Foundations of Geometry* than along those of a formal first-order theory. In particular no deep facts of logic are used, or presupposed. A knowledge of the meaning of the connectives and some experience in mathematics suffices. Nonetheless some non-trivial subjects are treated, such as the reflection principle, measurable cardinals, and models of set theory. Also, the theory of ordinals is put on a firm base. In particular the reader may benefit from the treatment of the cumulative hierarchy (due to Von Neumann), which provides extra insight into the set-theoretical universe. Throughout, the axiom of regularity (foundation) is assumed, and its role is discussed in Section II.1.

The last chapter contains a number of topics, such as Borel sets, inductive definitions, Boolean algebras, applications of the axiom of choice, infinite games and the axiom of determinateness, which illustrate the actual use of set theory in mathematics. Many of these topics go unnoticed in an ordinary mathematics text, so the authors thought it worthwhile to include them in the book as a demonstration of the usefulness of set theory in everyday mathematics.

From the above it may be clear that the book should be useful for mathematicians, teachers and students. In addition it offers a number of topics of interest to other scientists, such as linguists and biologists. Philosophers, too, will find quite a lot of material, that has traditionally called their attention. For example, the introduction of the fundamental number systems, cardinals (Frege-Russell), the notions of finite and infinite, inductive definitions (the notion of predicativity) and the paradoxes of Cantor and Russell.

No attention is paid to deeper metamathematical topics, such as Gödel's constructible sets or Cohen's forcing. For those topics the reader is referred to the existing literature (however, cf. exercise 2 of Section II.12).

Many exercises are added and the reader is strongly advised to try his hand at them.

For the convenience of the casual reader the book contains a number of redundancies. Definitions, for example, will be repeated when the notion turns up in a different context.

In the text several terms, such as "property", "operation", "class", etc., are used. This may seem confusing at first, but, when used judiciously, it makes life much more pleasant. In the Appendix the use of these terms is explained. The reader is asked to turn to the Appendix whenever he feels uneasy in handling these terms. Rather than reading the Appendix last, he should every now and then consult it. Also, the Appendix deals with some peculiarities that stem from the linguistic limitations of the theory.

How to use the book. Those interested in a guided introduction into set theory can read Chapter I (possibly skipping Sections 12 and 21) and in addition Chapter II, Sections 9, 10. Sections of Chapter II and Chapter III can be added if desired. The reader who is already familiar with naïve set theory can read Chapter I, Section 20 and go on to Chapter II. If he wants to restrict himself to the bare minimum, he can, after reading Section 2, stick to the Sections 3, 8, 9, 10, which are of central importance. Next, he can add to this Chapter II, Sections 4, 5 and 6.

Those interested in the foundations of mathematics should not skip anything (with the possible exception of Chapter II, Section 15).

The purely mathematically oriented reader can already on the basis of Chapter I turn to the applications in Chapter III. In particular, Chapter III, Section 3 will give him some idea how to handle ordinals.

References and cross-references. We refer to the Bibliografy by a name, followed by a date, in square brackets, e.g. [Gödel, 1940]. Cross references to a section (Theorem, Lemma, Definition, etc.) in the same chapter are made without mentioning the chapter, e.g. see Section 5 (see Section 5.3). Cross references to other chapters give the number of the chapter, e.g. see Section II.5 (see Section II.5.3).

Problems. The book contains a number of exercises of various degrees of difficulty. We do not claim any originality for the exercises. Some problems are really minor theorems and since they are quite often very useful, we strongly advise the reader to pay attention to them. Although it may seem superfluous we want to remind him that without solving problems (i.e. proving theorems) he himself will probably not get beyond the stage of collecting keywords.

Proofs. Not all proofs will be given in full detail and some proofs will altogether be left to the reader. The end of a proof is indicated by the sign □. It is hardly necessary to point out that only a fraction of the present corpus of set theory is treated in this book. Spectacular results, such as the independence of the axiom of choice and the continuum hypothesis in Zermelo-Fraenkel's system, require essential use of more sophisticated logic machinery. For these and other topics the reader is referred to [Cohen, 1966], [Gödel, 1940], [Fraenkel, Bar-Hillel, and Levy, 1973].
For those interested in the historical development of set theory it is worthwhile to trace it back to its sources. For this a reader should consult [Cantor, 1966], [Van Dalen, and Monna, 1972].

CHAPTER 1

NAÏVE SET THEORY

1 SOME IMPORTANT SETS AND NOTATIONS

We all know lots of sets, for example: the set of all people in Holland, the set of all triangles in a plane, the set of the numbers 1, 2, 3, 4 and 5. The latter is usually denoted by $\{1, 2, 3, 4, 5\}$.

Instead of "3 is an element of the set $\{1, 2, 3, 4, 5\}$" we write "$3 \in \{1, 2, 3, 4, 5\}$"; and instead of "7 is not an element of the set $\{1, 2, 3, 4, 5\}$" we write "$7 \notin \{1, 2, 3, 4, 5\}$".

The numbers 0, 1, 2, 3,... are called *natural numbers*. By $\mathbb{N}$ we mean the set of all natural numbers. So, for example, $3 \in \mathbb{N}$, $5 \in \mathbb{N}$ and $1024 \in \mathbb{N}$, while $-3 \notin \mathbb{N}$, $2/3 \notin \mathbb{N}$ and $\sqrt{2} \notin \mathbb{N}$.

The numbers ... -3, -2, -1, 0, 1, 2, 3, ... are called *integers*. By $\mathbb{Z}$ we mean the set of all integers. Note that each natural number is an integer, but not conversely. Examples: $2 \in \mathbb{Z}$, $-2 \in \mathbb{Z}$, $0 \in \mathbb{Z}$, $3 \in \mathbb{Z}$, $-3 \in \mathbb{Z}$, $2/3 \notin \mathbb{Z}$ and $\sqrt{2} \notin \mathbb{Z}$.

The numbers of the form p/q, where $p \in \mathbb{Z}$, $q \in \mathbb{N}$, $q \neq 0$ (and p and q relatively prime) are called *rational numbers*. By $\mathbb{Q}$ we mean the set of all rational numbers. Examples: $1/4 \in \mathbb{Q}$, $-1/4 \in \mathbb{Q}$, $2 \in \mathbb{Q}$, $-2 \in \mathbb{Q}$, $0 \in \mathbb{Q}$, $3/5 \in \mathbb{Q}$, $-3/5 \in \mathbb{Q}$, $\sqrt{2} \notin \mathbb{Q}$, $\sqrt{5} \notin \mathbb{Q}$, $\pi \notin \mathbb{Q}$. Note that all integers and hence also all natural numbers are rational numbers.

All rational numbers and numbers such as $\sqrt{2}$, $\sqrt{5}$, log 2, $\sqrt[4]{7}$, π, and so on, are called *real numbers*. By $\mathbb{R}$ we mean the set of all real numbers (a precise definition follows in Section 12).

In our examples, sets consisted of concrete and familiar objects, but once we have sets, we can form sets of sets, e.g. the set of all football teams. It is important to realize that the elements of a set may themselves be sets. Mathematics is full of examples of sets of sets. A straight line, for example, can be conceived as a set of points; the set of all straight lines in a plane is a natural example of a set of sets (of points). What may be astonishing, is not so much that sets occur as elements, but that for the greater part of mathematical practice we can confine ourselves to elements which are themselves sets. In this book we will study sets, sets of sets and similar towers, sometimes of great height and complexity.

In order to avoid excessive use of the word "set", we sometimes use the words "collection", "family", and "class", instead of "set". (In Chapter II we make a distinction, see also the Appendix).

We will use small Latin letters: a, b, c, x, y, z, u, v, w, ... and capital Latin letters: A, B, C, X, Y, Z, U, V, W, ... for sets.

The principal relation in set theory is the relation "is an element of" (" is a member of", "belongs to"). This is a primitive (that is, an undefined) relation. Instead of "x is an element of A" we write "$x \in A$". And instead of "x is not an element of A" we write "$x \notin A$".

Finite sets can easily be designated by listing all their elements: $\{a_1, \ldots, a_n\}$ is the set containing precisely the elements $a_1, \ldots, a_n$. Stated differently: $x \in \{a_1, \ldots, a_n\}$ if and only if $x = a_1$ or $x = a_2$ ... or $x = a_n$. In particular, $\{a\}$ is the set which has precisely a as its only element.

$\{a\}$ is called "the *singleton* (set)of a". Do not confuse a and $\{a\}$! They are quite different.

Examples:

(i) $\{0,1\}$ is a set with two elements, namely 0 and 1.
$\{\{0,1\}\}$ is a set with one element, namely $\{0,1\}$.

(ii) $\underset{\sim}{N}$ is a set with infinitely many elements.
$\{\underset{\sim}{N}\}$ is a set with one element, namely $\underset{\sim}{N}$.

If $a \in b$ and $b \in c$, then not necessarily $a \in c$!

Example: $2 \in \{2\}$ and $\{2\} \in \{\{2\}, 3\}$, but $2 \notin \{\{2\}, 3\}$; $\{\{2\}, 3\}$ is a set with two elements, namely $\{2\}$ and 3, and 2 is different from both.

2 EQUALITY OF SETS

Sets are, just like triangles and Riemann surfaces, legitimate mathematical objects. So it makes perfectly good sense to ask whether two sets are identical or not. If two sets A and B are identical (equal), we write $A=B$, if not, $A \neq B$. Identical sets have exactly the same properties, so, if $A = B$, then everything in A is in B and vice versa. One may wonder if, conversely, sets with exactly the same elements, are identical. Consider, for example, the set A of all even numbers and the set B of all sums of pairs of odd numbers. There is some reason to distinguish A and B: they are given in different ways. On the other hand, we feel (and mathematical practice confirms this) that definitions do not matter so much, it is rather content that counts.
So we make the explicit choice to consider sets as merely being determined by their elements. Hence, "having the same elements" means "being equal". We sum this up by the

Axiom of extensionality: $A = B$ if and only if A and B have the same elements.

As observed above, the "only if" holds trivially.

Now, if we express "for all x" by the notation "$\forall x$" and "if and only if" by the symbol "$\leftrightarrow$", then we can formulate our axiom as follows:
Axiom of extensionality: $A = B \leftrightarrow \forall x\, [x \in A \leftrightarrow x \in B]$.

By this axiom the following hold:

$\{3,4,5\} = \{4,3,5\}$, $\{2,3\} \neq \{3,4\}$

$\{3,3,7\} = \{3,7\}$, $\{0,1\} \neq \{1,2\}$

$\{2,3\} = \{2,3,3\}$, $\{2,\{3,4\}\} \neq \{\{2,3\}, 4\}$.

In the literature one uses, instead of "$\forall x$", also "Λx" and "(x)", and instead of "$\leftrightarrow$", also "$\rightleftarrows$" and "$\equiv$".

Since we will have to present a great number of definitions in the remainder of the text, it is convenient to have some notations at hand. We will use ":=" or "iff (if and only if)" for definitional equality (or equivalence). For convenience we sometimes use "if" instead of "iff".

Examples: $2 := 1+1$; x is even iff for some y, $x = 2y$; x is odd if for some y, $x = 2y + 1$.

Exercises

i Check that $\{0, \{1\}, \{0,1\}\} = \{\{1\}, \{0,1\}, 0\}$
and that $\{0, \{1\}, \{1\}\} \neq \{\{1\}, \{0,1\}\}$.

ii Check that $\{0, \{1\}, \{0,1\}\} \neq \{\{1\}, \{0,1\}\}$
and that $\{\{0,1\}, \{\{0,1\}\}\} \neq \{\{0,1\}\}$.

3 SUBSETS

When dealing with sets it is often important to know that one set is contained in another one. This is formulated as follows.

Definition 3.1: A is a *subset* of B iff each element of A is also an element of B. In other words: for all x, if $x \in A$, then $x \in B$.

If A is a subset of B, we write: $A \subseteq B$, or $B \supseteq A$. If A is not a subset of B, we write: $A \not\subseteq B$.

Now, if again we express "for all x" by the notation "$\forall x$", and "if..., then" by the symbol "$\rightarrow$", then we can formulate our definition as follows:
$A \subseteq B := \forall x\, [x \in A \rightarrow x \in B]$.

The definition of $A \subseteq B$ is illustrated in Fig. 1:

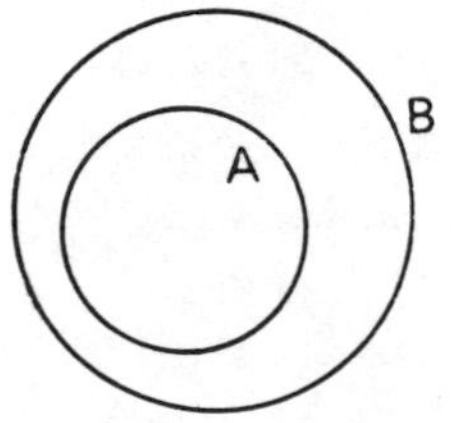

Fig. 1

Examples:

{2,3}	$\subseteq$	{1,2,3},	$\mathbb{Z} \subseteq \mathbb{Q}$,		
{2,3}	$\subseteq$	{2,3},	$\mathbb{Q} \subseteq \mathbb{R}$,		
{{2}}	$\subseteq$	{{2}, 4},	$\mathbb{N} \subseteq \mathbb{R}$,		
{2,3}	$\not\subseteq$	{1,3},	{{1}, {2}}	$\not\subseteq$	{{1},2},
{2}	$\not\subseteq$	{{2}, 3},	{{1},2}	$\not\subseteq$	{{1},{2}}.

Warning: Do not confuse $\in$ and $\subseteq$!

Examples:

(i) {2} $\in$ {{2}, 3}, but {2} $\not\subseteq$ {{2}, 3}.

(ii) {2,3} $\subseteq$ {1,2,3}, but {2,3} $\notin$ {1,2,3} .

<u>Theorem 3.2:</u> For all sets A,B and C:

(i) $A \subseteq A$,

(ii) if $A \subseteq B$ and $B \subseteq A$, then $A = B$,

(iii) if $A \subseteq B$ and $B \subseteq C$, then $A \subseteq C$.

The proof follows immediately from the definition of "$A \subseteq B$". For (ii), in addition, the axiom of extensionality is needed.
Instead of (i) one can say that $\subseteq$ is reflexive;
instead of (ii) that $\subseteq$ is antisymmetric and
instead of (iii) that $\subseteq$ is transitive.

Now, if we extend our symbolism by writing for "and" the symbol "$\wedge$", then we can formulate Theorem 3.2 as follows:
For all sets A,B and C

(i) $A \subseteq A$,

(ii) $(A \subseteq B \wedge B \subseteq A) \rightarrow A = B$,

(iii) $(A \subseteq B \wedge B \subseteq C) \rightarrow A \subseteq C$.

<u>Definition 3.3:</u> A is a *proper subset* of B iff $A \subseteq B$ and $A \neq B$.

If A is a proper subset of B, we write: $A \subset B$. If A is not a proper subset of B, we write: $A \not\subset B$. (Instead of "$A \subset B$" also "$A \subsetneqq B$" and "$A \underset{\neq}{\subset} B$" are used.)

Examples:

$\{1, \{2\}\} \subset \{1, \{2\}, 3\}$,

$\{4,5\} \subset \{4,5,6\}$,

$\{1, \{2\}\} \not\subset \{1, \{2\}\}$,

$\{1, \{2\}\} \not\subset \{1,2\}$

We extend our symbolism once again by writing for "not" the symbol "$\neg$".

Then "$x \notin A$" is equivalent to "$\neg\ (x \in A)$", "$A \not\subseteq B$" is equivalent to "$\neg(A \subseteq B)$" and "$A \not\subset B$" is equivalent to "$\neg\ (A \subset B)$".

4 THE NAÏVE COMPREHENSION PRINCIPLE AND THE EMPTY SET

One can define (or specify) sets by writing down all the elements, as for example "$\{2, 7, 43\}$", but this procedure has two drawbacks. In the first place it is clumsy. The set defined as "$\{9261, 10648, 13824, 12167, 15625, 21952, 17576, 24389, 19683\}$" is, for most of us, an unrecognizable list of numbers. If we indicate the same set by "the set of all x^3, with x between 20 and 30", it is already much more manageable. In the second place we can only indicate finite sets in this way.

We therefore usually specify a set by a property. Following Cantor, we will formulate explicitly the hidden assumptions behind this practice:

Let Φ be a property of objects (sets). Then there is a set whose elements are exactly those objects which have the property Φ.

Eventually this *naïve comprehension principle* turned out to be untenable (paradox of Cantor, 1895; Russell, 1902). We will use the principle with caution and formulate a better principle in due course. For the time being we confine ourselves to the remark that it can do no harm to define a subset of a given set (by the comprehension principle). It is only in the definition of "large" sets that paradoxes may arise.

As the reader can check in the following sections, we shall almost always apply the comprehension principle to define a subset of a given set with a certain property.

Let Φ be a property of objects (sets).
The naive comprehension principle says:

I. There is *at least* one set, whose elements are exactly those objects, which have the property Φ.

Now suppose that there are two sets, whose elements are exactly those objects which have the property Φ. Then those two sets have the same elements and hence, by the axiom of extensionality, they are equal. So

II. There is *at most* one set, whose elements are exactly those objects which have the property Φ.

Combining I and II: There is *precisely* one set, whose elements are exactly those objects which have the property Φ.
We shall use the comprehension principle in this convenient form.
Hence we can give a name to this set: *the* set of all objects with the property Φ.
We denote this set by

$$\{x \mid \Phi\ (x)\}$$

where $\Phi\ (x)$ is an abbreviation for "x has the property Φ".

Hence, if Φ indicates a property, then
(i) $\{x \mid \Phi\ (x)\}$ is a set, and
(ii) for all y, $y \in \{x \mid \Phi\ (x)\} \leftrightarrow \Phi\ (y)$.

Examples:
1. $\{x|x$ is a natural number$\}$ is a set, namely $\underset{\sim}{N}$.
2. $\{x|x = 1$ or $x = 2\}$ is a set, namely $\{1,2\}$. Now, if we extend our symbolism by writing the symbol "v" for the connective "or", then we can write the foregoing as follows:
$\{x|x = 1 \vee x = 2\}$ is a set, namely $\{1,2\}$.
3. $\{x|x=2 \vee x = 3 \vee x = 6\}$ is a set, namely $\{2,3,6\}$.
4. $\{x|x = 3\}$ is a set, namely $\{3\}$.

Let $A = \{x|\Phi(x)\}$ and $B = \{x|\Psi(x)\}$. Check that the following holds (no matter what Φ and Ψ are):

$$A = B \leftrightarrow \forall x\,[\Phi(x) \leftrightarrow \Psi(x)],$$
$$A \subseteq B \leftrightarrow \forall x\,[\Phi(x) \rightarrow \Psi(x)].$$

Now, take for $\Phi(z)$: $z \neq z$, and consider $\{z|z \neq z\}$.
This set has no elements, in other words:
$\forall y[\, y \notin \{z|z \neq z\}]$. So we have shown:

There is *at least* one set which has no elements (namely $\{z|z \neq z\}$). (*)

Actually, by the improved comprehension principle, we have

Theorem 4.1: There is exactly one set which has no elements.

Proof: By the improved comprehension principle there is exactly one set, containing all elements z, such that $z \neq z$. Since every object is identical to itself, this set contains no elements. □

We call this set *the empty set* , which we denote by $\emptyset$.

If we extend our symbolism once again by using the symbols "$\exists x$" for "there is at least one x" and "$\exists!x$" for "there is precisely one x", then we can formulate (*) as follows: $\exists x \forall y[\, y \notin x]$. And Theorem 4.1 can be written as: $\exists!x\forall y[\, y \notin x]$. (Instead of "$\exists x$" the symbols "$\bigvee x$" and "$(Ex)$" are also used in the literature.)

Theorem 4.2: For all sets a, $\emptyset \subseteq a$. In symbols: $\forall a[\,\emptyset \subseteq a]$.

Proof: Let a be a set and suppose $\emptyset \not\subseteq a$. Then there is an element which is in $\emptyset$ and not in a. This contradicts the fact that $\emptyset$ contains no elements. □

$\emptyset$ (we repeat) is a set without elements. $\{\emptyset\}$, on the other hand, is a set with one element, namely $\emptyset$. Hence $\emptyset \neq \{\emptyset\}$. $\{\{\emptyset\}\}$ is the set with $\{\emptyset\}$ as its only element. Hence $\{\{\emptyset\}\} \neq \{\emptyset\}$. $\{\emptyset, \{\emptyset\}\}$ is the set with $\emptyset$ and $\{\emptyset\}$ as its only elements.

Check that

$\emptyset \subseteq \emptyset$,	$\{\emptyset\} \subseteq \{\emptyset\}$,	$\{\emptyset\} \not\subseteq \{\{\emptyset\}\}$,	$\emptyset \in \{\emptyset, \{\emptyset\}\}$,
$\emptyset \notin \emptyset$,	$\{\emptyset\} \notin \{\emptyset\}$,	$\{\{\emptyset\}\} \subseteq \{\{\emptyset\}\}$,	$\{\emptyset\} \in \{\emptyset, \{\emptyset\}\}$,
$\emptyset \subseteq \{\emptyset\}$,	$\emptyset \subseteq \{\{\emptyset\}\}$,	$\{\emptyset\} \in \{\{\emptyset\}\}$,	$\{\emptyset\} \subseteq \{\emptyset, \{\emptyset\}\}$,
$\emptyset \in \{\emptyset\}$,	$\emptyset \notin \{\{\emptyset\}\}$,	$\emptyset \subseteq \{\{\emptyset\}\}$,	$\{\{\emptyset\}\} \subseteq \{\emptyset, \{\emptyset\}\}$.

5 UNION, INTERSECTION AND RELATIVE COMPLEMENT

After a bit of experimenting with sets one readily discovers that there is a number of operations on sets that bear a striking resemblance to the familiar operations of elementary algebra. One can, for example, join sets together, take away sets, construct the common part, etc. These operations have been studied extensively since (and even slightly before) the birth of set theory. For historical reasons (cf. also III.2) these operations are called Boolean operations.

In order to get acquainted with the operations and their properties, it is often convenient to make use of geometrical representations, called Venn diagrams (or diagrams, for short). The reader is urged to make diagrams whenever the need arises, it may be very helpful in visualizing theorems or in solving problems.

Let A and B be sets. We define a new set $A \cup B$, *the union of A and B*, as the set of all objects which belong to at least one of the sets A and B (see Fig. 2).

<u>Definition 5.1</u>: $A \cup B := \{x | x \in A \vee x \in B\}$.

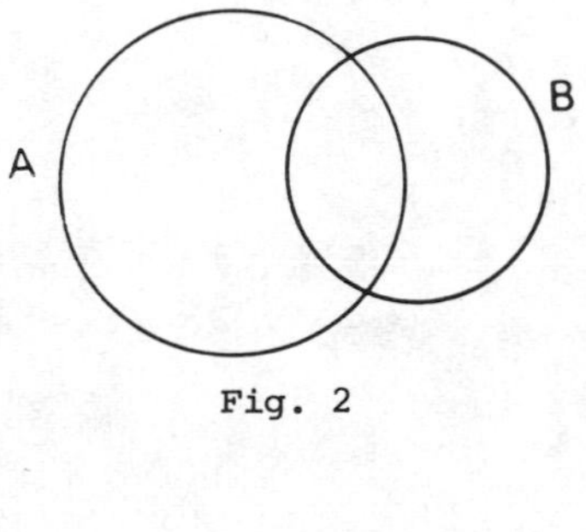

Fig. 2

Note that $A \cup B$ is a set by our naive comprehension principle and that for all y,

$y \in A \cup B \leftrightarrow y \in A \vee y \in B.$

Examples:

$\{1,2\} \cup \{5,6\} = \{1,2,5,6\}$, $\{1,2\} \cup \{2\} = \{1,2\}$,
$\{1,2\} \cup \{2,6\} = \{1,2,6\}$, $\{1,2\} \cup \emptyset = \{1,2\}$,
$\{1,2\} \cup \{1,2\} = \{1,2\}$.

The elementary notions in set theory, such as "subset", "union", "intersection", and so on, have many elementary properties associated with them. The proofs of these properties are usually quite trivial; nevertheless they should all be carried out. We shall give a few of these proofs here and urge the reader to go through the other proofs himself.

<u>Theorem 5.2:</u> $A \cup A = A$ ($\cup$ is idempotent),
$A \cup B = B \cup A$ ($\cup$ is commutative),
$A \cup (B \cup C) = (A \cup B) \cup C$ ($\cup$ is associative).

Proof: $x \in A \cup A \leftrightarrow x \in A \vee x \in A$
$\leftrightarrow x \in A.$

$$\begin{aligned} x \in A \cup B &\leftrightarrow x \in A \vee x \in B \\ &\leftrightarrow x \in B \vee x \in A \\ &\leftrightarrow x \in B \cup A \end{aligned}$$

$$\begin{aligned} x \in A \cup (B \cup C) &\leftrightarrow x \in A \vee x \in B \cup C \\ &\leftrightarrow x \in A \vee (x \in B \vee x \in C) \\ &\leftrightarrow (x \in A \vee x \in B) \vee x \in C \\ &\leftrightarrow x \in A \cup B \vee x \in C \\ &\leftrightarrow x \in (A \cup B) \cup C. \end{aligned}$$ □

<u>Theorem 5.3:</u> $A \cup \emptyset = A$ $A \subseteq B \to A \cup C \subseteq B \cup C$
$A \subseteq A \cup B$ $(A \subseteq C \wedge B \subseteq C) \leftrightarrow A \cup B \subseteq C.$
$B \subseteq A \cup B$

We shall prove here the last assertion of this theorem:
$(A \subseteq C \wedge B \subseteq C) \leftrightarrow A \cup B \subseteq C$.

Proof of $\rightarrow$: Suppose $A \subseteq C$ and $B \subseteq C$. We have to show that $A \cup B \subseteq C$; in other words: each element x of $A \cup B$ is also an element of C. So let $x \in A \cup B$. Then $x \in A$ or $x \in B$. If $x \in A$, then, because of $A \subseteq C$, also $x \in C$ and if $x \in B$, then, because of $B \subseteq C$, also $x \in C$ (see Figure 3).

Proof of $\leftarrow$: Suppose $A \cup B \subseteq C$. We have to show that $A \subseteq C$ and $B \subseteq C$; in other words: each element x of A is also an element of C and each element x of B is also an element of C. So let $x \in A$; then also $x \in A \cup B$; hence, because of $A \cup B \subseteq C$, $x \in C$. Let $x \in B$; then also $x \in A \cup B$; hence, because of $A \cup B \subseteq C$, also $x \in C$. □

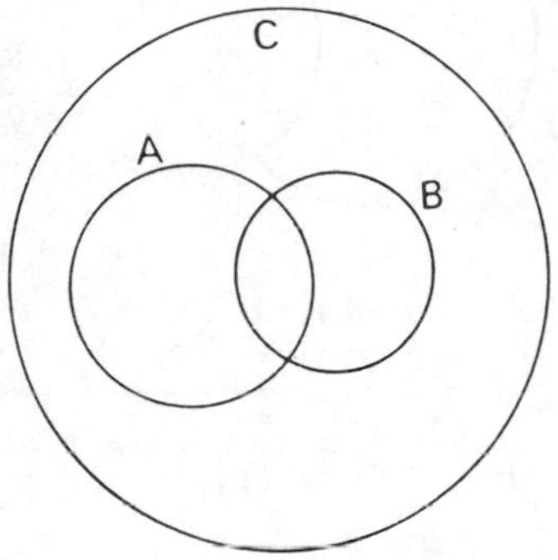

Fig. 3

Theorem 5.4: $A \subseteq B \leftrightarrow A \cup B = B$

Proof: By Theorem 5.3 we know $A \subseteq A \cup B$. From this it follows immediately that $A \cup B = B \rightarrow A \subseteq B$. We have to show next that $A \subseteq B \rightarrow A \cup B = B$. So suppose $A \subseteq B$, i.e. that each element of A is also an element of B. Then also each element of $A \cup B$ is an element of B, in other words: $A \cup B \subseteq B$. By Theorem 5.3 we also know that $B \subseteq A \cup B$. Hence $A \cup B = B$ (using 3.2 ii). □

Theorem 5.5: $\{a\} \cup \{b\} = \{a,b\}$

$\{a\} \cup \{b\} \cup \{c\} = \{a,b,c\}$.

We leave the proof as an exercise to the reader. Note that $\{a\} \cup \{b\} = \{a,b\}$ also holds in case $b = a$ (i.e. $\{a,a\} = \{a\}$).

The *intersection* of two sets A and B, $A \cap B$, is defined as the set of all objects, which are both elements of A and elements of B.

Definition 5.6: $A \cap B := \{x | x \in A \wedge x \in B\}$.

$A \cap B$ is a set by our naive comprehension principle, and for all y, $y \in A \cap B \leftrightarrow y \in A \wedge y \in B$ (see Fig. 4).

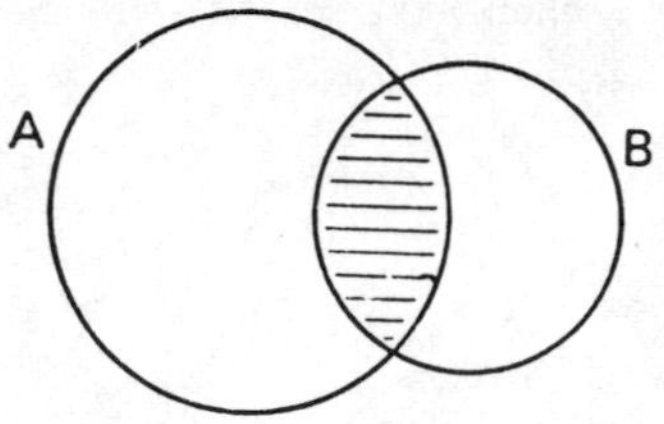

Fig. 4

Examples:

$\{1,2,3\} \cap \underset{\sim}{N} = \{1,2,3\}$, $\{1,2\} \cap \{1,2\} = \{1,2\}$

$\{1,2,3\} \cap \{2,9\} = \{2\}$, $\{1,2\} \cap \emptyset = \emptyset$

$\{1,2,3\} \cap \{6,9\} = \emptyset$.

Theorem 5.7: $A \cap A = A$ ($\cap$ is idempotent)

$A \cap B = B \cap A$ ($\cap$ is commutative)

$A \cap (B \cap C) = (A \cap B) \cap C$ ($\cap$ is associative).

We leave the proof as an exercise to the reader.

Theorem 5.8: $A \cap \emptyset = \emptyset$

$A \cap B \subseteq A$

$A \cap B \subseteq B$

$A \subseteq B \rightarrow A \cap C \subseteq B \cap C$

$(C \subseteq A \wedge C \subseteq B) \leftrightarrow C \subseteq A \cap B$.

We leave the proof as an exercise to the reader.

Theorem 5.9: $A \subseteq B \leftrightarrow A \cap B = A$.

We leave the proof as an exercise to the reader.

The following theorem concerns unions and intersections.

Theorem 5.10: Absorption laws: $A \cap (A \cup B) = A$

$$A \cup (A \cap B) = A.$$

Distributive laws: $A \cap (B \cup C) = (A \cap B) \cup (A \cap C)$

$$A \cup (B \cap C) = (A \cup B) \cap (A \cup C).$$

Proof: We prove two of these assertions.

(i) $A \cap (A \cup B) = A$. By Theorem 5.8 we know that $A \cap (A \cup B) \subseteq A$.
We also know $A \subseteq A$ and $A \subseteq A \cup B$;
hence, again by 5.8, $A \subseteq A \cap (A \cup B)$.
By Theorem 3.2 (ii) it follows that $A \cap (A \cup B) = A$.

(ii) $A \cup (B \cap C) = (A \cup B) \cap (A \cup C)$.

$$\begin{aligned} x \in A \cup (B \cap C) &\leftrightarrow x \in A \vee x \in B \cap C \\ &\leftrightarrow x \in A \vee (x \in B \wedge x \in C) \\ &\leftrightarrow (x \in A \vee x \in B) \wedge (x \in A \vee x \in C) \\ &\leftrightarrow x \in A \cup B \wedge x \in A \cup C \\ &\leftrightarrow x \in (A \cup B) \cap (A \cup C). \end{aligned}$$ □

If B and A are sets, we define $B - A$, the (set theoretical) *difference* of B and A, as follows:

Definition 5.11: $B - A := \{x | x \in B \wedge x \notin A\}$.

$B - A$ is again a set by our naive comprehension principle, and for all y, $y \in B - A \leftrightarrow y \in B \wedge y \notin A$ (see Fig. 5).

It often happens in mathematics that we work (for a time) within a certain

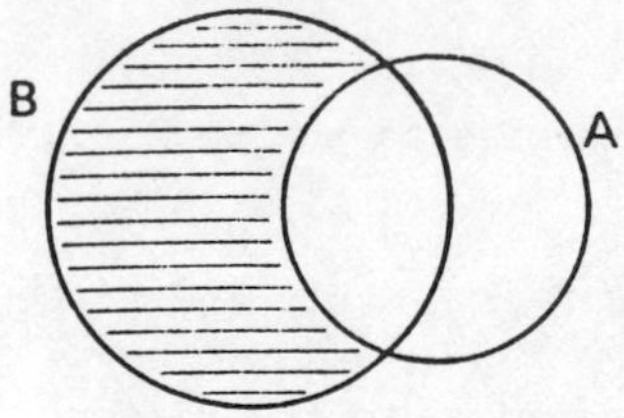

Fig. 5

set V; i.e. all the (mathematical) objects that we consider are elements of V, and all the sets we consider are subsets of V (for example, V may be the set of all real numbers, or the set of all complex numbers). We can think of V as our (temporary) "universe" or "universal set".

If A is a subset of a set V, we call $V - A$ the *relative complement* of A in V. In this case the notation $C_V(A)$ is often used instead of $V - A$ (see Fig. 6).

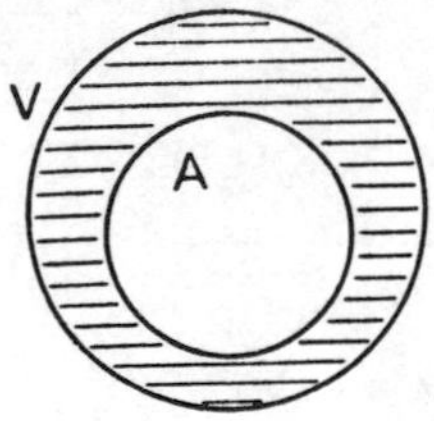

Fig. 6

Examples:

$\{1,2,3\} - \{1,3,7\} = \{2\}$

$\{\{1\}, \{1,2\}\} - \{1,2,9\} = \{\{1\},\{1,2\}\}$

$\{\{1\}, \{1,2\}\} - \{\{1,2\},1\} = \{\{1\}\}$.

$V := \{1,\{2\},\{1,3\}\}$; $C_V(\{\{2\}\}) = \{1,\{1,3\}\}$

$C_V(\{1\}) = \{\{2\},\{1,3\}\}$

$C_V(\{1,\{2\}\}) = \{\{1,3\}\}$.

Theorem 5.12: For all sets V, $C_V(\emptyset) = V$

$C_V(V) = \emptyset$.

We leave the proof as an exercise to the reader.

Theorem 5.13: Let V be a set. Then for all A with $A \subseteq V$,

$$A \cup C_V(A) = V$$
$$A \cap C_V(A) = \emptyset$$
$$C_V(C_V(A)) = A.$$

We leave the proof as an exercise to the reader.

If it is clear from the context what the "universal set" V is, then we often simply write "A^c" instead of "$C_V(A)$" and we simply speak of "the *complement* of A". (Instead of "A^c", "$-A$" and "$\bar{A}$" are also used in the literature.)

Theorem 5.14: Given sets $A \subseteq V$ and $B \subseteq V$, we have:

$A \subseteq B \leftrightarrow A^c \supseteq B^c$ (first law of reciprocity)

$(A \cup B)^c = A^c \cap B^c$ (second law of reciprocity)

$(A \cap B)^c = A^c \cup B^c$ (third law of reciprocity).

The second and the third law of reciprocity are also known as the laws of *De Morgan*.

As an illustration, we shall prove here the first law of reciprocity: $A \subseteq B \leftrightarrow A^c \supseteq B^c$. The proofs of the other laws are left as exercises to the reader.

$A \subseteq B \leftrightarrow$

for all x, if $x \in A$, then $x \in B$ $\leftrightarrow$

for all x, if $x \notin B$, then $x \notin A$ $\leftrightarrow$

for all x, if $x \in B^c$, then $x \in A^c$ $\leftrightarrow$

$B^c \subseteq A^c$ (see Fig. 7). □

The three laws of reciprocity can also be formulated as follows: in passing to the relative complement in V, "$\subseteq$" and "$\supseteq$" are interchanged and likewise "$\cup$" and "$\cap$".

Using this, one can derive the commutative law for "$\cup$" from the one for "$\cap$", and conversely. Likewise, the associative law for "$\cup$" can be derived from the one for "$\cap$", and conversely; and finally, each distributive law can be

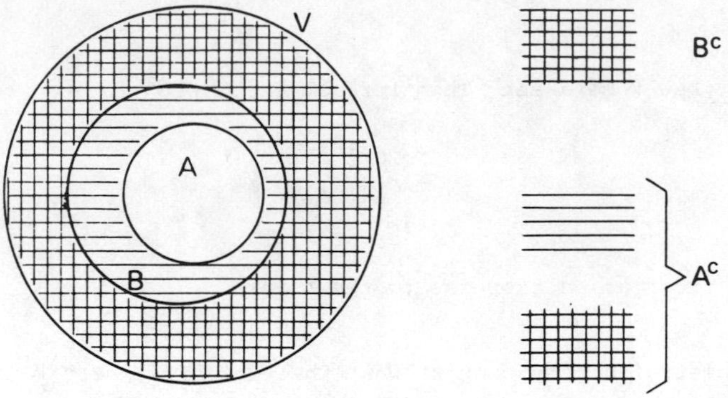

Fig. 7

derived from the other.

As an example, we will prove the one distributive law, $(A \cap B) \cup C = (A \cup C) \cap (B \cup C)$, from the other, $(X \cup Y) \cap Z = (X \cap Z) \cup (Y \cap Z)$, where A, B, C, X, Y and Z are all subsets of V: taking A^c, B^c and C^c for X, Y and Z respectively, the second distributive law gives $(A^c \cup B^c) \cap C^c = (A^c \cap C^c) \cup (B^c \cap C^c)$; taking the complement on both sides, we get: $(A^c \cup B^c)^c \cup (C^c)^c = (A^c \cap C^c)^c \cap (B^c \cap C^c)^c$; hence $((A^c)^c \cap (B^c)^c) \cup (C^c)^c = ((A^c)^c \cup (C^c)^c) \cap ((B^c)^c \cup (C^c)^c)$; and hence $(A \cap B) \cup C = (A \cup C) \cap (B \cup C)$.

Exercises

1. Let $A \subseteq B$ and $C \subseteq D$. Prove that
   (a) $A \cup C \subseteq B \cup D$ (b) $A \cap C \subseteq B \cap D$.

2. Prove that for all sets A, B and C
   (i) $(A \subset B \wedge B \subseteq C) \rightarrow A \subset C$
   (ii) $(A \subseteq B \wedge B \subset C) \rightarrow A \subset C$.

3. Let $A \subseteq V$ and $B \subseteq V$. Prove that
   (a) $A \cap (A^c \cup B) = A \cap B$ (b) $A - B = B^c - A^c$.

4. Prove that for all sets A, B and C
   (i) $A \cap C = \emptyset \rightarrow A \cap (B \cup C) = A \cap B$
   (ii) $A \cap B = \emptyset \rightarrow A - B = A$
   (iii) $(A \cap B = \emptyset \wedge A \cup B = C) \rightarrow A = C - B$.

5. Prove that for all sets A, B and C

(i) $A \cap (B - C) = (A \cap B) - C$

(ii) $(A \cup B) - C = (A - C) \cup (B - C)$

(iii) $A - (B \cup C) = (A - B) \cap (A - C)$

(iv) $A - (B \cap C) = (A - B) \cup (A - C)$

(v) $(A - B)^c = B \cup A^c$

(vi) $A - (A - B) = A \cap B$

(vii) $A \cup B = A \cup (B - A)$.

6. We define the operation "+" on sets (the "*symmetric difference*") as follows: $A + B := (A - B) \cup (B - A)$.

Prove:

(i) $A + B = B + A$

(ii) $A + (B + C) = (A + B) + C$

(iii) $A \cap (B + C) = (A \cap B) + (A \cap C)$

(iv) $A + A = \emptyset$

(v) $A + B = (A \cup B) - (A \cap B)$

(vi) $A + \emptyset = A$

(vii) $A^c + B^c = A + B$.

7. Let $A \subseteq V$ and $B \subseteq V$. Prove:

(i) $A \cup B = \emptyset \rightarrow (A = \emptyset \wedge B = \emptyset)$

(ii) $A \cap B^c = \emptyset \leftrightarrow A \subseteq B$

(iii) $A + B = \emptyset \leftrightarrow A = B$.

8. Prove: (i) $A \cup C = B \cup C \leftrightarrow A + B \subseteq C$

(ii) $(A \cup C) + (B \cup C) = (A + B) - C$.

9. Prove: $(A \subseteq B \wedge C = B - A) \rightarrow A = B - C$.

10. Prove: $(A \cup B) \cap (B \cup C) \cap (C \cup A) =$
$(A \cap B) \cup (B \cap C) \cup (C \cap A)$.

Generalize this result to more than three sets.

11. Prove: (i) $A - B = A \leftrightarrow B - A = B$

(ii) $A - B \subseteq C \leftrightarrow B - C \subseteq A$.

12. Prove: $A \subseteq B \cup C \leftrightarrow A \cap C^c \subseteq B$.

6 POWER SET

As we have said before, sets may themselves be elements of sets. In particular, we can conceive of all the subsets of a given set V as together forming a set, which we denote by "$\mathcal{P}(V)$".

<u>Definition 6.1:</u> $\mathcal{P}(V) := \{A \mid A \subseteq V\}$.

$\mathcal{P}(V)$ is called the *power set* of V.

$\mathcal{P}(V)$ is a set by our naive comprehension principle, and for all y: $y \in \mathcal{P}(V) \leftrightarrow y \subseteq V$.

Examples:

(i) $\emptyset$ has only one subset, namely $\emptyset$ itself; hence $\mathcal{P}(\emptyset) = \{\emptyset\}$.

(ii) $\{a\}$ has two subsets, namely $\emptyset$ and $\{a\}$; hence $\mathcal{P}(\{a\}) = \{\emptyset,\{a\}\}$.

(iii) $\{a,b\}$ has four subsets, namely $\emptyset$, $\{a\}$, $\{b\}$ and $\{a,b\}$; hence $\mathcal{P}(\{a,b\}) = \{\emptyset,\{a\}, \{b\}, \{a,b\}\}$.

(iv) $\{a,b,c\}$ has eight subsets, namely $\emptyset$, $\{a\}$, $\{b\}$, $\{c\}$, $\{a,b\}$, $\{a,c\}$, $\{b,c\}$ and $\{a,b,c\}$; hence $\mathcal{P}(\{a,b,c\}) =$
$\{\emptyset,\{a\},\{b\},\{c\}, \{a,b\}, \{a,c\}, \{b,c\}, \{a,b,c\}\}$.

These examples suggest that a set with n (n a natural number) elements has 2^n subsets. We shall prove that this actually is the case.

<u>Lemma 6.2:</u> Let W be a finite set, $a \notin W$ and $V = W \cup \{a\}$. Then V has exactly twice as many subsets as W has.

Proof: To each subset of W one can add the element a, or not. By this process one obtains all subsets of V, since from each subset A of W there arise two subsets of V: A and $A \cup \{a\}$. Hence V has exactly twice as many subsets as W has. □

<u>Theorem 6.3:</u> For all natural numbers n and for all sets V: if V has n elements, then $\mathcal{P}(V)$ has 2^n elements.

We will prove the theorem by mathematical induction on the number of elements of V. Since the reader may not be familiar with application of this principle in the context of set theory we will give a detailed proof. In the future we shall trust the reader to supply the details.

For a start we reformulate the theorem in a somewhat pedantic way.

For $n \in \underset{\sim}{N}$ let $\Phi(n)$ be the proposition: for all sets V, if V has n elements then $\mathcal{P}(V)$ has 2^n elements.

<u>Theorem 6.3</u> (re-formulated): $\forall n \in \underset{\sim}{N}\,[\Phi(n)]$.

Proof: It is sufficient to prove

1. $\Phi(0)$ and
2. $\forall k \in \underset{\sim}{N}\,[\Phi(k) \rightarrow \Phi(k+1)]$.

Proof of $\Phi(0)$: Let V be a set and suppose V has zero elements. Then $V = \emptyset$. Hence $\mathcal{P}(V) = \{\emptyset\}$. So $\mathcal{P}(V)$ has 1 element, and $2^0 = 1$.

Proof of $\forall k \in \underset{\sim}{N}\,[\Phi(k) \rightarrow \Phi(k+1)]$: Let $k \in \underset{\sim}{N}$ and suppose $\Phi(k)$, i.e. for all sets W, if W has k elements, then $\mathcal{P}(W)$ has 2^k elements. We now have to prove $\Phi(k+1)$. So suppose V is a set with $k+1$ elements. Let $a \in V$ and let $W := V - \{a\}$. Then W has k elements and hence W has 2^k subsets. Hence, by Lemma 6.2, V has $2.2^k = 2^{k+1}$ subsets. □

Remark: We have proved Theorem 6.3 by means of the following principle:

$$(\Phi(0) \wedge \forall k \in \underset{\sim}{N}\,[\Phi(k) \rightarrow \Phi(k+1)]) \rightarrow \forall n \in \underset{\sim}{N}\,[\Phi(n)].$$

This principle is called the principle of *mathematical induction*. In $\forall k \in N\,[\Phi(k) \rightarrow \Phi(k+1)]$, $\Phi(k)$ is called the *induction hypothesis*.

The principle of mathematical induction says: if 0 has the property Φ, and if, further, for all $k \in \underset{\sim}{N}$, whenever k has the property Φ then so does $k+1$, then all natural numbers have the property Φ.

Remark: If, for example, V has 10 elements, then $\mathcal{P}(V)$ has $2^{10} = 1024$ elements. If V has 20 elements, then $\mathcal{P}(V)$ has $2^{10}.2^{10} = 1024.1024$ (more than a million) elements. For finite sets V, $\mathcal{P}(V)$ is in general much larger than V. What if V has infinitely many elements? Is, for example, $\mathcal{P}(\underset{\sim}{N})$ also "larger" than $\underset{\sim}{N}$?

In order to be able to investigate this kind of question (cf. Section 15), we will first have to develop some more machinery.

Theorem 6.4: (i) $A \subseteq B \rightarrow \mathcal{P}(A) \subseteq \mathcal{P}(B)$.

(ii) $\mathcal{P}(A) \subseteq \mathcal{P}(B) \rightarrow A \subseteq B$.

(iii) $\mathcal{P}(A) = \mathcal{P}(B) \rightarrow A = B$.

(iv) $\mathcal{P}(A) \in \mathcal{P}(B) \rightarrow A \in B$.

Proof: (i) Suppose that $A \subseteq B$. Then $\forall x\, [x \subseteq A \rightarrow x \subseteq B]$, in other words $\forall x\, [x \in \mathcal{P}(A) \rightarrow x \in \mathcal{P}(B)]$ and this means precisely that $\mathcal{P}(A) \subseteq \mathcal{P}(B)$.

(ii) Suppose $\mathcal{P}(A) \subseteq \mathcal{P}(B)$, i.e. $\forall x\, [x \in \mathcal{P}(A) \rightarrow x \in \mathcal{P}(B)]$, in other words $\forall x\, [x \subseteq A \rightarrow x \subseteq B]$. Now we know $A \subseteq A$. Hence also $A \subseteq B$.

(iii) Suppose $\mathcal{P}(A) = \mathcal{P}(B)$. Then $\mathcal{P}(A) \subseteq \mathcal{P}(B)$ and $\mathcal{P}(B) \subseteq \mathcal{P}(A)$. Hence, applying (ii) twice, $A \subseteq B$ and $B \subseteq A$. Hence $A = B$.

(iv) Suppose $\mathcal{P}(A) \in \mathcal{P}(B)$, i.e. $\mathcal{P}(A) \subseteq B$. Now $A \in \mathcal{P}(A)$ (since $A \subseteq A$), and so $A \in B$. □

Warning: The converse of Theorem 6.4 (iv): $A \in B \rightarrow \mathcal{P}(A) \in \mathcal{P}(B)$ does not hold! Counterexample: let $A := \{\emptyset\}$ and $B := \{\{\emptyset\}\}$. Then $\mathcal{P}(A) = \{\emptyset, \{\emptyset\}\}$ and $\mathcal{P}(B) = \{\emptyset, \{\{\emptyset\}\}\}$. So $\mathcal{P}(A) \notin \mathcal{P}(B)$, while $A \in B$.

Check that:

I (i) $\{2,3\} \in \mathcal{P}(\underset{\sim}{N})$ and $\{4\} \in \mathcal{P}(\underset{\sim}{N})$.

(ii) $\{5,8,9\} \in \mathcal{P}(\underset{\sim}{N})$ and $\{7\} \in \mathcal{P}(\underset{\sim}{N})$.

II (i) $\{\{2,3\},\{4\}\} \in \mathcal{P}(\mathcal{P}(\underset{\sim}{N}))$.

(ii) $\{\{5,8,9\}, \{7\}\} \in \mathcal{P}(\mathcal{P}(\underset{\sim}{N}))$.

III $\{\{\{2,3\}, \{4\}\}, \{\{5,8,9\}, \{7\}\}\} \in \mathcal{P}(\mathcal{P}(\mathcal{P}(\underset{\sim}{N})))$.

For convenience we will sometimes omit parentheses in the notation for power sets, writing, e.g. " $\mathcal{P}\mathcal{P}(\underset{\sim}{N})$ " for "$\mathcal{P}(\mathcal{P}(\underset{\sim}{N}))$", etc.

Exercises

1. (i) Determine $\mathcal{P}(\emptyset)$, $\mathcal{P}\mathcal{P}(\emptyset)$, $\mathcal{P}\mathcal{P}\mathcal{P}(\emptyset)$ and $\mathcal{P}\mathcal{P}\mathcal{P}\mathcal{P}(\emptyset)$.

(ii) Determine $\mathcal{P}(\{a\})$, $\mathcal{P}\mathcal{P}(\{a\})$ and $\mathcal{P}\mathcal{P}\mathcal{P}(\{a\})$.

(iii) Determine $\mathcal{P}(\{a,b\})$ and $\mathcal{P}\mathcal{P}(\{a,b\})$.

2. (i) $\mathcal{P}(A) \cap \mathcal{P}(B) = \mathcal{P}(A \cap B)$.

(ii) $\mathcal{P}(A) \cup \mathcal{P}(B) \subseteq \mathcal{P}(A \cup B)$.

(iii) Show that it is not true that for all sets A and B, $\mathcal{P}(A) \cup \mathcal{P}(B) = \mathcal{P}(A \cup B)$.

3. $A \cap B = \emptyset \leftrightarrow \mathcal{P}(A) \cap \mathcal{P}(B) = \{\emptyset\}$.

4. Prove that $\{x \mid x \in A \wedge x \notin x\} \in \mathcal{P}(A) - A$.
Hence there is no set A such that $\mathcal{P}(A) \subseteq A$.

7 UNIONS AND INTERSECTIONS OF FAMILIES

Although, up to now, we have introduced the notions of union and intersection only for pairs of sets, the extension of these notions to a finite collection of sets does not present any problems.

In elementary algebra there is an analogous phenomenon. Although the sum $a + b$ is defined for pairs a,b, one invariably writes $a+b+c$, $a+b+c+d$, etc. This custom is justified by the fact that addition is associative, i.e. $a+(b+c) = (a+b)+c$. So, no matter how we restore the parentheses in $a+b+c$, the outcome will be the same. Formulated in a slightly pedantic way: we have introduced a ternary operation $\langle a,b,c \rangle \mapsto a+b+c$.

In exactly the same way we can introduce $A \cup B \cup C$, $A \cap B \cap C$, etc. , that is to say we can define finite unions and intersections by iterating the ordinary union and intersection.

Now we will go further and define the union and intersection of an arbitrary collection of sets, whether the collection is finite or not.

Let A be an arbitrary collection of sets.
We define the *union* of A, $\bigcup A$, as the set of all objects, which belong to at least one set in the collection A.

Definition 7.1: $\bigcup A := \{x \mid \exists y[y \in A \wedge x \in y]\}$.

For $A = \{c_1, c_2, c_3\}$ we indicate $\bigcup A$ in Fig. 8.

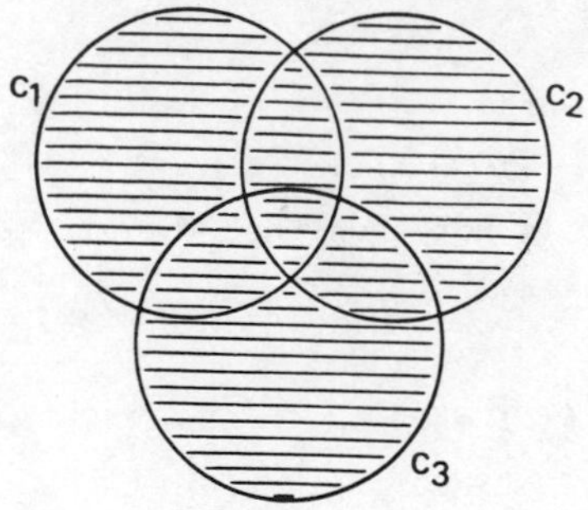

Fig. 8

Instead of $\bigcup A$ one also writes $\bigcup\{c \mid c \in A\}$ or $\bigcup_{c \in A} c$.

We define the *intersection* of A, $\bigcap A$, as the set of all objects which belong to every set in the collection A. (We assume for the time being that $A \neq \emptyset$.)

Definition 7.2: $\bigcap A := \{x \mid \forall y[\, y \in A \to x \in y]\,\}$.

For $A = \{c_1, c_2, c_3\}$ we indicate $\bigcap A$ in Fig. 9.

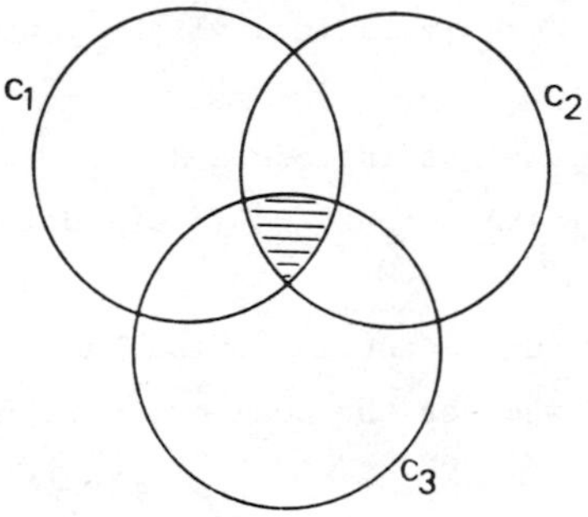

Fig. 9

Instead of $\bigcap A$ one also writes $\bigcap\{c \mid c \in A\}$ or $\bigcap_{c \in A} c$.

Examples:

$\bigcup\{\{3,4\},\{4,5\}\} = \{3,4\} \cup \{4,5\} = \{3,4,5\}$.

$\bigcup\{\{3,4\},\{4,5\},\emptyset\} = \{3,4\} \cup \{4,5\} \cup \emptyset = \{3,4,5\}$.

$\bigcup\{\{3,4\},\{4,5\},\{\emptyset\}\} = \{3,4\} \cup \{4,5\} \cup \{\emptyset\} = \{3,4,5,\emptyset\}$.

$\bigcap \{\{3,4\}, \{4,5\}\} = \{3,4\} \cap \{4,5\} = \{4\}.$

$\bigcap \{\{3,4\}, \{4,5\}, \emptyset\} = \{3,4\} \cap \{4,5\} \cap \emptyset = \{4\} \cap \emptyset = \emptyset.$

$\bigcap \{\{\{3,4\}\},\{3,4\}, \{3\}\} = \{\{3,4\}\} \cap \{3,4\} \cap \{3\} = \{\{3,4\}\} \cap \{3\} = \emptyset.$

"There is some $y \in A$ such that $\Phi(y)$" can be formulated in our symbolism by "$\exists y[y \in A \wedge \Phi(y)]$". Instead of "$\exists y[y \in A \wedge \Phi(y)]$" we will simply write "$\exists y \in A[\Phi(y)]$". For instance:

$$\bigcup A = \{x \mid \exists y \in A[x \in y]\}.$$

We can formulate "for all $y \in A$, $\Phi(y)$" in our symbolism by "$\forall y[y \in A \rightarrow \Phi(y)]$". Instead of "$\forall y[y \in A \rightarrow \Phi(y)]$" we just write "$\forall y \in A[\Phi(y)]$". For instance:

$$\bigcap A = \{x \mid \forall y \in A[x \in y]\}.$$

From the diagrams we can see that a union of a collection includes each of its elements and that an intersection of a collection is contained in each of its elements. We formulate this more precisely as follows.

<u>Theorem 7.3:</u> (i) $c \in A \rightarrow c \subseteq \bigcup A$,

(ii) $c \in A \rightarrow \bigcap A \subseteq c$,

(iii) $\bigcap A \subseteq \bigcup A$ (if $A \neq \emptyset$).

Proof: (i) Suppose $c \in A$ and $x \in c$. Then there is some y, namely c, such that $x \in y$ and $y \in A$. Hence $x \in \bigcup A$.

We leave the proof of (ii) as an exercise to the reader.

(iii) Suppose $A \neq \emptyset$. Then there is some c with $c \in A$. By (ii) $\bigcap A \subseteq c$ and by (i) $c \subseteq \bigcup A$. Hence $\bigcap A \subseteq \bigcup A$. □

<u>Theorem 7.4:</u> $\bigcup \emptyset = \emptyset$

$\bigcup\{B\} = B$ $\quad$ $\bigcap \{B\} = B$

$\bigcup\{B,C\} = B \cup C$ $\quad$ $\bigcap \{B,C\} = B \cap C.$

We leave the proofs as an exercise to the reader.

<u>Theorem 7.5:</u> $A \subseteq B \rightarrow \bigcup A \subseteq \bigcup B$,

$A \subseteq B \rightarrow \bigcap A \supseteq \bigcap B.$

Again we leave the proofs as an exercise to the reader. Theorem 7.5 has the following simple formulation: if you take the union (intersection) of more sets, you get a larger (smaller) set.

If we examine the definition of the intersection of a collection A more closely, we can see why A is taken to be non-empty. For we have

$$x \in \bigcap A \leftrightarrow \forall y[\, y \in A \rightarrow x \in y] .$$

Now, if $A = \emptyset$, then "$y \in A \rightarrow x \in y$" is true for all y and for all x (for suppose "$y \in \emptyset \rightarrow x \in y$" were false, then "$y \in \emptyset$" would be true and "$x \in y$" false, while we know that "$y \in \emptyset$" is false for any y). Hence, each x would be an element of $\bigcap \emptyset$, and so $\bigcap \emptyset$ would be the set of all sets. Such a set leads to a paradox, as we will see in Section 19. Hence the condition "$A \neq \emptyset$".

However, we can consider $\bigcap \emptyset$ if we restrict ourselves beforehand to a fixed universe V; for in that case $\bigcap A = \{x \in V \mid \forall y[\, y \in A \rightarrow x \in y]\,\}$ and hence $\bigcap \emptyset = V$.

In future we will use "$\bigcap A$" without scruples; if it is clear from the context that we are working in a fixed universe, we omit the condition "$A \neq \emptyset$".

Unfortunately we are forced by the difficulties mentioned above, to treat "$\bigcap \emptyset$" as an exceptional case. For example, $\bigcap A \subseteq \bigcup A$ holds in general only if $A \neq \emptyset$ (since $V = \bigcap \emptyset \not\subseteq \bigcup \emptyset = \emptyset$).

<u>Theorem 7.6:</u> (i) $\forall c \in A[\, c \subseteq V] \rightarrow \bigcup A \subseteq V$.

In words: if each element of A is a subset of V, then $\bigcup A$ is also a subset of V.

(ii) $\forall c \in A[\, V \subseteq c] \rightarrow V \subseteq \bigcap A$.

In words: if V is a subset of each element of A, then V is also a subset of $\bigcap A$.

We leave the proof as an exercise to the reader.
(The theorem can easily be illustrated by taking, say, $A = \{c_1, c_2, c_3\}$.)

The laws of De Morgan (second and third law of reciprocity) are also valid for unions and intersections of arbitrary collections of sets: the complement of the union is the intersection of the complements and the complement of the intersection is the union of the complements. More precisely:

Theorem 7.7: Let A be a collection in a universe V (i.e. a collection of subsets of V). Then $(\bigcup_{x \in A} x)^c = \bigcap_{x \in A} x^c$ and $(\bigcap_{x \in A} x)^c = \bigcup_{x \in A} x^c$.

Proof: We shall prove the first proposition (and leave proof of the second as an exercise):

$$\begin{aligned} y \in (\bigcup_{x \in A} x)^c &\leftrightarrow y \notin \bigcup_{x \in A} x \\ &\leftrightarrow \neg\exists x[x \in A \wedge y \in x] \\ &\leftrightarrow \forall x[\neg(x \in A \wedge y \in x)] \\ &\leftrightarrow \forall x[x \in A \rightarrow y \notin x] \\ &\leftrightarrow \forall x[x \in A \rightarrow y \in x^c] \\ &\leftrightarrow y \in \bigcap_{x \in A} x^c . \end{aligned}$$

□

In set theory we frequently speak of the *greatest set X with the property* Φ. By that we mean a set X which satisfies the following two conditions:
(1) X has the property Φ, and
(2) if Y is any set with the property Φ, then $Y \subseteq X$.
One can easily prove that there can be at most one such set. This justifies our use of the phrase "*the* greatest set with the property Φ" (when such a set exists).

Examples:

1. The set E of the even numbers is the greatest subset of $\underset{\sim}{N}$, which does not contain odd numbers.
2. E (as in 1) is the greatest subset of $\underset{\sim}{N}$, which does not contain odd numbers and which is closed under multiplication.
3. Let $\Phi(X)$:= X is a finite subset of $\underset{\sim}{N}$. Then there is clearly no greatest set with the property Φ.

Analogously we say that X is the *smallest set with the property* Φ, if
(1) X has the property Φ, and
(2) if Y is any set with the property Φ, then $X \subseteq Y$.

Example: The set of the Fibonacci-numbers is the smallest set which contains

1 and 2, and is closed under addition of two successive numbers.

With the notions "greatest set, which ..." and "smallest set, which ..." we can characterize intersection and union in a handy way:

Theorem 7.8: (i) $A \cap B$ is the greatest set which is included in both A and B (i.e. which is a subset of both A and B).

(ii) $A \cup B$ is the smallest set which includes both A and B (i.e. which has both A and B as subsets).

Proof:

(i) 1. $A \cap B \subseteq A$ and $A \cap B \subseteq B$ (Theorem 5.8).

2. $C \subseteq A \wedge C \subseteq B \rightarrow C \subseteq A \cap B$ (Theorem 5.8).

(ii) 1. $A \subseteq A \cup B$ and $B \subseteq A \cup B$ (Theorem 5.3).

2. $A \subseteq C \wedge B \subseteq C \rightarrow A \cup B \subseteq C$ (Theorem 5.3). □

Note that we can use Theorem 7.8 to define intersection and union exclusively in terms of "subset of". Conversely, we can define $A \subseteq B$ by means of intersection (union respectively), as is shown by the Theorems 5.9 and 5.4.

In almost all branches of mathematics it is convenient to use indices for the purpose of listing sets. As a matter of fact it was a great improvement on the older notation of the sort "let the points $A,B,C,\ldots,K,L$ be given". Instead we say now "consider the set $\{A_i \mid i \leqslant n\}$ of points"; this allows for greater precision and flexibility in notation.

Let us analyse what "indexing" means or what "an indexed set" is. A set I of indices is given and to each $i \in I$ we associate an element A_i of a certain family. So, in the language of set theory, an *indexing* is a mapping (we anticipate the definition of "mapping" or "function", cf. Section 13) A with domain I and the *indexed set* is the range of A: $\{A(i) \mid i \in I\}$. In conformity with tradition we write "A_i" instead of "$A(i)$" and we call i the index of A_i. I is called the index set of $\{A_i \mid i \in I\}$.

Trivially each set allows an indexing (by the identity function), so it is not an exciting feature. Rather it is a convenient tool.
We can then write "$\bigcup_{i \in I} A_i$" and "$\bigcap_{i \in I} A_i$" instead of "$\bigcup A$" and "$\bigcap A$" respectively.

So: $\bigcup_{i \in I} A_i = \{x \mid \exists i[i \in I \wedge x \in A_i]\}$ and

$\bigcap_{i \in I} A_i = \{x \mid \forall i[i \in I \rightarrow x \in A_i]\}$.

Example: $\bigcup\{A_i \mid i \in \{1,2,3\}\} = A_1 \cup A_2 \cup A_3$

$\bigcap\{A_i \mid i \in \{4,6\}\} = A_4 \cap A_6$.

In the special case when the index set is $\underset{\sim}{N}$, the family $\{A_i \mid i \in \underset{\sim}{N}\}$ is called a *sequence* of sets.

Definition 7.9: Two sets V and W are called *disjoint* if $V \cap W = \emptyset$; in other words, if V and W have no elements in common.

A collection of sets is called *pairwise disjoint* (or simply "disjoint") if any two distinct sets in it are disjoint.

Exercises

1. The open interval (a,b), with $a,b \in \underset{\sim}{R}$, is by definition the set $\{x \mid x \in \underset{\sim}{R} \wedge a < x < b\}$ and the closed interval $[a,b]$ is by definition the set $\{x \mid x \in \underset{\sim}{R} \wedge a \leqslant x \leqslant b\}$. $\underset{\sim}{N}^+$ is the set $\underset{\sim}{N} - \{0\}$.

   (a) If $A_i = (-\frac{1}{i}, \frac{1}{i})$, for $i = 1,2, \ldots$, what is the set $\bigcap_{i \in \underset{\sim}{N}^+} A_i$? Is the result the same if all intervals A_i

   (i) are all closed (i.e. $A_i = [-\frac{1}{i}, \frac{1}{i}]$),

   (ii) are alternately open and closed?

   (b) Repeat the problem (a) for the sequence of intervals $B_1 = (\frac{1}{2},1)$, $B_2 = (\frac{1}{2},\frac{3}{4})$, $B_3 = (\frac{5}{8},\frac{3}{4})$, $B_4 = (\frac{5}{8},\frac{11}{16})$, ..., obtained by repeated bisection of $(0,1)$ with alternate choice of the right and the left half.

2. If $X_0 \supseteq X_1 \supseteq X_2 \supseteq \ldots$ and $Y_0 \supseteq Y_1 \supseteq Y_2 \supseteq \ldots$, show that $\bigcap_{i \in \underset{\sim}{N}}(X_i \cup Y_i) = \bigcap_{i \in \underset{\sim}{N}} X_i \cup \bigcap_{i \in \underset{\sim}{N}} Y_i$.

   Does this equality hold for two arbitrary families $\{X_i \mid i \in \underset{\sim}{N}\}$ and $\{Y_i \mid i \in \underset{\sim}{N}\}$?

3. Given two collections A and B of sets, prove that $\bigcup(A \cup B) = \bigcup A \cup \bigcup B$ and $\bigcup(A \cap B) \subseteq \bigcup A \cap \bigcup B$. Show that $\bigcup(A \cap B) = \bigcup A \cap \bigcup B$ if the elements of $A \cup B$

50360

are pairwise disjoint.

4. (a) Let A be a collection of sets and let A^* be the collection of all sets $X - Y$ with $X, Y \in A$. Prove that $A^* \subseteq (A^*)^*$. Give an example to show that the inclusion $\subseteq$ in this proposition cannot be replaced by the equality $=$.
   (b) Does $A^* \subseteq (A^*)^*$ hold, if
   (i) $A^* = \{X + Y \mid X, Y \in A\}$ (see Exercise 6, Section 5),
   (ii) $A^* = \{X \cup Y \mid X, Y \in A\}$,
   (iii) $A^* = \{X \cap Y \mid X, Y \in A\}$?

5. Let $\{A_i \mid i \in I\}$ and $\{B_i \mid i \in I\}$ be two families of sets with the same index set I. Suppose that for all $i \in I$, $A_i \subseteq B_i$.
   Prove that (a) $\bigcup_{i \in I} A_i \subseteq \bigcup_{i \in I} B_i$ and (b) $\bigcap_{i \in I} A_i \subseteq \bigcap_{i \in I} B_i$.

6. Let $\{A_i \mid i \in I\}$ and $\{B_j \mid j \in J\}$ be two families of sets. Suppose $\forall i \in I \ \exists j \in J \ [B_j \subseteq A_i]$.
   Prove that $\bigcap_{j \in J} B_j \subseteq \bigcap_{i \in I} A_i$.

7. (a) $(\bigcup_{i \in I} A_i) - (\bigcup_{j \in J} B_j) = \bigcup_{i \in I} (\bigcap_{j \in J} [A_i - B_j])$,
   (b) $(\bigcap_{i \in I} A_i) - (\bigcap_{j \in J} B_j) = \bigcap_{i \in I} (\bigcup_{j \in J} [A_i - B_j])$.

8. Prove that $\bigcap(A \cup B) = (\bigcap A) \cap (\bigcap B)$.

9. (i) If $A \in B$, then $A \subseteq \bigcup B$ and $\bigcap B \subseteq A$,
   (ii) $A \subseteq B \leftrightarrow \bigcup A \subseteq \bigcup B$,
   (iii) $\emptyset \in A \rightarrow \bigcap A = \emptyset$.

10. Let $\{A_i \mid i \in \underset{\sim}{N}\}$ be a sequence of sets.
    We define the upper and lower limits of this sequence as follows:
    $\limsup A_i := \bigcap_{n \in \underset{\sim}{N}} \bigcup_{k \in \underset{\sim}{N}} A_{n+k}$ (limes superior),
    $\liminf A_i := \bigcup_{n \in \underset{\sim}{N}} \bigcap_{k \in \underset{\sim}{N}} A_{n+k}$ (limes inferior).

    (a) $x \in \limsup A_i \leftrightarrow x$ is an element of infinitely many A_i.
    $x \in \liminf A_i \leftrightarrow x$ is an element of all A_i, except finitely many.
    $\bigcap_{i \in \underset{\sim}{N}} A_i \subseteq \liminf A_i \subseteq \limsup A_i \subseteq \bigcup_{i \in \underset{\sim}{N}} A_i$.

(b) $\liminf A_i^c = (\limsup A_i)^c$,

$\liminf (A_i \cap B_i) = (\liminf A_i) \cap (\liminf B_i)$,

$\limsup (A_i \cup B_i) = (\limsup A_i) \cup (\limsup B_i)$.

11. If $\liminf A_i = \limsup A_i$ (= A say), then we say that the sequence $\{A_i \mid i \in \underset{\sim}{N}\}$ converges to the common value A of these two limits and we write then $\lim A_i = A$.

(a) If $A_i \subseteq A_{i+1}$ for all $i \in \underset{\sim}{N}$, then $\lim A_i$ exists and is equal to $\bigcup_{i \in \underset{\sim}{N}} A_i$.

(b) If $A_i \supseteq A_{i+1}$ for all $i \in \underset{\sim}{N}$, then $\lim A_i$ exists and is equal to $\bigcap_{i \in \underset{\sim}{N}} A_i$.

8 ORDERED PAIRS

In the plane the pairs <4,2> and <2,4> indicate different points (Fig. 10).

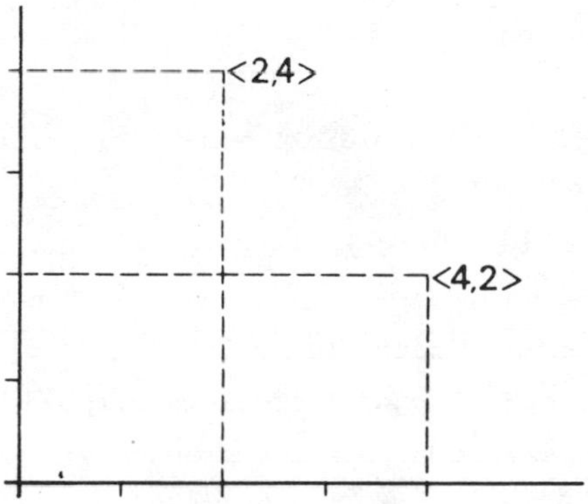

Fig. 10

The order of the numbers 2 and 4 is here of importance, in the same way that the order of letters is of importance in constructing words: "pin" and "nip" contain the same letters, but in a different order.

A pair of objects, say "a" and "b", in which their order is relevant, is called the *ordered pair* of a and b, written "$\langle a,b \rangle$". (The notation "(a,b)" is sometimes used.)

This is different from the ordinary (unordered) pair $\{a,b\}$, which is the same as $\{b,a\}$. Ordered pairs have the characteristic property

$$(*)\quad \langle a,b\rangle = \langle c,d\rangle \leftrightarrow a = c \wedge b = d.$$

Unordered pairs do not have this property (since $\{a,b\} = \{b,a\}$ even for $a \neq b$).

We can introduce the notion of ordered pair as a primitive notion (i.e. undefined) and introduce the above-mentioned property (*) as an axiom.

However, it is a wise rule not to introduce more primitive notions than necessary ("Ockham's razor") and hence we shall define a set, which behaves as an ordered pair, i.e. which satisfies the desired property (*).

<u>Definition 8.1:</u> $\langle a,b\rangle := \{\{a\},\{a,b\}\}$.

(This is not the only definition which will work: see the exercises.)

We must now show that this is indeed a good definition of the ordered pair, i.e. that (*) is satisfied.

<u>Theorem 8.2:</u> $\langle a,b\rangle = \langle c,d\rangle \leftrightarrow a = c \wedge b = d$.

Proof: The implication from right to left is trivial. So suppose $\langle a,b\rangle = \langle c,d\rangle$, i.e. $\{\{a\},\{a,b\}\} = \{\{c\},\{c,d\}\}$. If two sets are equal, then they have the same elements; hence $\{a\} = \{c\}$ and $\{a,b\} = \{c,d\}$ or $\{a\} = \{c,d\}$ and $\{a,b\} = \{c\}$. In the first case it follows simply that $a = c$ and $b = d$. In the second case we can conclude: $a = c = d$ and $a = b = c$; so, also in this case, $a = c$ and $b = d$. □

For this definition of $\langle a,b\rangle$ (Definition 8.1) the following theorem holds:

<u>Theorem 8.3:</u> If $a \in A$ and $b \in B$, then $\langle a,b\rangle \in \mathcal{P}\mathcal{P}(A \cup B)$.

Proof: Suppose $a \in A$ and $b \in B$. Then

(i) $a \in A \cup B$, so $\{a\} \subseteq A \cup B$, in other words $\{a\} \in \mathcal{P}(A \cup B)$, and

(ii) $b \in A \cup B$, so $\{a,b\} \subseteq A \cup B$, in other words $\{a,b\} \in \mathcal{P}(A \cup B)$.

From (i) and (ii) it follows that $\{\{a\},\{a,b\}\} \subseteq \mathcal{P}(A \cup B)$, in other words, $\{\{a\},\{a,b\}\} \in \mathcal{P}\mathcal{P}(A \cup B)$. □

We can generalize the notion of ordered pair to the notion of ordered n-tuple ($n \in \mathbb{N}$, $n \geqslant 1$):

Definition 8.4: $\langle a \rangle := a$

$\langle a_1,\ldots,a_n,a_{n+1}\rangle := \langle\langle a_1,\ldots,a_n\rangle, a_{n+1}\rangle$.

By means of mathematical induction one easily verifies that the object $\langle a_1,\ldots,a_n\rangle$ ($n \in \mathbb{N}$, $n \geqslant 1$), defined above, indeed behaves as an ordered n-tuple:

Theorem 8.5: For all $n \in \mathbb{N}$, $n \geqslant 1$:

$$\langle a_1,\ldots,a_n\rangle = \langle b_1,\ldots,b_n\rangle \leftrightarrow a_1 = b_1 \wedge \ldots \wedge a_n = b_n.$$

Proof: For $n = 1$, $\langle a_1\rangle = a_1$ and $\langle b_1\rangle = b_1$, so the proposition holds for $n = 1$.

Now suppose (induction hypothesis) that the proposition holds for n, i.e. $\langle a_1,\ldots,a_n\rangle = \langle b_1,\ldots,b_n\rangle \leftrightarrow a_1 = b_1 \wedge \ldots \wedge a_n = b_n$. And suppose that $\langle a_1,\ldots,a_n,a_{n+1}\rangle = \langle b_1,\ldots,b_n,b_{n+1}\rangle$, i.e. $\langle\langle a_1,\ldots,a_n\rangle, a_{n+1}\rangle = \langle\langle b_1,\ldots,b_n\rangle, b_{n+1}\rangle$.
Then, by Theorem 8.2, $\langle a_1,\ldots,a_n\rangle = \langle b_1,\ldots,b_n\rangle$ and $a_{n+1} = b_{n+1}$. Hence, by the induction hypothesis, $a_1 = b_1 \wedge \ldots \wedge a_n = b_n \wedge a_{n+1} = b_{n+1}$. □

Exercises

1. We provide an alternative notion of ordered pair: $\langle a,b\rangle = \{\{a,\emptyset\},\{b,\{\emptyset\}\}\}$. Prove, making use of this definition, that $\langle a,b\rangle = \langle c,d\rangle \leftrightarrow a = c \wedge b = d$.

2. The same as in 1, but now for $\{\{a,\emptyset\},\{b\}\}$.

3. Check whether for $\langle a,b\rangle$, defined as $\{\{a,\emptyset\},b\}$, the equivalence $\langle a,b\rangle = \langle c,d\rangle \leftrightarrow a = c \wedge b = d$ holds.

4. Prove: (i) $\bigcap\bigcap \langle a,b\rangle = a$,

   (ii) $\bigcap(\bigcup\langle a,b\rangle - \bigcap \langle a,b\rangle) = b$, if $a \neq b$.

9 CARTESIAN PRODUCT

The *Cartesian product* $A \times B$ of two sets A and B is by definition the set of all ordered pairs $\langle a,b \rangle$ with $a \in A$ and $b \in B$ (Fig. 11).

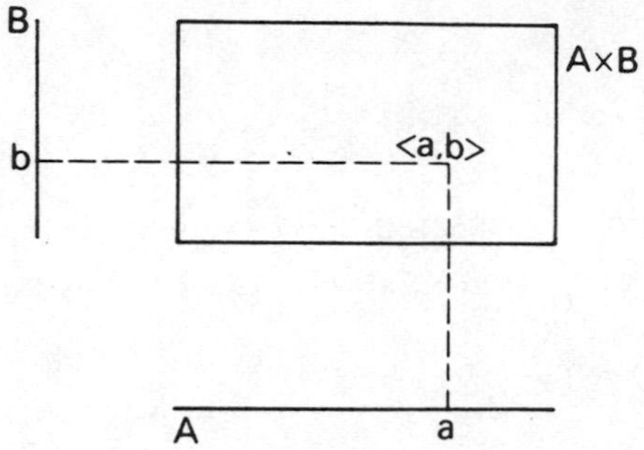

Fig. 11

Definition 9.1: $A \times B := \{x \mid$ there is some $a \in A$ and there is some $b \in B$ such that $x = \langle a,b \rangle\}$, in other words, $A \times B := \{\langle a,b \rangle \mid a \in A \wedge b \in B\}$.

Examples: $\{2,3\} \times \{4\} = \{\langle 2,4 \rangle , \langle 3,4 \rangle\}$

$\{1\} \times \{4,5\} = \{\langle 1,4 \rangle , \langle 1,5 \rangle\}$

$\{2,3\} \times \{4,5\} = \{\langle 2,4 \rangle , \langle 3,4 \rangle , \langle 2,5 \rangle , \langle 3,5 \rangle\}$

$\mathbb{R} \times \mathbb{R} := \{\langle x,y \rangle \mid x \in \mathbb{R} \wedge y \in \mathbb{R}\}$

i.e. $\mathbb{R} \times \mathbb{R}$ corresponds to the set of all points in the Euclidean plane (Fig . 12).

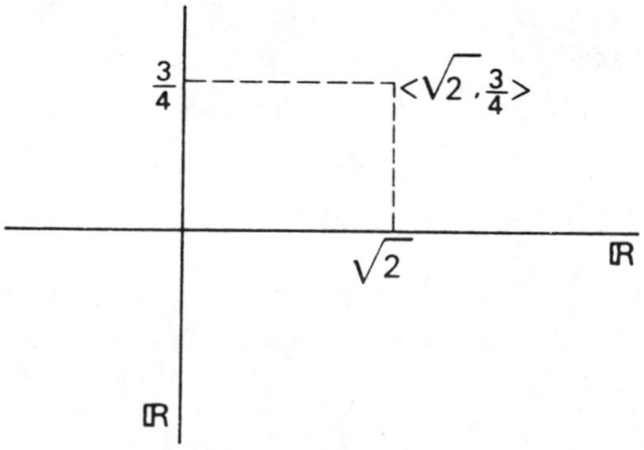

Fig. 12

"There is some $a \in A$ and there is some $b \in B$ such that $x = \langle a,b\rangle$" can be formulated in our symbolism as follows: $\exists a \in A\, \exists b \in B[x = \langle a,b\rangle]$.
So $A \times B = \{x | \exists a \in A\, \exists b \in B\, [x = \langle a,b\rangle]\}$.

From definition 9.1 and theorem 8.3 we immediately conclude

Corollary 9.2:
$A \times B = \{x | x \in \mathcal{P}\,\mathcal{P}(A \cup B) \wedge \exists a \in A\, \exists b \in B\, [x = \langle a,b\rangle]\}$.

Notation: Instead of "$\{x | x \in C \wedge \Phi(x)\}$" we usually write "$\{x \in C | \Phi(x)\}$". Some authors use $\{x \in C : \Phi(x)\}$ or $\{x | x \in C | \Phi(x)\}$. In this notation, Corollary 9.2 is formulated as follows: $A \times B = \{x \in \mathcal{P}\mathcal{P}(A \cup B) | \exists a \in A\, \exists b \in B\, [x = \langle a,b\rangle]\}$, or simply $A \times B = \{\langle a,b\rangle \in \mathcal{P}\mathcal{P}(A \cup B) | a \in A \wedge b \in B\}$.

$\{2\} \times \{4\} = \{\langle 2,4\rangle\}$, but $\{4\} \times \{2\} = \{\langle 4,2\rangle\}$. So it is not true that for all sets A and B, $A \times B = B \times A$; in other words, the operation $\times$ is not commutative (see also Exercise 13).

Theorem 9.3: The operation $\times$ is distributive with respect to union and intersection, i.e. $A \times (B \cup C) = (A \times B) \cup (A \times C)$ and
$A \times (B \cap C) = (A \times B) \cap (A \times C)$.

Proof: We shall prove the first of these two identities and leave the proof of the second to the reader. By the axiom of extensionality, it is sufficient to prove that $A \times (B \cup C)$ and $(A \times B) \cup (A \times C)$ have the same elements, in other words that for all y, $y \in A \times (B \cup C) \leftrightarrow y \in (A \times B) \cup (A \times C)$.

Proof of $\rightarrow$: Suppose $y \in A \times (B \cup C)$, i.e. $y = \langle a,z\rangle$ for some $a \in A$ and for some $z \in B \cup C$; say $y = \langle a,z\rangle$ with $a \in A$ and $z \in B \cup C$; so $z \in B$ or $z \in C$.
If $z \in B$, then $y \in A \times B \subseteq (A \times B) \cup (A \times C)$.
If $z \in C$, then $y \in A \times C \subseteq (A \times B) \cup (A \times C)$.
Hence $y \in (A \times B) \cup (A \times C)$.
Proof of $\leftarrow$: Suppose $y \in (A \times B) \cup (A \times C)$, i.e.
$y \in A \times B$ or $y \in A \times C$, so
$y = \langle a,b\rangle$ for some $a \in A$ and some $b \in B$, or
$y = \langle a,c\rangle$ for some $a \in A$ and some $c \in C$.

Hence, in both cases there is some $a \in A$ and some $z \in B \cup C$ such that $y = \langle a,z \rangle$. Hence $y \in A \times (B \cup C)$. □

Instead of "$A \times A$" we usually write "A^2".

Example: $\{3,4\}^2 = \{3,4\} \times \{3,4\} = \{\langle 3,3\rangle, \langle 3,4\rangle, \langle 4,3\rangle, \langle 4,4\rangle\}$.

More generally, we define A^n ($n \in \underset{\sim}{N}$, $n \geqslant 1$) inductively by

<u>Definition 9.4:</u>

$$A^1 := A$$

$$A^{n+1} := A^n \times A$$

<u>*Example*:</u> $\{3,4\}^3 := \{3,4\}^2 \times \{3,4\} = \{\langle 3,3\rangle, \langle 3,4\rangle, \langle 4,3\rangle, \langle 4,4\rangle\} \times \{3,4\} =$

$\{\langle\langle 3,3\rangle,3\rangle, \langle\langle 3,3\rangle,4\rangle, \langle\langle 3,4\rangle,3\rangle, \langle\langle 3,4\rangle,4\rangle, \langle\langle 4,3\rangle,3\rangle, \langle\langle 4,3\rangle,4\rangle,$

$\langle\langle 4,4\rangle,3\rangle, \langle\langle 4,4\rangle,4\rangle\}$.

Analogous to Definition 9.4 one can define the cartesian product with finitely many factors:

<u>Definition 9.5:</u>

$$\mathop{\mathrm{X}}\limits_{i=1}^{1} A_i := A_1$$

$$\mathop{\mathrm{X}}\limits_{i=1}^{n+1} A_i := \left(\mathop{\mathrm{X}}\limits_{i=1}^{n} A_i\right) \times A_{n+1}$$

Example: Let $A_1 = \{1,2\}$, $A_2 = \{3,4\}$ and $A_3 = \{7,8,9\}$. Then
$\mathop{\mathrm{X}}\limits_{i=1}^{3} A_i = (A_1 \times A_2) \times A_3 = (\{1,2\} \times \{3,4\}) \times \{7,8,9\}$.

Finally we mention that $\times$ is not associative; see Exercise 1.

<u>Exercises</u>

1. Give a simple example to show that the operation $\times$ is not associative, i.e. that $A \times (B \times C) = (A \times B) \times C$ does not hold for all sets A, B and C.

2. Prove that $\{a\} \times \{a\} = \{\{\{a\}\}\}$.

3. Prove that $A \times (B \cap C) = (A \times B) \cap (A \times C)$ and generalize this identity to: $A \times \bigcap_{i \in I} B_i = \bigcap_{i \in I} (A \times B_i)$.

4. (a) $A \times (B - C) = (A \times B) - (A \times C)$
(b) $B \subseteq C \rightarrow A \times B \subseteq A \times C$
(c) $(A \subseteq C \wedge B \subseteq D) \rightarrow A \times B \subseteq C \times D$
(d) $(A \times B \subseteq A \times C \wedge A \neq \emptyset) \rightarrow B \subseteq C$
(e) $(A \cap B) \times (C \cap D) = (A \times C) \cap (B \times D)$
(f) $A \times A = B \times B \rightarrow A = B$

5. Give an example to show that $\cup$ is not distributive with respect to $\times$, i.e. that $A \cup (B \times C) = (A \cup B) \times (A \cup C)$ does not hold for all sets A, B and C.

6. Let $A = \{1,2,3,4\}$, $B = \{5,6,7\}$, $C = \{8,9,10\}$.
Determine $A \times B$, $B \times A$, $C \times (B \times A)$, $(A \cup B) \times C$, $(A \times C) \cup (B \times C)$, $(A \cup B) \times (B \cup C)$.

7. $(A \times B) \cap (C \times D) = (A \times D) \cap (C \times B)$

8. $A \cap B = \emptyset \leftrightarrow (A \times C) \cap (B \times C) = \emptyset$ for all non-empty sets C.

9. Let $A \neq \emptyset$ and $C \neq \emptyset$. Prove:
$(A \subseteq B \wedge C \subseteq D) \leftrightarrow A \times C \subseteq B \times D$.

10. Let A, B, C and D be non-empty sets. Prove:
$A \times B = C \times D \leftrightarrow (A = C \wedge B = D)$.

11. $(A \times B) \cap (A^c \times C) = \emptyset$
$(B \times A) \cap (C \times A^c) = \emptyset$

12. $A \times B = \emptyset \leftrightarrow (A = \emptyset \vee B = \emptyset)$

13. $A \times B = B \times A \leftrightarrow A = B \vee A = \emptyset \vee B = \emptyset$

14. $(\bigcap_{i \in I} A_i) \times (\bigcap_{j \in J} B_j) = \bigcap_{<i,j> \in I \times J} (A_i \times B_j)$

15. The family $\{B_i \mid i \in I\}$ is defined to be a *cover* of A if and only if $A \subseteq \bigcup_{i \in I} B_i$.
Suppose that $\{B_i \mid i \in I\}$ and $\{C_j \mid j \in J\}$ are two covers of A.
Prove that $\{B_i \cap C_j \mid <i,j> \in I \times J\}$ is also a cover of A.

16. Prove that $A^n - B^n$ is the union of n cartesian products.

10 RELATIONS

We start with a few examples of binary relations R between the elements of a set X and the elements of a set Y (or: "between X and Y").

Examples:

1	X = M(en)	Y = W(omen)	$xRy := x$ is the son of y
2	$X = \underset{\sim}{N}$	$Y = \underset{\sim}{N}$	$xRy := y = x + 1$
3	$X = \underset{\sim}{N}$	$Y = \underset{\sim}{R}$	$xRy := y = \sqrt{x}$
4	$X = \underset{\sim}{N}^2$	$Y = \underset{\sim}{N}^2$	$<a,b>R<c,d> := a + d = b + c$
5	$X = \underset{\sim}{N} \times (\underset{\sim}{Z} - \{0\})$	$Y = X$	$<a,b>R<c,d> := a \cdot d = b \cdot c$
6	$X = \underset{\sim}{N}$	$Y = \mathcal{P}(\underset{\sim}{N})$	$xRy := x \in y$

Some examples of a ternary relation R between the elements of a set X, the elements of a set Y and the elements of a set Z:

X = M(en), Y = W(omen), Z = P(eople);

$R(x,y,z) := x$ and y are the parents of z.

$X = Y = Z = \underset{\sim}{N}$; $R(x,y,z) := x + y = z$.

In order not to make this story too tedious, we will discuss here only binary relations.

The following sets correspond to the relations of the examples 1,..., 6 :

1 $\{<x,y> \in M \times W \mid x$ is the son of $y\}$

2 $\{<x,y> \in \underset{\sim}{N} \times \underset{\sim}{N} \mid y = x + 1\}$

3 $\{<x,y> \in \underset{\sim}{N} \times \underset{\sim}{R} \mid y = \sqrt{x}\}$

4 $\{<<a,b>, <c,d>> \in \underset{\sim}{N}^2 \times \underset{\sim}{N}^2 \mid a + d = b + c\}$

5 $\{\langle\langle a,b\rangle, \langle c,d\rangle\rangle \in (\mathbb{N} \times (\mathbb{Z} - \{0\}))^2 \mid a.d = b.c\}$

6 $\{\langle x,y\rangle \in \mathbb{N} \times \mathcal{P}(\mathbb{N}) \mid x \in y\}$

It will be clear that we can represent the mathematical notion "relation" by a set: each binary relation R between the elements of a set X and the elements of a set Y determines a subset of $X \times Y$; and, conversely, each subset of $X \times Y$ determines a binary relation between the elements of X and of Y. Hence we have the following definition.

Definition 10.1: R is a (binary) *relation* between X and Y if and only if $R \subseteq X \times Y$.

We use the following notation for binary relations R:

$$xRy := \langle x,y\rangle \in R.$$

One sometimes uses "$R(x,y)$" for "xRy".

Definition 10.2: For $R \subseteq X \times Y$ we define:

$\mathrm{Dom}(R) := \{x \in X \mid \exists y \in Y\,[xRy]\}$ (*domain* of R)

$\mathrm{Ran}(R) := \{y \in Y \mid \exists x \in X\,[xRy]\}$ (*range* of R).

For the relations of examples 1,..., 6 Dom(R) and Ran (R) are respectively:

	Dom(R)	Ran(R)
1.	the set of all men	the set of all mothers
2.	$\mathbb{N}$	$\mathbb{N} - \{0\}$
3.	$\mathbb{N}$	$\{y \in \mathbb{R} \mid \exists x \in \mathbb{N}\,[y = \sqrt{x}]\}$
4.	$\mathbb{N}^2$	$\mathbb{N}^2$
5.	$\mathbb{N} \times (\mathbb{Z} - \{0\})$	$\mathbb{N} \times (\mathbb{Z} - \{0\})$
6.	$\mathbb{N}$	$\mathcal{P}(\mathbb{N}) - \{\emptyset\}$

If $R \subseteq X \times X$, then we call R simply a relation *on* X. Example 2 gives a relation on $\mathbb{N}$, example 4 a relation on $\mathbb{N}^2$ and example 5 a relation on $\mathbb{N} \times (\mathbb{Z} - \{0\})$.

Let $R \subseteq X \times Y$.

If R is a relation between X and Y, then the *converse* relation $\breve{R}$ is the relation between Y and X, defined as follows.

Definition 10.3: $\breve{R} := \{<u,v> \in Y \times X | <v,u> \in R\}$

For the relations of examples 1,...,6 the converse relations are respectively:

1. $\{<u,v> \in W \times M \mid u \text{ is the mother of } v\}$
2. $\{<u,v> \in \underset{\sim}{N} \times \underset{\sim}{N} \mid v = u-1\}$
3. $\{<u,v> \in \underset{\sim}{R} \times \underset{\sim}{N} \mid v = u^2\}$
4. $\{<<u_1,u_2>, <v_1,v_2>> \in \underset{\sim}{N}^2 \times \underset{\sim}{N}^2 \mid u_1 - v_1 = u_2 - v_2\}$
5. $\{<<u_1,u_2>, <v_1,v_2>> \in (\underset{\sim}{N} \times (\underset{\sim}{Z} - \{0\}))^2 \mid u_1 . v_2 = u_2 . v_1\}$
6. $\{<u,v> \in \mathcal{P}(\underset{\sim}{N}) \times \underset{\sim}{N} \mid v \in u\}$

Note that in examples 4 and 5, $\breve{R} = R$.

Let $R \subseteq U \times V$ and $S \subseteq V \times W$.

The *composite relation* of S and R, or the *composition* of S and R, is the relation $S \circ R$ between U and W defined as follows.

Definition 10.4: $S \circ R := \{<x,z> \in U \times W | \exists y \in V [<x,y> \in R \wedge <y,z> \in S]\}$ (Figure 13).

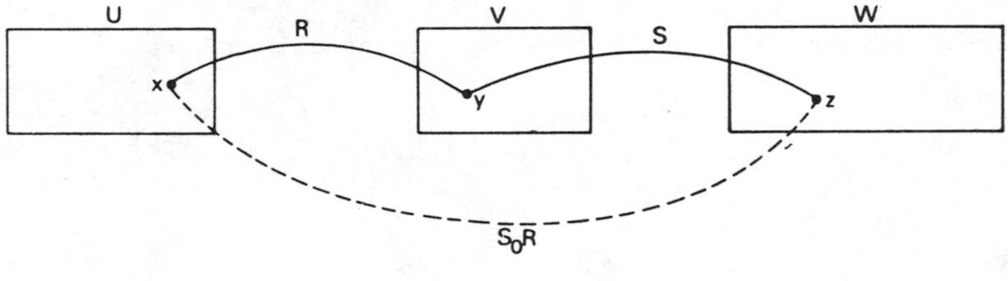

Fig. 13

Example I: Let R be the relation of example 2, $R \subseteq \underset{\sim}{N} \times \underset{\sim}{N}$, and let S be the relation of example 3, $S \subseteq \underset{\sim}{N} \times \underset{\sim}{R}$. Then

$$S \circ R := \{<x,z> \in \underset{\sim}{N} \times \underset{\sim}{R} | \exists y \in \underset{\sim}{N} [<x,y> \in R \wedge <y,z> \in S]\}$$
$$= \{<x,z> \in \underset{\sim}{N} \times \underset{\sim}{R} | \exists y \in \underset{\sim}{N} [y = x + 1 \wedge z = \sqrt{y}]\}$$
$$= \{<x,z> \in \underset{\sim}{N} \times \underset{\sim}{R} | z = \sqrt{(x+1)}\}.$$

Example II: Let M be the set of all men and xRy := y is the father of x.

$R \circ R := \{\langle x,z\rangle \in M \times M \mid \exists y \in M [\langle x,y\rangle \in R \wedge \langle y,z\rangle \in R]\}$

$= \{\langle x,z\rangle \in M \times M \mid \exists y \in M [$ y is the father of x and z is the father of $y]\}$

$= \{\langle x,z\rangle \in M \times M \mid z$ is the grandfather of $x\}$.

In other words: $x(R \circ R)z$:= z is the grandfather of x.

<u>Definition 10.5:</u> Let V be a set. The relation $I_V := \{\langle x,y\rangle \in V \times V \mid x = y\}$ is called the *identity relation* on V (or the "diagonal" of $V \times V$).

Note that $u \in \bigcup Z \leftrightarrow \exists z[z \in Z \wedge u \in z]$.

Hence: if $\{a,b\} \in Z$, then $a \in \bigcup Z$ and $b \in \bigcup Z$. (*)

From this follows immediately:

<u>Theorem 10.6:</u> If $\langle a,b\rangle \in R$, then $a \in \bigcup(\bigcup R)$ and $b \in \bigcup(\bigcup R)$.

Proof: Suppose $\langle a,b\rangle = \{\{a\}, \{a,b\}\} \in R$. By (*), with R substituted for Z: $\{a\} \in \bigcup R$ and $\{a,b\} \in \bigcup R$. Again by (*), but now with $\bigcup R$ substituted for Z: $a \in \bigcup(\bigcup R)$ and $b \in \bigcup(\bigcup R)$. □

In Definition 10.2 we have defined Dom(R) and Ran(R) for the case that $R \subseteq X \times Y$. We can generalize this definition for arbitrary sets R:

<u>Definition 10.7:</u> $\text{Dom}(R) := \{x \mid \exists y [\langle x,y\rangle \in R]\}$

$\text{Ran}(R) := \{y \mid \exists x [\langle x,y\rangle \in R]\}$

Note that if R does not contain ordered pairs, then $\text{Dom}(R) = \emptyset$ and $\text{Ran}(R) = \emptyset$.

From definition 10.7 and theorem 10.6 we immediately have:

<u>Corollary 10.8:</u> $\text{Dom}(R) = \{x \in \bigcup(\bigcup R) \mid \exists y [\langle x,y\rangle \in R]\}$

$\text{Ran}(R) = \{y \in \bigcup(\bigcup R) \mid \exists x [\langle x,y\rangle \in R]\}$

Definition 10.9: Let R be a relation on a set X (i.e. $R \subseteq X \times X$).
R is *reflexive* := for all $x \in X$, $x \, R \, x$.
R is *symmetric* := for all $x, y \in X$, $x \, R \, y \rightarrow y \, R \, x$.
R is *transitive* := for all $x,y,z \in X$, $x \, R \, y \wedge y \, R \, z \rightarrow x \, R \, z$.

Examples:

i) The relation $\leqslant$ on $\underset{\sim}{N}$ is reflexive and transitive, but not symmetric.
ii) The relation $<$ on $\underset{\sim}{N}$ is transitive, but not reflexive or symmetric.
iii) The relation $=$ on $\underset{\sim}{N}$ is reflexive, symmetric and transitive.
iv) The relation "is parallel to" on the set of all straight lines in the Euclidean plane is reflexive, symmetric and transitive. (Note that we define that each line is parallel to itself.)

Exercises

1. Let $R = \{<0,1>,<0,3>,<0,4>,<2,1>,<1,2>,<4,7>\}$.
Determine $\text{Dom}(R)$, $\text{Ran}(R)$ and $\breve{R}$. Let $S = \{<1,4>,<3,2>,<5,0>\}$. Determine $R \circ S$ and $S \circ R$.

2. Let R be a relation on a set X.
Prove: R is reflexive $\leftrightarrow I_X \subseteq R \cap \breve{R}$.
R is symmetric $\leftrightarrow R = \breve{R}$.
R is transitive $\leftrightarrow R \circ R \subseteq R$.

3. (i) Let R be a relation on a set E and let $*X$ and $X*$ be defined for each subset X of E as follows: $*X := \{x \in E \,|\, \forall y \in X \, [x \, R \, y]\}$
$X* := \{x \in E \,|\, \forall y \in X \, [y \, R \, x]\}$.

Prove that:
(a) $A \subseteq B \rightarrow (*A \supseteq *B \wedge A* \supseteq B*)$
(b) $A \subseteq (*A)*$ and $A \subseteq *(A*)$
(c) $(*(A*))* = A*$ and $*((*A)*) = *A$.

(ii) Note that if R is symmetric, then $*X = X*$. Show in this case that $A = A** \wedge B = B** \rightarrow A \cap B = (A \cap B)**$. Give an example, which shows that we cannot replace $\cap$ by $\cup$.

(iii) E is the collection of subsets of a given set and $x\ R\ y :=$ $x \cap y \neq \emptyset$. Prove that, if $\{x_i \mid i \in I\}$ is a collection of elements of E, then $\bigcup_{i \in I} x_i = (\bigcup_{i \in I} x_i)^{**}$. Does the same result hold if $x\ R\ y := x \cap y = \emptyset$?

4. Let R be a relation on A. Prove:
   (a) $\forall x \in A\ [\neg\ x\ R\ x] \leftrightarrow R \cap I_A = \emptyset$
   (b) $\forall x,y \in A\ [x\ R\ y \rightarrow \neg\ y\ R\ x] \leftrightarrow R \cap \breve{R} = \emptyset$
   (c) $\forall x,y,z \in A\ [x\ R\ y \wedge y\ R\ z \rightarrow \neg\ x\ R\ z] \leftrightarrow (R \circ R) \cap R = \emptyset$.

5. Let R be a reflexive relation on A. Prove that for each relation S on A: $S \subseteq R \circ S$ and $S \subseteq S \circ R$.

6. Suppose R and S are relations on A. Suppose that R is reflexive, S is reflexive and that S is transitive. Prove: $R \subseteq S \leftrightarrow R \circ S = S$.

7. Prove that for given relations R,S,T and U :
   (a) $\breve{R} \cup \breve{S} = (R \cup S)\breve{}$ $\qquad \breve{R} \cap \breve{S} = (R \cap S)\breve{}$
   (b) $(S \circ R)\breve{} = \breve{R} \circ \breve{S}$
   (c) $T \circ (S \circ R) = (T \circ S) \circ R$
   (d) $(R \subseteq S \wedge T \subseteq U) \rightarrow T \circ R \subseteq U \circ S$
   (e) $(S \cup T) \circ R = (S \circ R) \cup (T \circ R)$
   (f) $(S \cap T) \circ R \subseteq (S \circ R) \cap (T \circ R)$
   Give in case (f) an example, which shows that $\subseteq$ cannot be replaced by $=$.

11 EQUIVALENCE RELATIONS

We now give five examples of a relation R on a set X, which all have the following three properties:

1. for all $x \in X$, $x\ R\ x$ (reflexivity)
2. for all x and $y \in X$, $x\ R\ y \rightarrow y\ R\ x$ (symmetry)
3. for all x,y and $z \in X$, $x\ R\ y \wedge y\ R\ z \rightarrow x\ R\ z$ (transitivity)

Examples:

I $\quad X = \mathbb{N} \qquad x\ R\ y := x = y$

II $\quad X = \mathbb{N} \qquad x\ R\ y := \exists k \in \mathbb{Z}\ [x - y = 3k]$ (x is congruent to y modulo 3)

III $X = \underset{\sim}{N}^2$ $<a,b>\ R\ <c,d>\ := a+d = b+c$

IV $X = \underset{\sim}{N} \times (\underset{\sim}{Z} - \{0\})$ $<a,b>\ R\ <c,d>\ := a.d = b.c$

V X is the set of all straight lines in the Euclidean space and $x\ R\ y$:= x is parallel to y.

<u>Definition 11.1:</u> A relation R on a set X is an *equivalence relation* on X if R is reflexive, symmetric and transitive.

Check that for examples I,..,V R is an equivalence relation on X.

It is a well-established custom to denote equivalence relations by suggestive symbols, such as $\sim$, $\approx$, $\simeq$. Whenever convenient we will adopt this notation.

<u>Definition 11.2:</u> Let $\sim$ be an equivalence relation on a set X. The *equivalence class* $[x]_\sim$ of an element x of X with respect to $\sim$ (or *modulo* $\sim$) is by definition the subset of X, consisting of all those elements y in X for which $x \sim y$.

$$[x]_\sim := \{y \in X \mid x \sim y\}$$

x is called a *representative* of the class $[x]_\sim$.(Instead of "$[x]_\sim$" one sometimes writes "$x/\sim$".)

Note that if $\sim$ is an equivalence relation on X, then for all $x,y \in X$,

$$x \sim y \leftrightarrow [x]_\sim = [y]_\sim.$$

For the equivalence relations $\sim$ on X from I,...,IV we now give a number of examples of equivalence classes $[x]_\sim$:

I $[0]_\sim := \{y \in \underset{\sim}{N} \mid y = 0\}$, hence $[0]_\sim = \{0\}$.
$[1]_\sim := \{y \in \underset{\sim}{N} \mid y = 1\}$, hence $[1]_\sim = \{1\}$.

II $[0]_\sim := \{y \in \underset{\sim}{N} \mid -y \text{ is divisible by } 3\} = \{0,3,6,\ldots\}$.
$[1]_\sim := \{y \in \underset{\sim}{N} \mid 1-y \text{ is divisible by } 3\} = \{1,4,7,\ldots\}$.
$[2]_\sim := \{y \in \underset{\sim}{N} \mid 2-y \text{ is divisible by } 3\} = \{2,5,8,\ldots\}$.
$[3]_\sim := \{y \in \underset{\sim}{N} \mid 3-y \text{ is divisible by } 3\} = \{0,3,6,\ldots\}$. Hence $[3]_\sim = [0]_\sim$.
Check that $[4]_\sim = [1]_\sim$, $[5]_\sim = [2]_\sim$, $[6]_\sim = [3]_\sim = [0]_\sim$, and so on.

III $[\langle 0,0\rangle]_\sim := \{\langle c,d\rangle \in \mathbb{N}^2 \mid c = d\}$, hence

$[\langle 0,0\rangle]_\sim = \{\langle 0,0\rangle , \langle 1,1\rangle , \langle 2,2\rangle, \ldots \}$.

Check that $[\langle 1,1\rangle]_\sim = [\langle 0,0\rangle]_\sim$, $[\langle 2,2\rangle]_\sim = [\langle 0,0\rangle]_\sim$, and so on.

$[\langle 0,1\rangle]_\sim := \{\langle c,d\rangle \in \mathbb{N}^2 \mid d = c+1\}$, hence

$[\langle 0,1\rangle]_\sim = \{\langle 0,1\rangle , \langle 1,2\rangle , \langle 2,3\rangle , \ldots\}$.

Check that $[\langle 1,2\rangle]_\sim = [\langle 0,1\rangle]_\sim$, $[\langle 2,3\rangle]_\sim = [\langle 0,1\rangle]_\sim$, and so on.

In words: $[\langle 0,1\rangle]_\sim$ is the set of all pairs $\langle c,d\rangle$ of natural numbers, for which $c-d = -1$ (or $d = c+1$). This suggests the possibility of defining the integer -1 in terms of natural numbers and the addition operation on them, namely as the set $\{\langle c,d\rangle \in \mathbb{N}^2 \mid d = c+1\}$.

$[\langle 0,2\rangle]_\sim := \{\langle c,d\rangle \in \mathbb{N}^2 \mid d = c+2\}$, hence $[\langle 0,2\rangle]_\sim = \{\langle 0,2\rangle,\langle 1,3\rangle,\langle 2,4\rangle,\ldots\}$.

In words: $[\langle 0,2\rangle]_\sim$ is the set of all pairs of natural numbers $\langle c,d\rangle$, for which $c - d = -2$ (or $d = c+2$). This suggests the possibility of defining the integer -2 in terms of natural numbers and the addition operation on them, namely as the set $\{\langle c,d\rangle \in \mathbb{N}^2 \mid d = c+2\}$.

It should be clear that in this way we can define each integer in terms of natural numbers and the addition operation on them (by taking appropriate equivalence classes). Further we shall see that we can define in this way the set $\mathbb{Z}$ of all integers in terms of $\mathbb{N}$, and that we can then define, on this set $\mathbb{Z}$, operations of "addition" and "multiplication" (in terms of the addition and multiplication operations on $\mathbb{N}$) which will satisfy all the characteristic properties of addition and multiplication on the integers.

We shall also see that not only the integers, but also the rational and real numbers can be defined in terms of natural numbers, and hence not only $\mathbb{Z}$, but also $\mathbb{Q}$ and $\mathbb{R}$, are definable in terms of $\mathbb{N}$, and, further, the corresponding operations of addition and multiplication on these structures are definable in terms of the addition and multiplication operations on $\mathbb{N}$.

One may express this very roughly by saying that the natural numbers form the basis of mathematics.

IV $X = \mathbb{N} \times (\mathbb{Z} - \{0\})$. $\qquad \langle a,b\rangle \sim \langle c,d\rangle := a.d = b.c$

$[\langle 1,2\rangle]_\sim := \{\langle c,d\rangle \in \mathbb{N} \times (\mathbb{Z} - \{0\}) \mid d = 2.c\}$,

hence $[\langle 1,2\rangle]_\sim = \{\langle 1,2\rangle, \langle 2,4\rangle, \langle 3,6\rangle, \ldots\}$.

In words: $[<1,2>]_\sim$ is the set of all pairs $<c,d> \in \underset{\sim}{N} \times (\underset{\sim}{Z} - \{0\})$, for which $\frac{c}{d} = \frac{1}{2}$ (or $d = 2.c$). This suggests the possibility of defining the rational number $\frac{1}{2}$ in terms of natural numbers and integers, and hence in terms of natural numbers, and the multiplication operation on them, namely as the set $\{<c,d> \in \underset{\sim}{N} \times (\underset{\sim}{Z} - \{0\}) \mid d = 2.c\}$.

$[<2,3>]_\sim := \{<c,d> \in \underset{\sim}{N} \times (\underset{\sim}{Z} - \{0\}) \mid 2.d = 3.c\}$, hence
$[<2,3>]_\sim = \{<2,3>, <4,6>, <6,9>, \ldots\}$. This again suggests the possibility of defining the rational number $\frac{2}{3}$ as the set $[<2,3>]_\sim$.

Equivalence classes have, either nothing, or everything in common, as detailed by the following

Lemma 11.3: $[a]_\sim \cap [b]_\sim \neq \emptyset \rightarrow [a]_\sim = [b]_\sim$.

Proof: Suppose $[a]_\sim \cap [b]_\sim \neq \emptyset$. Then there is some c with $c \in [a]_\sim$ and $c \in [b]_\sim$, hence $c \sim a$ and $c \sim b$. Now if $x \in [a]_\sim$, then $x \sim a$, $a \sim c$ (because of $c \sim a$) and $c \sim b$. Hence, because $\sim$ is transitive, $x \sim b$.
So we have shown: $x \in [a]_\sim \rightarrow x \in [b]_\sim$. Hence $[a]_\sim \subseteq [b]_\sim$. In the same way we can prove $[b]_\sim \subseteq [a]_\sim$. Hence $[a]_\sim = [b]_\sim$. □

Definition 11.4: A collection $\mathbf{P}$ of subsets of X is called a *partition* of X if

i) $X = \bigcup\{A \mid A \in \mathbf{P}\}$, and
ii) for all $A,B \in \mathbf{P}$, $A = B$ or $A \cap B = \emptyset$.

An example of a partition is shown in Figure 14.

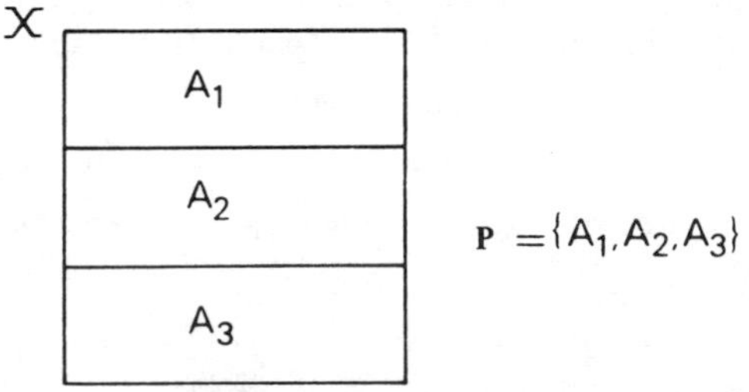

Fig. 14

Theorem 11.5: Let $\sim$ be an equivalence relation on X. Then $\{[x]_\sim \mid x \in X\}$ is a partition of X.

Proof: For each $x \in X$, $x \in [x]_\sim$. Hence $X \subseteq \bigcup \{[x]_\sim \mid x \in X\}$. For each $x \in X$, $[x]_\sim \subseteq X$. Hence $\bigcup\{[x]_\sim \mid x \in X\} \subseteq X$ (see 7.6). We have proved part i) of definition 11.4. The preceding lemma 11.3 proves ii). □

Definition 11.6: Let $\sim$ be an equivalence relation on X. The *quotient set* $X/\sim$ of X *modulo* $\sim$ is the set of equivalence classes $[x]_\sim$ for all $x \in X$. In other words: $X/\sim \; := \{[x]_\sim \mid x \in X\}$.

As examples let us consider the quotient sets from I, ..., IV:

I $\quad \underset{\sim}{N}/\sim \; = \{\{x\} \mid x \in \underset{\sim}{N}\}$, where $x \sim y \; := x = y$.

II $\quad \underset{\sim}{N}/\sim \; = \{[0]_\sim, [1]_\sim, [2]_\sim\}$, where $x \sim y := x - y$ is divisible by 3.

III $\quad \underset{\sim}{N}^2/\sim \; = \{\ldots, [\langle 0,2\rangle]_\sim, [\langle 0,1\rangle]_\sim, [\langle 0,0\rangle]_\sim, [\langle 1,0\rangle]_\sim, [\langle 2,0\rangle]_\sim, \ldots\}$, where $\langle a,b\rangle \sim \langle c,d\rangle := a + d = b + c \; (a - b = c - d)$.

Now, if we define -1 as $[\langle 0,1\rangle]_\sim$, -2 as $[\langle 0,2\rangle]_\sim$, and so on, and if we identify the natural numbers $0,1,2, \ldots$ with $[\langle 0,0\rangle]_\sim$, $[\langle 1,0\rangle]_\sim$, $[\langle 2,0\rangle]_\sim, \ldots$, then we can write: $\underset{\sim}{N}^2/\sim \; = \{\ldots, -2, -1, 0, 1, 2, \ldots\}$ (see Figure 15).

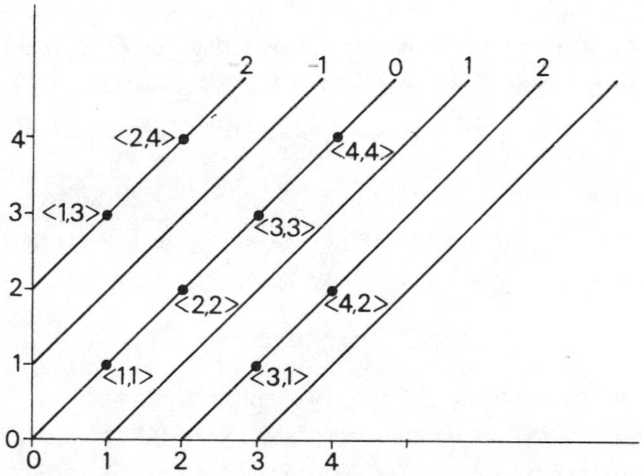

Fig. 15

On $\underset{\sim}{N}^2/\sim$ we can define operations of "addition" and "multiplication" in terms of the addition and multiplication of natural numbers:

$[\langle a,b\rangle]_\sim + [\langle c,d\rangle]_\sim := [\langle a+c,\ b+d\rangle]_\sim$

and

$[\langle a,b\rangle]_\sim \,.\, [\langle c,d\rangle]_\sim := [\langle a.c + b.d,\ b.c + a.d\rangle]_\sim$.

The reader can easily verify that these are good definitions, i.e. that the defined operations are independent of the choices of the representatives, in other words:

if $[\langle a,b\rangle]_\sim = [\langle a',b'\rangle]_\sim$ and $[\langle c,d\rangle]_\sim = [\langle c',d'\rangle]_\sim$,

then $[\langle a+c,\ b+d\rangle]_\sim = [\langle a'+c',\ b'+d'\rangle]_\sim$ and

$[\langle a.c + b.d,\ b.c + a.d\rangle]_\sim = [\langle a'.c' + b'.d',\ b'.c' + a'.d'\rangle]_\sim$.

Moreover, the reader can easily verify that the operations of addition and multiplication on $\underset{\sim}{N}^2/\sim$, defined above, have all properties which we should expect:

i) For all $A,B,C \in \underset{\sim}{N}^2/\sim$,
$A + B = B + A$ (+ is commutative)
$A + (B + C) = (A + B) + C$ (+ is associative)
$A + [\langle 0,0\rangle]_\sim = A$ ($[\langle 0,0\rangle]_\sim$ is neutral element of the addition).

ii) For each $A \in \underset{\sim}{N}^2/\sim$ there is precisely one $B \in \underset{\sim}{N}^2/\sim$ such that $A + B = [\langle 0,0\rangle]_\sim$. This B is called "minus A" and is indicated by $-A$.

iii) For all $A,B,C \in \underset{\sim}{N}^2/\sim$, $A.(B+C) = A.B + A.C$ (. is distributive with respect to +).

iv) For each $A \in \underset{\sim}{N}^2/\sim$, $A.[\langle 1,0\rangle]_\sim = A$ ($[\langle 1,0\rangle]_\sim$ is the identity element of the multiplication).

This shows that we can indeed define the set $\underset{\sim}{Z}$ of the integers in terms of the set $\underset{\sim}{N}$ of the natural numbers, namely by

$$\underset{\sim}{Z} := \underset{\sim}{N}^2/\sim \ .$$

IV A closer examination of $\underset{\sim}{N} \times \underset{\sim}{Z} - \{0\}/\sim$, with $\langle a,b\rangle \sim \langle c,d\rangle := a.d = b.c$ ($\frac{a}{b} = \frac{c}{d}$), would show, in an analogous way, that $\underset{\sim}{N} \times \underset{\sim}{Z} - \{0\}/\sim$ has exactly the same structure as $\underset{\sim}{Q}$, so that we can in fact define $\underset{\sim}{Q}$ by

$$\mathbb{Q} := \mathbb{N} \times \mathbb{Z} - \{0\}/\sim$$

(see Figure 16).

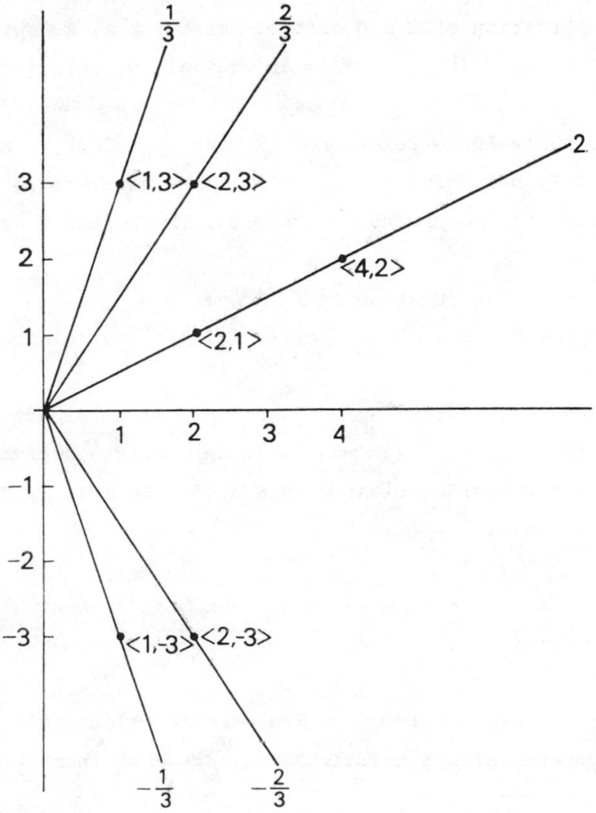

Fig. 16

The notion of "equivalence" can be used to clarify a popular practice in mathematics, that goes by the name of "identification".

One identifies objects which are "the same" in some way. E.g. Cauchy sequences that converge to the same limit, or angles that are equal modulo 2π. This process can be made precise in the following way: let $\sim$ be an equivalence relation on a set, then the *identification under* $\sim$ is the canonical mapping $a \mapsto [a]_\sim$.

Metaphorically speaking: equivalent elements are identified and the result is the equivalence class under $\sim$. Although this manner of expression is rather loose, we will occasionally use it, relying on the above clarification.

Exercises

1. Let **P** be a partition of X and define, for $x,y \in X$, $x \approx y :=$ $\exists A \in \mathbf{P}\, [x \in A \wedge y \in A]$. Then $\approx$ is an equivalence relation on X.

2. Let $\sim$ be an equivalence relation on X. Let $\mathbf{P} = \{[x]_\sim \mid x \in X\}$, i.e. the partition of X, belonging to $\sim$. Let $\approx$ be the equivalence relation belonging to **P** according to the preceding exercise. Then $\sim$ and $\approx$ are identical.

3. Let R be a reflexive relation on X. Prove:
$\forall a,b,c \in X\, [a\, R\, b \wedge b\, R\, c \rightarrow c\, R\, a] \leftrightarrow R$ is an equivalence relation on X.

4. Show that the factor group G/H of a group G with respect to a normal subgroup H is a quotient set modulo an appropriate equivalence relation. The same for the residue-class ring R/N of a commutative ring R with respect to an ideal N.

5. Prove: if $R \subseteq A \times A$ and Dom $(R) = A$, then R is an equivalence relation if and only if $R \circ \breve{R} = R$.

6. Let $\{R_i \mid i \in I\}$ be a family of equivalence relations on A. Prove that $\bigcap_{i \in I} R_i$ is an equivalence relation on A. Check whether $\bigcup_{i \in I} R_i$ is also an equivalence relation on A.

7. Let R and S be equivalence relations on A.
Prove:
i) $R \circ S$ is an equivalence relation on $A \leftrightarrow R \circ S = S \circ R$
ii) $R \cup S$ is an equivalence relation on $A \leftrightarrow$
$(R \circ S \subseteq R \cup S \wedge S \circ R \subseteq R \cup S)$.

8. For each of the following relations on $\mathbb{Z}$, check whether it is an equivalence relation or not.

a) $R = \{\langle x,y\rangle \in \mathbb{Z}^2 \mid x + y < 3\}$

b) $R = \{\langle x,y\rangle \in \mathbb{Z}^2 \mid x \text{ is a divisor of } y\}$

c) $R = \{\langle x,y\rangle \in \mathbb{Z}^2 \mid x \text{ and } y \text{ are relatively prime}\}$

d) $R = \{\langle x,y\rangle \in \mathbb{Z}^2 \mid x + y \text{ is even}\}$

e) $R = \{\langle x,y\rangle \in \mathbb{Z}^2 \mid x = y \vee x = -y\}$

f) $R = \{\langle x,y\rangle \in \mathbb{Z}^2 \mid x + y \text{ is even and } y \text{ is a divisor of } x\}$

g) $R = \{\langle x,y\rangle \in \mathbb{Z}^2 \mid y = x+1\}$

9. For each $n \in \mathbb{Z}$ let $B_n = \{m \in \mathbb{Z} \mid \exists q \in \mathbb{Z}\ [m = n+5.q]\}$. Prove that $\{B_n \mid n \in \mathbb{Z}\}$ is a partition of $\mathbb{Z}$.

10. Prove that in each of the following cases $\{B_r \mid r \in \mathbb{R}\}$ is a partition of $\mathbb{R} \times \mathbb{R}$. Describe geometrically the members of this partition. Find the equivalence relations corresponding to the partitions (see exercises 1 and 2).

a) $B_r = \{\langle x,y\rangle \in \mathbb{R}^2 \mid y = x + r\}$

b) $B_r = \{\langle x,y\rangle \in \mathbb{R}^2 \mid x^2 + y^2 = r\}$

Hint: $y = x + r$ is the equation of a line and $x^2 + y^2 = r$ is the equation of a circle.

11. Prove that the following relations are equivalence relations on $\mathbb{R} \times \mathbb{R}$:

a) $G = \{\langle\langle a,b\rangle , \langle c,d\rangle\rangle \in \mathbb{R}^2 \times \mathbb{R}^2 \mid a^2 + b^2 = c^2 + d^2\}$

b) $H = \{\langle\langle a,b\rangle , \langle c,d\rangle\rangle \in \mathbb{R}^2 \times \mathbb{R}^2 \mid b - a = d - c\}$

c) $J = \{\langle\langle a,b\rangle , \langle c,d\rangle\rangle \in \mathbb{R}^2 \times \mathbb{R}^2 \mid a+b = c+d\}$.

Find the partitions corresponding to these equivalence relations and describe geometrically the members of this partition.

12. If H and J are the equivalence relations of exercise 11, describe the equivalence relation $H \cap J$. Describe the equivalence classes modulo $H \cap J$.

13. Let H and J be the equivalence relations of exercise 11. Prove that $H \circ J = J \circ H$. Conclude that $H \circ J$ is an equivalence relation (see exercise 7) and describe the equivalence classes modulo $H \circ J$.

14. Let L be the set of all straight lines in the plane. Let G and H be the following relations on L:
$G = \{\langle l_1, l_2\rangle \in L^2 \mid l_1 \text{ is parallel to } l_2\}$
$H = \{\langle l_1, l_2\rangle \in L^2 \mid l_1 \text{ is perpendicular to } l_2\}$.
Prove the following:
a) G is an equivalence relation on L
b) $H \circ G = H$ and $G \circ H = H$
c) $G \cup H$ is an equivalence relation; describe its equivalence classes.

15. Let A be a set. Prove that I_A and $A \times A$ are equivalence relations on A. Describe the partitions induced, respectively, by I_A and $A \times A$.

16. Let $\{A_i \mid i \in I\}$ be a partition of A and let $\{B_j \mid j \in J\}$ be a partition of B. Prove that $\{A_i \times B_j \mid \langle i,j\rangle \in I \times J\}$ is a partition of $A \times B$.

17. Let G and H be equivalence relations on A.
Prove that for each $x \in A$, $[x]_{G\cap H} = [x]_G \cap [x]_H$. (Cf. exercise 6)

18. Suppose that G and H are equivalence relations on A and that $G \cup H$ is also an equivalence relation on A. Prove that for each $x \in A$, $[x]_{G\cup H} = [x]_G \cup [x]_H$.

19. Prove that $\underset{\sim}{Z} := \underset{\sim}{N}^2/\sim$, with $\langle a,b\rangle \sim \langle c,d\rangle := a+d = b+c$, is a commutative ring with unit element and without zero-divisors.

20. Prove that $\underset{\sim}{Q} := \underset{\sim}{N} \times \underset{\sim}{Z} - \{0\}/\sim$, with $\langle a,b\rangle \sim \langle c,d\rangle := a.d = b.c$, is a field, ordered by $<_{\underset{\sim}{Q}}$, defined by $[a,b]_\sim <_{\underset{\sim}{Q}} [c,d]_\sim := a.d < b.c$.

21. Give an example of a relation, which is transitive and symmetric, but not reflexive.

22. Spot the flaw in the following argument: Let R be transitive and symmetric. Then $xRy \wedge yRz \rightarrow xRz$ for all x,y and z. Also $xRy \rightarrow yRx$ holds for all x and y. Now take any x and y such that xRy, then, by the preceding lines, xRx. Hence R is reflexive.

12 REAL NUMBERS

This section may be omitted by readers who are not primarily interested in the construction of $\underset{\sim}{R}$. The remaining part of the text is independent of the present introduction of the reals. We have inserted this topic to show that, indeed, set theory allows us to recover the fundamental structures of analysis.

It is a well-known fact that real numbers can be constructed, once the rationals are given. The construction is slightly more sophisticated than the constructions of $\underset{\sim}{Z}$ from $\underset{\sim}{N}$ or $\underset{\sim}{Q}$ from $\underset{\sim}{Z}$, since essentially transfinite objects must be used.

Traditionally there are three main processes for the introduction of reals, namely a construction via

chains of segments	(Cantor)
Dedekind-cuts	(Dedekind)
Cauchy-sequences	(Cauchy)

The resulting structures are identifiable in an obvious way.

In this section we will sketch the first two processes. In exercise 4 we shall treat Cauchy sequences.

We note that
$$\begin{array}{lcl} 1 & < \sqrt{2} < & 2 \\ 1.4 & < \sqrt{2} < & 1.5 \\ 1.41 & < \sqrt{2} < & 1.42 \\ 1.414 & < \sqrt{2} < & 1.415 \\ 1.4142 & < \sqrt{2} < & 1.4143 \end{array}$$

The numbers in the left-hand and right-hand column, by which $\sqrt{2}$ is bounded, are rational numbers, but $\sqrt{2}$ itself is *not* a rational number (as was first proved by Pythagoras).

We see also that $\sqrt{2}$ is determined by an infinite sequence of pairs of rational numbers, namely by the sequence $\langle 1,2\rangle$, $\langle 1.4, 1.5\rangle$, $\langle 1.41, 1.42\rangle$,.... Hence $\sqrt{2}$ is determined by an infinite process. We say that the sequence

$\langle 1,2 \rangle$, $\langle 1.4, 1.5 \rangle$, ... is a chain of segments in $\mathbb{Q}$.

Definition 12.1 (Cantor): $\langle a_n, b_n \rangle_{n \in \mathbb{N}}$ is a *chain of segments* in $\mathbb{Q}$ iff

1. for all $n \in \mathbb{N}$, $a_n \in \mathbb{Q}$ and $b_n \in \mathbb{Q}$
2. for all $n \in \mathbb{N}$, $a_n \leqslant a_{n+1} \leqslant b_{n+1} \leqslant b_n$
3. for all $n \in \mathbb{N}$, $b_n - a_n \leqslant 2^{-n}$.

$$\begin{matrix} \times & \circ & \cdot & \cdot & \circ & \times \\ a_0 & a_1 & a_2 & b_2 & b_1 & b_0 \end{matrix}$$

For convenience we will write $\langle a_n, b_n \rangle$ instead of $\langle a_n, b_n \rangle_{n \in \mathbb{N}}$.

Remark:

i) The object $\langle a_n, b_n \rangle_{n \in \mathbb{N}}$ is clearly not finite. As a matter of fact it is a special kind of function, defined on $\mathbb{N}$. A proper definition will be given in definition 13.3a.

ii) In 3 the "speed of convergence" of the chain has been fixed. This is not necessary, but some condition on convergence has to be given, and this one happens to be rather simple.

Note that not only $\langle 1,2 \rangle$, $\langle 1.4, 1.5 \rangle$, ... , but also (for example) $\langle 1.41, 1.42 \rangle$, $\langle 1.414, 1.415 \rangle$, ... is a chain of segments in $\mathbb{Q}$. Both chains of segments determine the same real number, namely $\sqrt{2}$. We say that these two chains of segments are equivalent.

Definition 12.2: Let $\langle a_n, b_n \rangle$ and $\langle c_n, d_n \rangle$ be chains of segments in $\mathbb{Q}$. $\langle a_n, b_n \rangle$ is *equivalent* to $\langle c_n, d_n \rangle$ iff for all $k \in \mathbb{N}$, $b_k \geqslant c_k$ and $d_k \geqslant a_k$.

Notation: $\langle a_n, b_n \rangle \sim \langle c_n, d_n \rangle$.

$$\begin{matrix} \circ & \times & \circ & \times \\ a_k & c_k & b_k & d_k \end{matrix} \qquad \qquad \begin{matrix} \times & \circ & \times & \circ \\ c_k & a_k & d_k & b_k \end{matrix}$$

Intuitively, $\langle a_n, b_n \rangle \sim \langle c_n, d_n \rangle$ means that for each $k \in \underset{\sim}{\mathrm{N}}$ the segment $[a_k, b_k]$ meets $[c_k, d_k]$.

We will prove now that $\sim$ is an equivalence relation.

<u>Theorem 12.3:</u> $\sim$ is an equivalence relation on the set of all chains of segments in $\underset{\sim}{\mathrm{Q}}$.

Proof: It is immediately obvious that $\sim$ is symmetric and reflexive. It is a bit more difficult to see that $\sim$ is transitive. Suppose that $\langle a_n, b_n \rangle \sim \langle c_n, d_n \rangle$ $\langle c_n, d_n \rangle \sim \langle e_n, f_n \rangle$ and not $\langle a_n, b_n \rangle \sim \langle e_n, f_n \rangle$, i.e. $\neg \forall n \in \underset{\sim}{\mathrm{N}}\, [b_n \geqslant e_n \wedge f_n \geqslant a_n]$. Then there is some $m \in \underset{\sim}{\mathrm{N}}$ such that $\neg(b_m \geqslant e_m \wedge f_m \geqslant a_m)$, in other words, $b_m < e_m \vee f_m < a_m$. So

$$a_m \leqslant b_m < e_m \leqslant f_m \qquad \text{or} \qquad e_m \leqslant f_m < a_m \leqslant b_m$$

Let us consider the case in which $b_m < e_m$. (0)

The other case, $f_m < a_m$, is treated analogously.

Take $k \in \underset{\sim}{\mathrm{N}}$ such that $2^{-k} < e_m - b_m$ and such that $k \geqslant m$. Now consider $\langle c_k, d_k \rangle$.

Since $k \geqslant m$ we know that $\left.\begin{array}{l} a_m \leqslant a_k \leqslant b_k \leqslant b_m \\ \text{and } e_m \leqslant e_k \leqslant f_k \leqslant f_m . \end{array}\right\}$ (1)

Since $\langle a_n, b_n \rangle \sim \langle c_n, d_n \rangle$ we know that $b_k \geqslant c_k \wedge d_k \geqslant a_k$. (2)

Since $\langle c_n, d_n \rangle \sim \langle e_n, f_n \rangle$ we know that $d_k \geqslant e_k \wedge f_k \geqslant c_k$. (3)

Finally we observe that $d_k - c_k \leqslant 2^{-k} < e_m - b_m$. (4)

(0),..., (4) together yield a contradiction, as can be seen from the following picture.

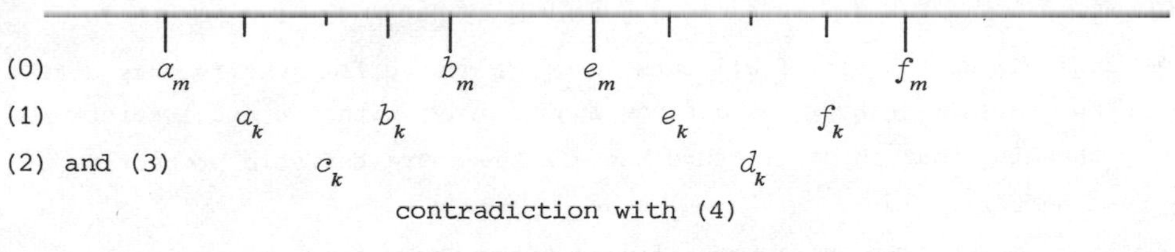

□

To return to our example: we can define $\sqrt{2}$ as the equivalence class of the chain of segments $\langle 1, 2 \rangle$, $\langle 1.4, 1.5 \rangle$, ... If we adopt for a moment the naive point of view that we already have got the continuum (cf. the status of the reals before the nineteenth century), then each real number determines a chain of segments, and exactly one equivalence class of such chains. One can check that in the set of all equivalence classes of chains of segments in $\underset{\sim}{Q}$, operations of "addition" and "multiplication" can be defined (satisfying all the characteristic properties of addition and multiplication on the real numbers). Thus the set of equivalence classes of chains of segments in $\underset{\sim}{Q}$, together with these two operations, has exactly the same (abstract) structure as the set of real numbers with addition and multiplication. This makes it plausible to adopt the standard mathematical practice: we simply forget the naive (intuitive) continuum and we *define* $\underset{\sim}{R}$ as the quotient set of the set of chains of segments in $\underset{\sim}{Q}$ modulo $\sim$, in other words, as the set of equivalence classes of chains of segments in $\underset{\sim}{Q}$.

Because $\underset{\sim}{Q}$ was definable in terms of $\underset{\sim}{Z}$ and hence in terms of $\underset{\sim}{N}$, this means that $\underset{\sim}{R}$ can also (ultimately) be defined in terms of $\underset{\sim}{N}$.

An alternative approach to the construction of the continuum $\underset{\sim}{R}$ links up with its so-called "continuity": if $A, B \subseteq \underset{\sim}{R}$ are not empty, while $A \cup B = \underset{\sim}{R}$ and $\forall a \in A\ \forall b \in B\ [a < b]$, then either A has a greatest or B has a smallest element.

Definition 12.4 (Dedekind): A *cut* in $\underset{\sim}{Q}$ is a subset $A \subseteq \underset{\sim}{Q}$, such that

i) $A \neq \emptyset$ and $A \neq \underset{\sim}{Q}$

ii) if $p \in A$ and $q < p$, then $q \in A$

iii) A does not have a greatest element.

Example: $\{x \in \underset{\sim}{Q} \mid x < 0 \vee x^2 < 2\}$ is a cut in $\underset{\sim}{Q}$.

We define $\underset{\sim}{R}$ as the set of all cuts in $\underset{\sim}{Q}$. We must still justify this definition by providing the set $\underset{\sim}{R}$, defined in this way, with the usual structure and by checking that this structure has all the characteristic properties of the real numbers.

For this purpose we introduce a linear order on $\underset{\sim}{R}$: this is simply the subset relation between cuts. That this yields a linear order on $\underset{\sim}{R}$, is a direct consequence of 12.4 ii) (linear orders are considered in Section 14).

For $p \in \mathbb{Q}$, let $p^* := \{q \in \mathbb{Q} | q < p\}$. Then p^* is clearly a cut in $\mathbb{Q}$ and $p \leqslant q \leftrightarrow p^* \subseteq q^*$. Hence we can identify $\{p^* | p \in \mathbb{Q}\} \subseteq \mathbb{R}$ with $\mathbb{Q}$, because these sets are isomorphic with respect to the order (the notion of "isomorphism" is considered in Section 13).

We will generally identify $\mathbb{Q}$ and $\mathbb{Q}^* = \{p^* | p \in \mathbb{Q}\}$; in other words we treat $\mathbb{Q}$ as a subset of $\mathbb{R}$. (We say that $\mathbb{Q}$ has been *embedded* in $\mathbb{R}$.)

Furthermore, $\mathbb{Q}$ (as a subset of $\mathbb{R}$) is *dense* in $\mathbb{R}$, i.e. between every two real numbers there is a rational number: for suppose $A, B \in \mathbb{R}$ and $A \neq B$, for instance $p \in A - B$, then $B \subset q^* \subset A$ if $p < q$ and $q \in A$.

Finally, $\mathbb{R}$, with this order, has the "continuity" property mentioned above: for if $X, Y \subseteq \mathbb{R}$ are not empty, while $X \cup Y = \mathbb{R}$ and $\forall A \in X \; \forall B \in Y \; [A \subset B]$, and if, further, X does not have a greatest element, then one easily checks that $\cup X$ is the smallest element of Y:

$\cup X$ is clearly a cut, while $\cup X \notin X$ (otherwise $\cup X$ would be the greatest element of X); hence $\cup X$ is an element of Y and hence also the smallest element of Y.

This "continuity" property, together with the properties of having a countable dense subset, and no least or greatest element, characterizes the order on $\mathbb{R}$ (see Chapter III, 3, exercise 4).

Note that, in defining the order on $\mathbb{R}$, we have (only) used the order on $\mathbb{Q}$.

The addition operation for real numbers can now be defined by $A + B := \{p + q | p \in A \wedge q \in B\}$ (essentially because "+" is monotone with respect to $<$: if $p < p'$, then $p + q < p' + q$).

For the multiplication operation some care is necessary. (We have to distinguish cases, according as each number is positive or negative.)

Exercises

1. Prove that for each of the two definitions of $\mathbb{R}$ (Cantor and Dedekind) $\mathbb{R}$ is an ordered field.

2. Prove Cantor's intersection theorem:
   If $\langle a_n, b_n \rangle$ is a chain of segments in $\mathbb{R}$ (see Definition 12.1, but now with $\mathbb{R}$ instead of $\mathbb{Q}$), then there is precisely one $c \in \mathbb{R}$, such that for all $n \in \mathbb{N}$, $a_n \leqslant c \leqslant b_n$.

3. Prove Dedekind's supremum theorem:

If A is a cut in $\underset{\sim}{R}$ (see Definition 12.4, but now with $\underset{\sim}{R}$ instead of $\underset{\sim}{Q}$), then there is precisely one $c \in \underset{\sim}{R}$ such that c = supremum (A), i.e.

i $\forall a \in A\,[a \leqslant c]$ (c is an upper bound of A)

ii $\forall b \in \underset{\sim}{R}\,[b < c \rightarrow \exists a \in A\,[b < a]]$ (c is the least upper bound of A, in other words: if $b < c$, then b is not an upper bound of A).

4. We define a *Cauchy sequence* in $\underset{\sim}{Q}$ as a sequence $\{a_n\}_{n \in \underset{\sim}{N}}$ such that:

1. for all $n \in \underset{\sim}{N}$, $a_n \in \underset{\sim}{Q}$ and
2. $\forall k \in \underset{\sim}{N}\; \exists p \in \underset{\sim}{N}\; \forall n,m \in \underset{\sim}{N}\,[n,m \geqslant p \rightarrow |a_n - a_m| \leqslant 2^{-k}]$.

Now we can identify $\sqrt{2}$, for example, with the Cauchy sequence: 1 , 1.4 , 1.41 , 1.414 , 1.4142, ... and also with the following Cauchy sequence: 1 , 1.2 , 1.3 , 1.4 , 1.41 , 1.414 , 1.4142 , ...

We can define $\underset{\sim}{R}$ as the set of all Cauchy sequences in $\underset{\sim}{Q}$ modulo an equivalence relation. Determine this equivalence relation and then prove Cauchy's theorem:

If $\{a_n\}_{n \in \underset{\sim}{N}}$ is a Cauchy sequence in $\underset{\sim}{R}$, then there is precisely one $c \in \underset{\sim}{R}$ such that $\lim\limits_{n \to \infty} a_n = c$.

13 FUNCTIONS (MAPPINGS)

Let X and Y be sets. "f is a *function* (mapping) from X to Y" means intuitively: f assigns to each $x \in X$ a uniquely determined $y \in Y$.

Notation: $f : X \rightarrow Y$.

For each $x \in X$, the uniquely determined $y \in Y$, which is assigned by f to x, is called the image (under f) of x.

Notation: $y = f(x)$.

Examples of functions $f{:}X \rightarrow Y$.

1. $X = \{1,2,3\}$, $Y = \{4,5,6,7\}$, $f(1) = 4$
$f(2) = 4$
$f(3) = 6$

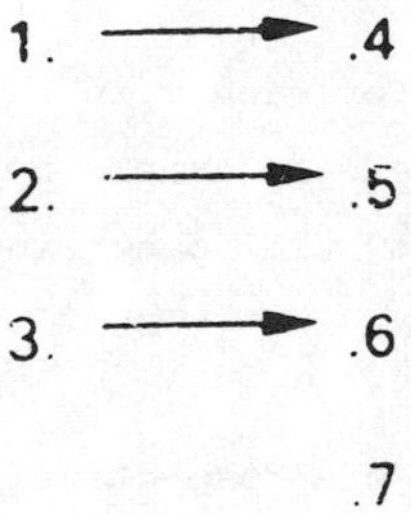

2. $X = \{1,2,3\}$, $Y = \{4,5,6,7\}$, $f(1) = 4$
$f(2) = 5$
$f(3) = 6$

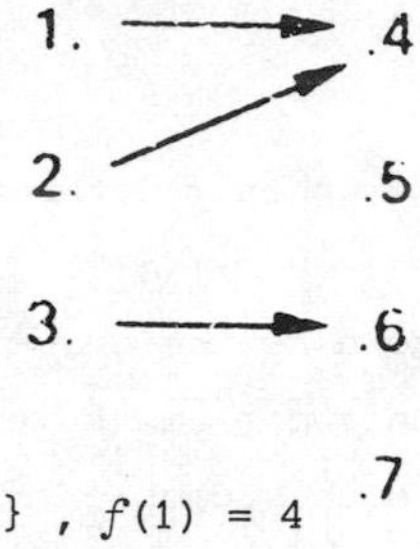

3. $X = \{1,2,3\}$, $Y = \{4,5\}$, $f(1) = 4$
$f(2) = 4$
$f(3) = 5$

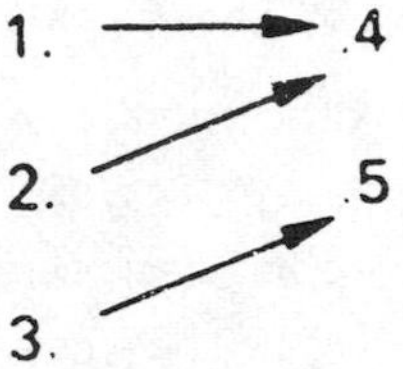

4. $X = \{1,2,3\}$, $Y = \{4,5,6\}$, $f(1) = 5$
$f(2) = 4$
$f(3) = 6$

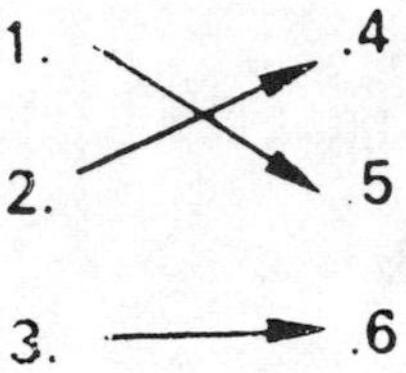

5. $X = \mathbb{N}$, $Y = \mathbb{N}$, $f(x) = 0$ if x is even
$f(x) = 1$ if x is odd.

6. $X = \mathbb{N}$, $Y = \mathcal{P}(\mathbb{N})$, $f(x) = \{x\}$.

7. $X = \mathbb{N}^2$, $Y = \mathbb{Z}$, $f(\langle n,m\rangle) = n-m$.

8. $X = \mathbb{R}_+$ with $\mathbb{R}_+ := \{x \in \mathbb{R} \mid x > 0\}$, $Y = \mathbb{R}$, $f(x) = \log(x)$ (see Figure 17).

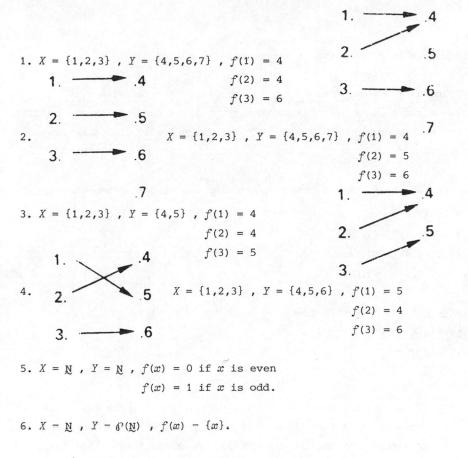

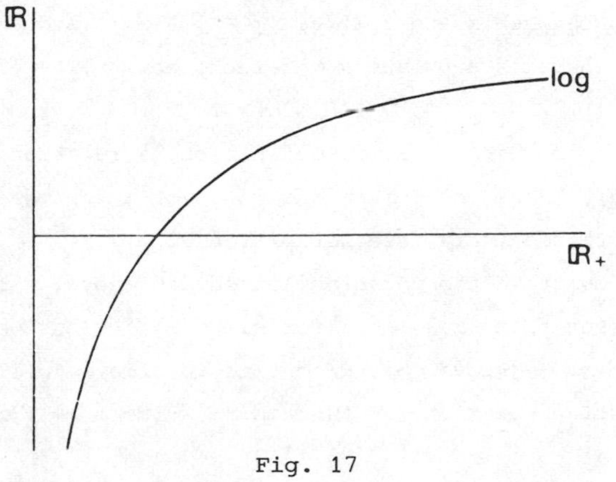

Fig. 17

If $f:X \to Y$, then f determines a set of ordered pairs, namely $\{\langle x,y\rangle \in X \times Y \mid y = f(x)\}$.
This set, known as the *graph* of f, has the property that for each x in X there is a unique element y in Y such that $\langle x,y\rangle$ is in the set (namely $y=f(x)$). Conversely, each subset of $X \times Y$ with this special property will determine a function $f : X \to Y$.

The graphs of the functions from the examples 1,..., 8 are respectively:

1. $\{\langle 1,4\rangle , \langle 2,4\rangle , \langle 3,6\rangle\}$
2. $\{\langle 1,4\rangle , \langle 2,5\rangle , \langle 3,6\rangle\}$
3. $\{\langle 1,4\rangle , \langle 2,4\rangle , \langle 3,5\rangle\}$
4. $\{\langle 1,5\rangle , \langle 2,4\rangle , \langle 3,6\rangle\}$

5. $\{\langle x,y\rangle \in \mathbb{N}^2 \mid (x \text{ is even} \wedge y = 0) \vee (x \text{ is odd} \wedge y = 1)\}$
6. $\{\langle x,y\rangle \in \mathbb{N} \times \mathcal{P}(\mathbb{N}) \mid y = \{x\}\}$
7. $\{\langle\langle n,m\rangle ,y\rangle \in \mathbb{N}^2 \times \mathbb{Z} \mid y = n-m\}$
8. $\{\langle x,y\rangle \in \mathbb{R}_+ \times \mathbb{R} \mid y = \log x\}$

Any function can thus be represented by a certain set, namely its graph. In fact, it is common in set theory to identify a function with its graph and thus reduce the notion of "function" to the notion of "set". This is what we will do.

Definition 13.1: f is a *function* from X to Y iff
1. f is a relation between X and Y (i.e., $f \subseteq X \times Y$), and
2. for each $x \in X$ there is a unique $y \in Y$ such that $\langle x,y\rangle \in f$.

So a function f from X to Y is, by definition, a relation between X and Y (of a special kind), hence the domain and range of f can be defined simply as Dom(f) and Ran(f) respectively (see Definition 10.2).

We shall still maintain the notation introduced above, namely "$f(x)$" for the unique $y \in Y$ such that $\langle x,y\rangle \in f$ (for all $x \in X$). Thus we have, for all $x \in X$, $y \in Y$: $y = f(x) \leftrightarrow \langle x,y\rangle \in f$. From time to time we will use a convenient and suggestive notation for functions: write $x \mapsto f(x)$ for $\langle x,f(x)\rangle \in f$.

Remark: This is not the only reasonable approach to the notion of "function". (In fact, in category theory the order of things is reversed: the notion of "function" is one of the primitive notions, and the notion of "set" is defined in terms of these!) However, the present approach is very convenient in the context of set theory.

Check that for $f: X \to Y$, Dom $(f) = X$ and Ran $(f) = \{y \in Y \mid \exists x \in X\,[y=f(x)]\}$.

Example: for the function f of example 1, Ran $(f) = \{4,6\}$; in example 2, Ran $(f) = \{4,5,6\}$.

Notation: If $f: X \to Y$ and $X' \subseteq X$, then "$f(X')$" and "$f''X'$ " are often used to denote the set $\{f(x) \mid x \in X'\}$, i.e., the set $\{y \in Y \mid \exists x \in X'[\,y = f(x)]\}$.

The notation $f(X')$ is ambiguous; for example, a subset of X can at the same time be an element ($X = \{0,1\ ,\{0\}\}$ and $f(0) = f(1) = f(\{0\}) = 0$). Nevertheless "$f(X')$" and "$f''X'$" will both be used if no confusion results.

If $f: X \to Y$ and $Y' \subseteq Y$, then " $\overset{-1}{f}(Y')$" is used to denote the set $\{x \in X \mid f(x) \in Y'\}$, i.e., the set $\{x \in X \mid \exists y \in Y'[\,f(x) = y]\}$.

Remark: Let Y be any set. Then $\emptyset \subseteq \emptyset \times Y$.
Further, because $\emptyset$ has no elements, it follows trivially that to each $x \in \emptyset$ there is a unique $y \in Y$ such that $\langle x,y\rangle \in \emptyset$. Hence, by Definition 13.1, $\emptyset$ is a function from $\emptyset$ to Y. Further, $\emptyset$ is the only function from $\emptyset$ to Y (since $\emptyset$ is the only relation with Dom $(\emptyset) = \emptyset$).

N.B. We cannot interchange $\emptyset$ and Y since for non-empty Y, dom $(\emptyset) \neq Y$ and by definition a function on Y has domain Y.(cf. exercise 30).

The set of all functions $f: X \to Y$

Note, if $f: X \to Y$, then $f \subseteq X \times Y$; hence $f \in \mathcal{P}(X \times Y)$.

<u>Definition 13.2</u>: Y^X is the set of all functions $f: X \to Y$.
In symbols, $Y^X := \{f \in \mathcal{P}(X \times Y) \mid f: X \to Y\}$.

Example: The set $\{1,2,3\}^{\{5,6\}}$ has $3^2 = 9$ elements $f_1, \ldots, f_9$, the functions $f_1, \ldots, f_9$ being defined by the following scheme:

	f_1	f_2	f_3	f_4	f_5	f_6	f_7	f_8	f_9
5	1	1	1	2	2	2	3	3	3
6	1	2	3	1	2	3	1	2	3

i.e. $f_1(5) = 1$ $f_2(5) = 1$ $f_7(5) = 3$
$f_1(6) = 1$ $f_2(6) = 2$ $f_7(6) = 1$ and so on.

Check that $\{5,6\}^{\{1,2,3\}}$ has $2^3 = 8$ elements.

Theorem 13.3: If Y is a set with m elements and X is a set with n elements ($m,n \in \underset{\sim}{N}$), then Y^X has m^n elements.

Remark: If Y is a set with 10 elements and X has 6 elements, then there are, by this theorem, one million functions $f: X \to Y$.

Proof of Theorem 13.3: Throughout the following argument, let $m \in \underset{\sim}{N}$ be fixed, and let Y be a fixed set with m elements.

Now let Φ be the following property of natural numbers: "if X is any set with n elements, Y^X has m^n elements".

We can re-formulate Theorem 13.3 as follows: $\forall n \in \underset{\sim}{N}\, [\Phi(n)]$.

Now it is sufficient to show:

1. $\Phi(0)$ and
2. $\forall k \in \underset{\sim}{N}\, [\Phi(k) \to \Phi(k+1)]$.

Proof of $\Phi(0)$: if X has 0 elements, i.e. $X = \emptyset$, then $\emptyset$ is the only function from X to Y; hence $Y^X = \{\emptyset\}$; so $Y^\emptyset$ has $m^0 = 1$ element.

Proof of $\forall k \in \underset{\sim}{N}\, [\Phi(k) \to \Phi(k+1)]$: Suppose $\Phi(k)$, i.e. if X is any set with k elements, then Y^X has m^k elements. We have to show now that $\Phi(k+1)$ holds. So let $\{x_1, \ldots, x_k, x_{k+1}\}$ be a set with $k+1$ elements. By the induction hypothesis $\Phi(k)$ there are m^k different functions from $\{x_1, \ldots, x_k\}$ to Y.

	f_1	f_2	...	f_{m^k}
x_1	*	*		*
x_2	*	*		*
⋮				
x_k	*	*		*
x_{k+1}				

Now for each i, $1 \leqslant i \leqslant m^k$, there are m different possible choices for $f_i(x_{k+1})$. Thus there are altogether $m.m^k = m^{k+1}$ different functions from $\{x_1, \ldots, x_k, x_{k+1}\}$ to Y. □

In mathematics (especially analysis) one frequently uses sequences of objects. We can now give an exact formulation of the notion of sequence.

Definition 13.3a: A *sequence* of elements of X is a function from $\underset{\sim}{N}$ to X (in other words, an element of $X^{\underset{\sim}{N}}$).

With this definition the missing link in the chain of notions required for the introduction of the real numbers (see Definition 12.1) has been supplied.

Injective, surjective and bijective functions.

The functions $f: X \rightarrow Y$ in examples 2,4,6 and 8 (at the beginning of this section) have the property that they assign distinct elements of Y to distinct elements of X; in other words: for all $x, x' \in X$, if $x \neq x'$, then $f(x) \neq f(x')$, or (equivalently): for all $x, x' \in X$, if $f(x) = f(x')$, then $x = x'$. We call such functions injective (one-to-one).

Definition 13.4: $f: X \rightarrow Y$ is *injective* (or an *injection*) :=
$\forall x \in X \; \forall x' \in X \, [x \neq x' \rightarrow f(x) \neq f(x')]$.

Notation: We write "$f: X \underset{1}{\leqslant} Y$" to indicate that $f: X \rightarrow Y$ is injective.

The functions $f: X \rightarrow Y$ in examples 3,4,7 and 8 have the property that each element $y \in Y$ is the image (under f) of an element $x \in X$; in symbols: $\forall y \in Y \; \exists x \in X \, [y = f(x)]$. We call such functions surjective (onto).

<u>Definition 13.5</u>: $f\colon X \to Y$ is *surjective* (or a *surjection*) := $\forall y \in Y\ \exists x \in X\ [y = f(x)]$.

Note that $f\colon X \to Y$ is surjective if and only if $\mathrm{Ran}(f) = Y$.

The functions in examples 1 and 5 are neither injective nor surjective. The functions in examples 4 and 8 have both properties. We call such functions bijective.

<u>Definition 13.6</u>: $f\colon X \to Y$ is *bijective* (or a *bijection*) if it is both injective and surjective.

Notation: One writes "$f\colon X \underset{1}{=} Y$" to indicate that $f\colon X \to Y$ is bijective.

A bijection $f\colon X \to Y$ gives a one-one correspondence between the elements of X and the elements of Y: for each $x \in X$ there is exactly one (f is a function) $y \in Y$ such that $y = f(x)$ and for each $y \in Y$ there is at least one (f is surjective) and precisely one (f is injective) $x \in X$ such that $y = f(x)$.

Some special functions

<u>Definition 13.7</u>: For $X \subseteq Y$ let $id_X\colon X \to Y$ be defined by $id_X(x) = x$.

id_X is called the *identity function* on X.

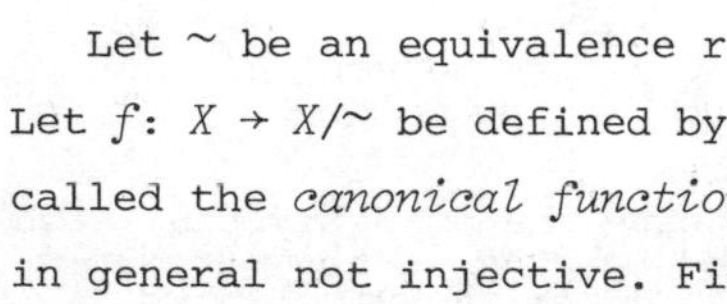

$id_X\colon X \to Y$ is of course injective, but surjective only if $X = Y$.

Let $\sim$ be an equivalence relation on X.

Let $f\colon X \to X/\!\sim$ be defined by $f(x) := [x]_\sim$. This function f is sometimes called the *canonical function* from X to $X/\!\sim$. It is of course surjective, but in general not injective. Find a condition (on $\sim$) which makes f injective.

Definition 13.8: For $U \subseteq V$ let $K_U: V \to \{0,1\}$ be defined by

$K_U(v) = 1$ if $v \in U$

$\quad\quad = 0$ if $v \notin U$.

Note that $K_U \in \{0,1\}^V$ (see Figure 18).

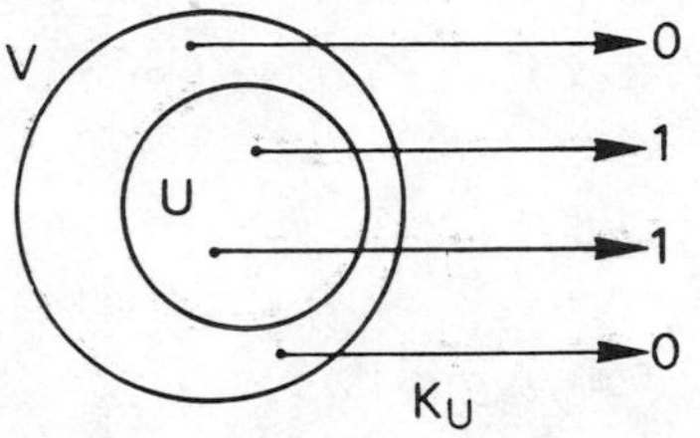

Fig. 18

$K_U: V \to \{0,1\}$ is called the *characteristic function* of U.

If $f: \mathbb{N} \to \{0,1\}$, then we can represent f schematically. Example:

n	0	1	2	3	4	5	6	...
$f(n)$	0	0	1	0	1	1	0	....

f is (by Definition 13.3a) a sequence. We can represent such a sequence by leaving out the top line in the above schema, thus:

0 0 1 0 1 1 0 ... ,

obtaining an infinite sequence of 0's and 1's.

Remark: In general, a sequence (as defined in 13.3a) can, in this way, be represented as an *infinite sequence* (in the ordinary sense). This is the motivation for Definition 13.3a.

For $U \subseteq \mathbb{N}$, $K_U: \mathbb{N} \to \{0,1\}$ and hence we can, if $U \subseteq \mathbb{N}$, represent K_U by an infinite sequence of 0's and 1's. For example, the sequence 1 0 1 1 0 ... represents the characteristic function of $\{0,2,3, \ldots\}$.

Equality of functions

Since we have defined functions as certain sets (of ordered pairs) the equality relation between functions is thereby determined. Let $f: X \to Y$ and $g: X \to Y$. Then, by the axiom of extensionality:

$f = g \leftrightarrow \forall x \in X\ \forall y \in Y\ [\langle x,y\rangle \in f \leftrightarrow \langle x,y\rangle \in g]$.

Hence, $f = g \leftrightarrow \forall x \in X\ \forall y \in Y\ [y = f(x) \leftrightarrow y = g(x)]$.

So;

$$f = g \leftrightarrow \forall x \in X\ [f(x) = g(x)].$$

Note:

$$f \neq g \leftrightarrow \exists x \in X\ [f(x) \neq g(x)].$$

Theorem 13.9: Let $K: \mathcal{P}(V) \to \{0,1\}^V$ be defined by $K(U) := K_U$. Then
i) K is injective, and
ii) K is surjective.

Proof of i): Suppose $U_1 \neq U_2$, i.e. there is some $v \in V$ such that $(v \in U_1 \wedge v \notin U_2) \vee (v \in U_2 \wedge v \notin U_1)$. But then $(K_{U_1}(v) = 1 \wedge K_{U_2}(v) = 0) \vee (K_{U_2}(v) = 1 \wedge K_{U_1}(v) = 0)$. So $\exists v \in V\ [K_{U_1}(v) \neq K_{U_2}(v)]$, and hence $K_{U_1} \neq K_{U_2}$.

Proof of ii): Suppose $f \in \{0,1\}^V$.

Let $U_f := \{v \in V \mid f(v) = 1\}$. Then for all $v \in V$, $K_{U_f}(v) = 1 \leftrightarrow v \in U_f$
$\leftrightarrow f(v) = 1.$

Hence for all $v \in V$, $K_{U_f}(v) = f(v)$. Therefore $K_{U_f} = f$. □

Composition of two functions; Inverse function; the restriction of a function.

Definition 13.10: Let $f: A \to B$ and $g: B \to C$.
The *composition* $g \circ f: A \to C$ is the function, defined by $g \circ f\ (x) := g(f(x))$. (see Figure 19).

Remark: This is in agreement with the definition (10.4) of the composition of two relations (since we are viewing functions are relations of a certain type).

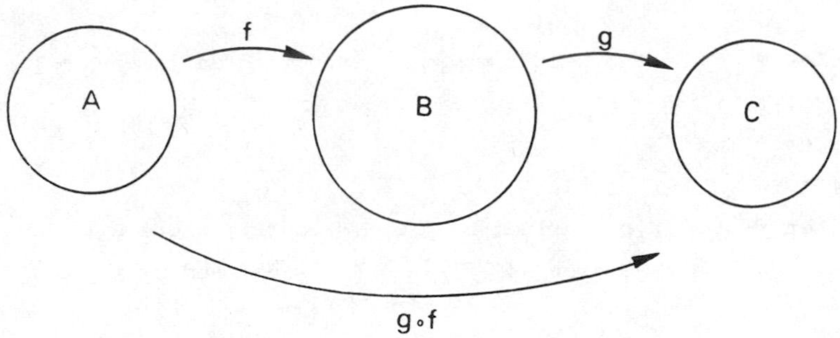

Fig. 19

Because in the composition of f and g first f and then g is applied, one might expect "$f \circ g$" as notation. However, we prefer the order to be determined by the fact that in $g(f(x))$, g precedes f.

Warning: Some authors (especially in algebra) denote what we call "$f(x)$" by "$(x)f$". In such a notation, "$g(f(x))$" is written as "$((x)f)g$", and then the composition of f and g (which we denote by "$g \circ f$") would be denoted by "$f \circ g$".

Example: $A = \underset{\sim}{N}$, $B = \underset{\sim}{Z}$ and $C = \underset{\sim}{Q}$.

Let f: $\underset{\sim}{N} \to \underset{\sim}{Z}$ be defined by $f(n) := -n$.

Let g: $\underset{\sim}{Z} \to \underset{\sim}{Q}$ be defined by $g(m) := \frac{1}{2}m$.

Then $g \circ f$: $\underset{\sim}{N} \to \underset{\sim}{Q}$ is defined by $g \circ f(n) = -\frac{1}{2}n$.

If $f{:}X \to Y$ is a bijection, then there is -because f is surjective- for each $y \in Y$ at least one $x \in X$ such that $y = f(x)$, and -because f is injective- there is for each $y \in Y$ at most one $x \in X$ such that $y = f(x)$. Hence, if $f{:}X \to Y$ is a bijection, then for each $y \in Y$ there is precisely one $x \in X$ such that $y = f(x)$.

Definition 13.11: Let $f{:}X \to Y$ be a bijection.
f^{-1}: $Y \to X$ is the function, defined by $f^{-1}(y)$:= the unique element x in X such that $y = f(x)$. f^{-1} is called the *inverse* function of f.

Note that $f^{-1} = \check{f}$ (see Definition 10.3).
If $f{:}X \to Y$ is a bijection, then $f^{-1}\ f$ is the identity function on X and $f\ f^{-1}$ is the identity function on Y.

Example: Let $\underset{\sim}{N}_{even}$ be the set $\{x \in \underset{\sim}{N} | x \text{ is even}\}$ and define $f:\underset{\sim}{N} \to \underset{\sim}{N}_{even}$ by $f(n) := 2n$. Then $f:\underset{\sim}{N} \to \underset{\sim}{N}_{even}$ is a bijection and $f^{-1} : \underset{\sim}{N}_{even} \to \underset{\sim}{N}$ is defined by $f^{-1}(m) := \frac{1}{2}m$.

Example: Let $f: \underset{\sim}{R}_+ \to \underset{\sim}{R}$ be defined by $f(x) := \log(x)$, where $\underset{\sim}{R}_+ := \{x \in \underset{\sim}{R} | x > 0\}$. Then $f: \underset{\sim}{R}_+ \to \underset{\sim}{R}$ is a bijection and $f^{-1}: \underset{\sim}{R} \to \underset{\sim}{R}_+$ is defined by $f^{-1}(x) := e^x$.

<u>Definition 13.12</u>: Let $f:X \to Y$ and $X_0 \subseteq X$.
$f \restriction X_0: X_0 \to Y$ is defined by $f \restriction X_0 \ (x) := f(x)$.
$f \restriction X_0$ is called the *restriction* of f to X_0 (see Figure 20).

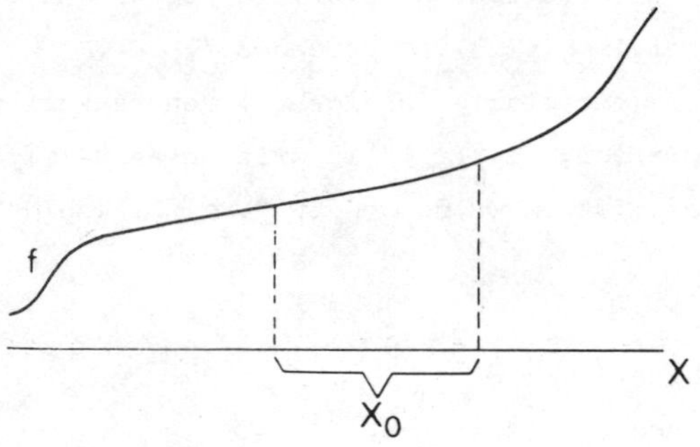

Fig. 20

Example: Let $f: \underset{\sim}{R} \to \underset{\sim}{R}$ be defined by $f(x) := \sin \pi x$.
$f \restriction \underset{\sim}{Z}: \underset{\sim}{Z} \to \underset{\sim}{R}$ is defined by $f \restriction \underset{\sim}{Z} \ (m) = \sin \pi m = 0$.

<u>Theorem 13.13</u>: Let U, V and W be sets.

i) There is a bijection from V to V.

ii) If there is a bijection from U to V, then there is also a bijection from V to U.

iii) If there are bijections from U to V and from V to W, then there is also a bijection from U to W.

Proof: i) $id_V : V \to V$ is a bijection (see 13.7).

ii) If $f: U \to V$ is a bijection, then $f^{-1}: V \to U$ is a bijection (see 13.11).

iii) If $f: U \to V$ and $g : V \to W$ are bijections, then $g \quad f: U \to W$ is a bijection (see 13.10). □

Structures and Isomorphisms

In algebra one is not interested in how the elements of a given set have been constructed, only in how they behave under certain given operations and relations on the set.

Thus, for example, the group of integers under addition does not differ algebraically from the additive group of even integers. Likewise there is no difference, as far as order properties are concerned, between the set of natural numbers and the set of prime numbers. More precisely: the elements of such a pair of sets can be made to correspond with each other in such a way that the operation of addition and the order relation respectively are preserved.

This brings us to the notions of structure and of isomorphism.

<u>Definition 13.14</u>: $\langle X,R_0,\dots,R_p\rangle$ is a (*relational*) *structure* iff X is a set and $R_0,\dots,R_p$ are relations on X.

Remark: A more general notion of structure is obtained by considering sets together with certain relations and operations on them (see Chapter III, Section 2).

Examples of (relational) structures:

1. $\langle A,\in\rangle$, A an arbitrary set.
2. $\langle \mathcal{P}(V), \subseteq\rangle$, V an arbitrary set.
3. $\langle \underset{\sim}{N}, \leqslant\rangle$, "$\leqslant$" being the "less than or equal" relation on $\underset{\sim}{N}$.
4. $\langle \underset{\sim}{N}, \leqslant, /\rangle$, where a/b iff a is a divisor of b.
5. $\langle \underset{\sim}{N}_{even}, \leqslant, /\rangle$, $\underset{\sim}{N}_{even}$ being the set of all even natural numbers.
6. $\langle \underset{\sim}{Z}, \leqslant\rangle$, where "$\leqslant$" is now the "less than or equal" relation on $\underset{\sim}{Z}$.
7. $\langle M$, is a brother of $\rangle$, M being the set of all men.

<u>Definition 13.15</u>: Let $\langle X,R_0,\dots,R_p\rangle$ and $\langle Y,S_0,\dots,S_p\rangle$ be two (relational) structures, such that for each i=0,...p R_i and S_i are both (n_i+1)-ary relations. Let $f\colon X\to Y$. f is an *isomorphism* from $\langle X,R_0,\dots,R_p\rangle$ to $\langle Y,S_0,\dots,S_p\rangle$ iff

1. f is a bijection from X to Y, and
2. for all $i=0,\ldots,p$ and for all $x_0,\ldots,x_{n_i} \in X$,

   $R_i\ (x_0,\ldots,x_{n_i}) \leftrightarrow S_i\ (f(x_0),\ldots, f(x_{n_i}))$.

Examples:

i) $f\colon \underset{\sim}{N} \to \underset{\sim}{N}_{even}$, defined by $f(n) = 2n$, is an isomorphism from $\langle \underset{\sim}{N},< \rangle$ to $\langle \underset{\sim}{N}_{even},< \rangle$ and likewise an isomorphism from $\langle \underset{\sim}{N},<,/ \rangle$ to $\langle \underset{\sim}{N}_{even},<,/ \rangle$, where "/" is the "divisibility" relation (see Figure 21).

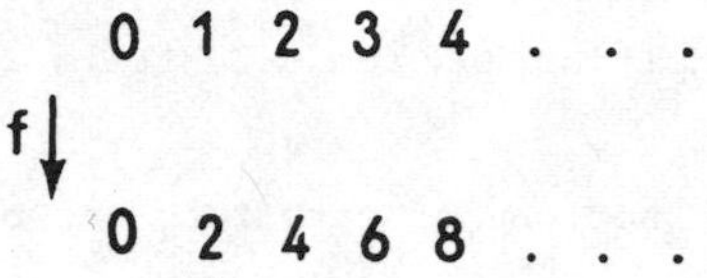

Fig. 21

ii) $f\colon \underset{\sim}{N} \to \underset{\sim}{Z}$, defined by $f(2n) = n$ and $f(2n-1)= -n$, is a bijection from $\underset{\sim}{N}$ to $\underset{\sim}{Z}$, but it is not an isomorphism from $\langle \underset{\sim}{N},< \rangle$ to $\langle \underset{\sim}{Z},< \rangle$ (see Figure 22).

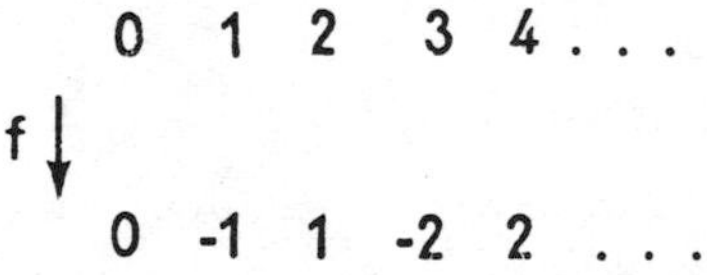

Fig. 22

Notation: If $\langle X,R_0,\ldots,R_p \rangle$ is mapped isomorphically by f to $\langle Y,S_0,\ldots,S_p \rangle$, we write

$f\colon \langle X,R_0,\ldots,R_p \rangle \simeq \langle Y,S_0,\ldots,S_p \rangle$.

If it is clear from the context what relations are involved, we may write more briefly

$$f\colon X \simeq Y$$

Definition 13.16: $\langle X,R_0,\ldots,R_p\rangle$ is *isomorphic* to $\langle Y,S_0,\ldots,S_p\rangle$ iff there is an isomorphism f from $\langle X,R_0,\ldots,R_p\rangle$ to $\langle Y,S_0,\ldots,S_p\rangle$.

Notation: $\langle X,R_0,\ldots,R_p\rangle \simeq \langle Y,S_0,\ldots,S_p\rangle$ or more briefly $X \simeq Y$, if it is clear from the context what relations are involved.

Example: f: $\underset{\sim}{N} \to \underset{\sim}{N}_{even}$, defined by $f(0)=2$, $f(1)=0$, $f(n)=2n$ for $n \geqslant 2$, is a bijection from $\underset{\sim}{N}$ to $\underset{\sim}{N}_{even}$, but it is not an isomorphism from $\langle \underset{\sim}{N}, <\rangle$ to $\langle \underset{\sim}{N}_{even}, <\rangle$ (see Figure 23).

$$\begin{array}{c} 0\ 1\ 2\ 3\ 4\ \ldots \\ f\downarrow \\ 2\ 0\ 4\ 6\ 8\ \ldots \end{array}$$

Fig. 23

However, f: $\underset{\sim}{N} \to \underset{\sim}{N}_{even}$, defined by $f(n) = 2n$, is an isomorphism from $\langle \underset{\sim}{N}, <\rangle$ to $\langle \underset{\sim}{N}_{even}, <\rangle$. Hence, $\langle \underset{\sim}{N}, <\rangle \simeq \langle \underset{\sim}{N}_{even}, <\rangle$ (see Figure 24).

$$\begin{array}{c} 0\ 1\ 2\ 3\ 4\ \ldots \\ f\downarrow \\ 0\ 2\ 4\ 6\ 8\ \ldots \end{array}$$

Fig. 24

Theorem 13.17: Let $\mathbf{A}$ be a relational structure $\langle X,R_0,\ldots,R_p\rangle$, $\mathbf{B}$ a relational structure $\langle Y,S_0,\ldots,S_p\rangle$ and $\mathbf{C}$ a relational structure $\langle Z,T_0,\ldots,T_p\rangle$, such that for each $i=0,\ldots,p$ R_i, S_i and T_i have the same number of arguments.

Then i) $\mathbf{A} \simeq \mathbf{A}$

ii) $\mathbf{A} \simeq \mathbf{B} \to \mathbf{B} \simeq \mathbf{A}$

iii) $(\mathbf{A} \simeq \mathbf{B} \wedge \mathbf{B} \simeq \mathbf{C}) \to \mathbf{A} \simeq \mathbf{C}$.

The proof is left as an exercise to the reader.

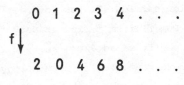

At this point we insert a word on the usage of terms in mathematical practice. As underlined at the beginning of this section, a function is thought of as a device that maps elements of a set on elements of another set. For example, $\exp : \mathbb{R} \to \mathbb{R}$, where exp is defined by $\exp(x) = e^x$. However, one could just as well write $\exp: \mathbb{R} \to \mathbb{R}_+$ (where $\mathbb{R}_+ = \{r \in \mathbb{R} | r > 0\}$).

Have we dealt with two different functions? By Definition 13.1 there was just one function exp (namely the graph). Nonetheless, one tends to distinguish somehow between the cases. This practice is codified in category theory, where mappings (morphisms) f are always presented together with X and Y, such that $f: X \to Y$, while Y may be distinct from $\text{Ran}(f)$ in our sense.

In set theory a function carries all its information, in particular one can construct $\text{Ran}(f)$ and $\text{Dom}(f)$ from f; in category theory (and in informal mathematics) one has to specify as an extra the set Y, called the codomain of f. As a result there is some confusion in terminology. The only point which might cause trouble concerns isomorphisms; by definition an isomorphism is a bijection, however, one sometimes wants to consider isomorphisms from $\mathbf{A}$ to a substructure of $\mathbf{B}$; on those occasions we will speak about "an isomorphism from $\mathbf{A}$ into (or in) $\mathbf{B}$", or "an isomorphic embedding from $\mathbf{A}$ in $\mathbf{B}$".

The cartesian product of a family of sets.

Now we can generalize the notion of "cartesian product", which we have defined in 9.1 and 9.5 for a finite number of sets only, to an arbitrary family of sets.

Definition 13.18: Let $A = \{A_i | i \in I\}$ be a family of sets.

$$\mathop{\times}_{i \in I} A_i := \{f: I \to \bigcup_{i \in I} A_i \,|\, \forall i \in I\, [f(i) \in A_i]\}$$

Notation: instead of $\mathop{\times}_{i \in I} A_i$ one also writes $\times A$.

We easily verify, that in the special case that $I = \{0,1\}$, $\mathop{\times}_{i \in \{0,1\}} A_i$ corresponds to $A_0 \times A_1$; more precisely:

Theorem 13.19: The function $g: \mathop{\times}_{i \in \{0,1\}} A_i \to A_0 \times A_1$, defined by $g(f) = \langle f(0), f(1) \rangle$, is a bijection from $\mathop{\times}_{i \in \{0,1\}} A_i$ to $A_0 \times A_1$.

The proof is left as an exercise to the reader.

Note that in the case that $A_0 = A_1 = A$, $\underset{i \in \{0, 1\}}{\times} A_i$ is the set $A^{\{0, 1\}}$ of all functions from $\{0,1\}$ to A. Check also: if $A_i = B$ for all $i \in I$, then $\underset{i \in I}{\times} A_i = B^I$.

Exercises

1. Examine, which of the following sets are relations, functions, injections, surjections or bijections from $\{1,2,3,4\}$ to $\{1,2,3,4\}$:

a) $R_1 = \{< 3,1> , < 4,2> , < 4,3 > , < 2,3 >\}$

b) $R_2 = \{< 2,3> , < 1,2> , < 3,2 > , < 4,3 >\}$

c) $R_3 = \{< 2,1> , < 1,2> , < 4,3 > , < 3,4 >\}$

d) $R_1 \circ R_2$

e) R_3

2. Let $f: A \to B$ and $g: B \to C$. Prove:

   a) if $g \circ f$ is injective, then f is injective.

   b) if $g \circ f$ is surjective, then g is surjective.

Let $f^*: \mathbb{N} \to \mathbb{N}$ be defined by $f^*(n) = n+1$ and let $g^*: \mathbb{N} \to \mathbb{N}$ be defined by $g^*(0) = 0$ and $g^*(n+1) = n$ (see figure 25).

Prove, using f^* and g^*, that not for all f and g:

   c) if $g \circ f$ is injective, then g is injective.

   d) if $g \circ f$ is surjective, then f is surjective.

   e) if $g \circ f$ is bijective, then f or g is bijective.

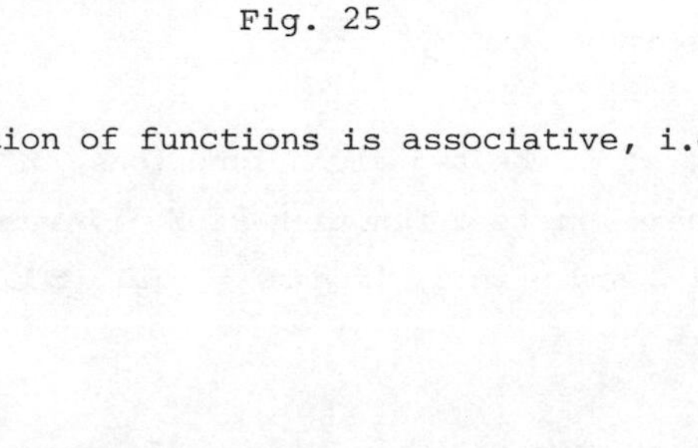

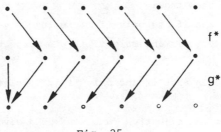

Fig. 25

3. Show that composition of functions is associative, i.e. $h \circ (g \circ f) = (h \circ g) \circ f$.

4. Given a function $f\colon X \to Y$ and sets $A,B \subseteq X$ and $C,D \subseteq Y$, prove that
   a) $f(A \cup B) = f(A) \cup f(B)$
   b) $f(A \cap B) \subseteq f(A) \cap f(B)$
   c) $\overset{-1}{f}(C \cup D) = \overset{-1}{f}(C) \cup \overset{-1}{f}(D)$
   d) $\overset{-1}{f}(C \cap D) = \overset{-1}{f}(C) \cap \overset{-1}{f}(D)$
   e) $\overset{-1}{f}(Y - C) = \overset{-1}{f}(Y) - \overset{-1}{f}(C)$

   In the case of b), show that proper inclusion can occur.

5. Given a function $f\colon X \to Y$ and sets A,B such that $A \subseteq X$ and $B \subseteq Y$, show that $\overset{-1}{f}(f(A)) \supseteq A$ and $f(\overset{-1}{f}(B)) \subseteq B$.

6. Show that, for a function $f\colon X \to Y$
   a) f is injective if and only if $f\ (A \cap B) = f(A) \cap f(B)$ for all $A,B \subseteq X$.
   b) f is surjective if and only if for all $A \subseteq X$, $f(X - A) \supseteq Y - f(A)$.
   c) f is bijective if and only if for all $A \subseteq X$, $f\ (X - A) = Y - f(A)$.

7. Prove that $f\colon \mathcal{P}(V) \to \mathcal{P}(V)$, defined by $f(A) = A^c$, is a bijection.

8. If $f\colon X \to Y$, $g\colon Y \to X$ and $h\colon Y \to X$ are functions such that $g \circ f = id_X$ and $f \circ h = id_Y$, then we say that g is a *left-inverse* of f and h is a *right-inverse* of f. Show that:
   a) f has a left-inverse if and only if it is injective.
   b) f has a right-inverse if and only if it is surjective.
      (cf. II 9, AC 3).
   c) f has a two-sided inverse if and only if it is bijective.

9. Show that the set of all bijective functions of a given set into itself is a group with respect to the operation $\circ$.

10. Let $f\colon E \to G$ and $g\colon E \to F$ be two given functions, of which g is surjective. Show that there exists a function $h\colon F \to G$ satisfying $f = h \circ g$ if and only if for all x and y in E, if $g(x) = g(y)$, then $f(x) = f(y)$.

11. Given three functions $f\colon A \to B$, $g\colon B \to C$ and $h\colon C \to D$, show that if $g \circ f$ and $h \circ g$ are both bijective, then f,g and h are bijective.

12. Let A be a set and $f = \{<x, <x,x>> \mid x \in A\}$. Prove that f is a bijection from A to I_A.

13. Let $f\colon A \to B$ and $g\colon A \to B$. Prove: if $f \subseteq g$, then $f = g$.

14. Let $f\colon A \to B$ and $g\colon C \to D$. Define the product of f and g as follows: $(f.g)(<x,y>) = <f(x),g(y)>$ for each $<x,y> \in A \times C$. Prove that $f.g$ is a function from $A \times C$ to $B \times D$.
Prove: if f and g are injective, then $f.g$ is injective, and, if f and g are surjective, then $f.g$ is surjective.
Prove that $\mathrm{Ran}(f.g) = \mathrm{Ran}(f) \times \mathrm{Ran}(g)$.

15. Prove: if $f\colon B \cup C \to A$, then $f = f \upharpoonright B \cup f \upharpoonright C$.

16. Let $f_1 : B \to A$, $f_2 : C \to A$ and $B \cap C = \emptyset$.
If $f = f_1 \cup f_2$, then
i) $f\colon B \cup C \to A$
ii) $f_1 = f \upharpoonright B$ and $f_2 = f \upharpoonright C$
iii) if $x \in B$, then $f(x) = f_1(x)$ and if $x \in C$, then $f(x) = f_2(x)$.

17. Let $f_1 : A \to B$ and $f_2 : C \to D$ be bijective, $A \cap C = \emptyset$ and $B \cap D = \emptyset$. Let $f = f_1 \cup f_2$. Prove that $f\colon A \cup C \to B \cup D$ is bijective.

18. Let $f\colon B \to A$, $g\colon C \to A$ and suppose that $f \upharpoonright B \cap C = g \upharpoonright B \cap C$. If $h = f \cup g$, then prove that $h\colon B \cup C \to A$ and that $f = h \upharpoonright B$ and $g = h \upharpoonright C$.

19. Let $f\colon A \to B$. Prove: $\breve{f}$ is a function from B to A if and only if f is bijective.

20. Let $f\colon A \to B$ and $g\colon B \to A$. If $g \circ f = id_A$ and $f \circ g = id_B$, then $f\colon A \to B$ is bijective and $g = f^{-1}$.

21. Let $f\colon A \to B$ and $g\colon B \to C$.
   i) If f and g are injective, then $g \circ f$ is injective.
   ii) If f and g are surjective, then $g \circ f$ is surjective.
   iii) If f and g are bijective, then $g \circ f$ is bijective.

22. Let $f\colon A \to B$. Prove that $id_B \circ f = f$ and $f \circ id_A = f$.

23. Let $f\colon A \to B$ and $g\colon B \to A$. Suppose that $y = f(x)$ if and only if $x = g(y)$. Prove that f is bijective and that $g = f^{-1}$.

24. Let $g\colon B \to C$ and $h\colon B \to C$. Let $A \neq \emptyset$. Suppose that for each function $f\colon A \to B$, $g \circ f = h \circ f$. Prove that $g = h$.

25. Let $g\colon A \to B$ and $h\colon A \to B$. Let C be a set with more than one element. Suppose that for each $f\colon B \to C$, $f \circ g = f \circ h$. Prove that $g = h$.

26. Let $f\colon B \to C$. Prove that f is injective if and only if for each pair of functions $g\colon A \to B$ and $h : A \to B$, if $f \circ g = f \circ h$, then $g = h$.

27. Let $f\colon B \to C$. Prove that f is surjective if and only if for each pair of functions $g\colon C \to D$ and $h\colon C \to D$, if $g \circ f = h \circ f$, then $g = h$.

28. Let $f\colon A \to C$ and $g\colon A \to B$. Prove that there is a function $h\colon B \to C$ such that $f = h \circ g$ if and only if for all $x,y \in A$, if $g(x)= g(y)$, then $f(x)= f(y)$. Prove that h is unique.

29. Let $f\colon C \to A$, $g\colon B \to A$ and suppose that g is bijective. Prove that there exists a function $h\colon C \to B$ such that $f = g \circ h$ if and only if $\mathrm{Ran}(f) \subseteq \mathrm{Ran}(g)$. Prove that h is unique.

30. i) $Y^{\emptyset}$ has precisely one element, namely $\emptyset$, whether Y is empty or not.
   ii) If X is not empty, then $\emptyset^X$ is empty.

31. Let $f\colon A \to B$ and G an equivalence relation on B .
   $\hat{f}(G) := \{\langle x,y\rangle \in A^2 \mid \langle f(x), f(y)\rangle \in G\}$.
   Prove that $\hat{f}(G)$ is an equivalence relation on A.

32. Let $f: A \to B$. We define a relation G on A as follows:
$G := \{\langle x,y \rangle \in A^2 \mid f(x) = f(y)\}$.
Prove that G is an equivalence relation on A. G is called the equivalence relation determined by f. Conversely, if G is an equivalence relation on A, then we define $f: A \to A/G$ as follows: $f(x) = [x]_G$; f is called the canonical function from A to A/G. Prove the following: if G is an equivalence relation on A and f is the canonical function from A to A/G, then G is the equivalence relation determined by f.

33. Let $f: A \to B$. Let G be the equivalence relation determined by f (see Exercise 32) and let r,s and t be the functions, defined as follows:
$r: A \to A/G$ is the canonical function from A to A/G (see Exercise 32).
$s: A/G \to f(A)$ is the function, defined by $s([x]_G) = f(x)$.
$t: f(A) \to B$ is defined by $t(y) = y$.
Prove that r is surjective, that s is bijective, that t is injective and that $f = t \circ s \circ r$. So the preceding results say that each function $f: A \to B$ can be expressed as a composition of three functions r,s and t, which are respectively surjective, bijective and injective. One speaks of the canonical decomposition of f, and uses the following diagram:

$$A \xrightarrow[surj]{r} A/G \xrightarrow[bij]{s} f(A) \xrightarrow[inj]{t} B$$

Here G is the equivalence relation determined by f.

34. Let $f: A \to B$ be surjective and let $\{B_i \mid i \in I\}$ be a partition of B. Prove that $\{\overset{-1}{f}(B_i) \mid i \in I\}$ is a partition of A.

35. Let $f: A \to B$ be injective and let $\{A_i \mid i \in I\}$ be a partition of A. Prove that $\{f(A_i) \mid i \in I\}$ is a partition of $f(A)$.

36. Prove that $(f \circ g) \restriction B = f \circ (g \restriction B)$.

37. Let $f: A \to A$ and for all x, $f \circ f(x) = x$. Then f is bijective and $f^{-1} = f$.

38. Prove that $f: \mathbb{N} \times \mathbb{N} \to \mathbb{N}$, defined by $f(n,m) = 2^m(2n+1)$, is bijective.

39. Let $f: A \to A$. Does there exist a set $C, C \subseteq A$, such that $f(C) = C$?

40. In category theory one starts with functions instead of sets, and in this way one produces most of the important mathematical objects. In the following exercises we give some examples.

i) Let $p_i : A_1 \times A_2 \to A_i$ (i=1,2) be defined by $p_i(\langle a_1, a_2\rangle) = a_i$. Then for each pair $f_1 : X \to A_1$, $f_2 : X \to A_2$ there is a unique $h: X \to A_1 \times A_2$, such that $f_i = p_i \circ h$ (i=1,2). We say that the f_i can be factorized through $A_1 \times A_2$ (see Figure 26).

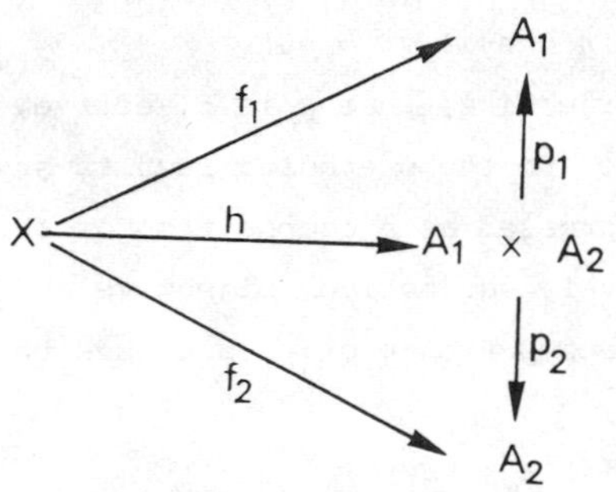

Fig. 26

ii) Let P be a set and $q_i : P \to A_i$ (i=1,2) such that for each X and for each pair $f_i : X \to A_i$ there is a unique $h: X \to P$ with $f_i = q_i \circ h$. Then $F: P \to A_1 \times A_2$, defined by $F(x) = \langle q_1(x), q_2(x)\rangle$, is a bijection (see Figure 27). So P can be taken as (a copy of) $A_1 \times A_2$, defined in "category-theoretical" terms.

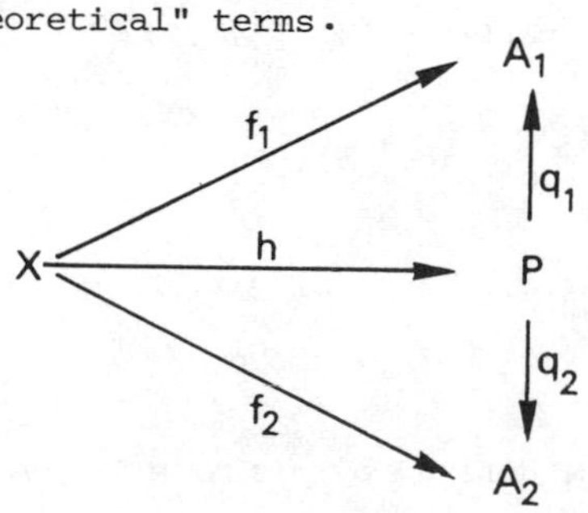

Fig. 27

Analogously, one can give a "category-theoretical" definition of the disjoint union $A_1 \dot{\cup} A_2$ of two sets A_1 and A_2 (as well as the "set-theoretical" definition $(A_1 \times \{1\}) \cup (A_2 \times \{2\})$).

iii) Let $j_i(x) = \langle x,i \rangle$, $i=1,2$. Then for each X and for each pair $f_i : A_i \to X$ there is a unique $h: A_1 \dot{\cup} A_2 \to X$ such that $f_i = h \circ j_i$ (see Figure 28).

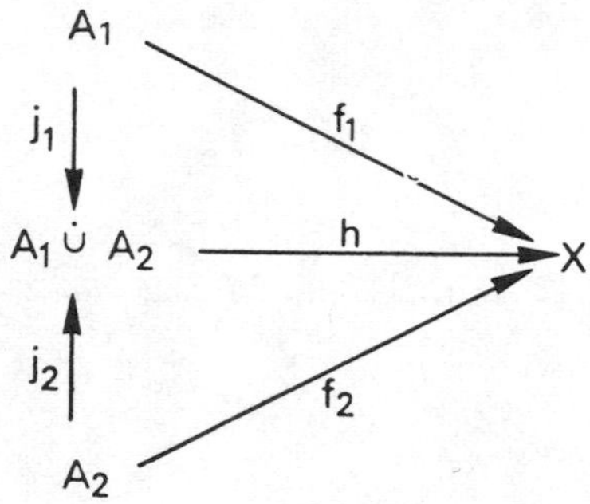

Fig. 28

iv) Let S be a set and $k_i : A_i \to S$ $(i=1,2)$ such that for each X and for each pair $f_i : A_i \to S$ $(i=1,2)$ there is a unique $h: S \to X$ with $f_i = h \circ k_i$. Then $G: A_1 \dot{\cup} A_2 \to S$, defined by $G(\langle x,i \rangle) = k_i(x)$, is a bijection (see Figure 29).

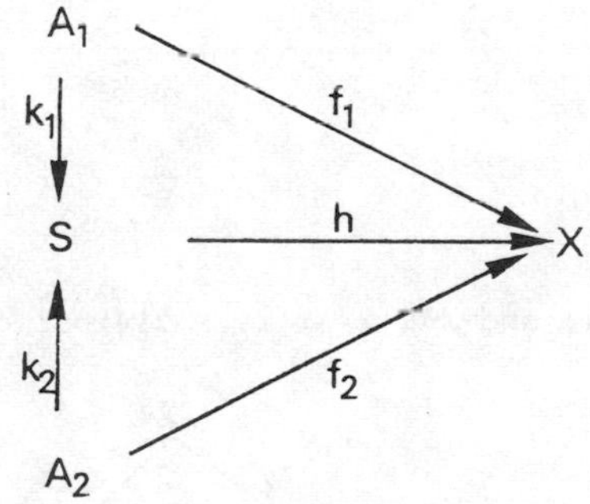

Fig. 29

v) Prove that $\{f \in A \cup B^{\{0,1\}} | f(0) \in A \wedge f(1) \in B\}$ satisfies the requirements for a product given in ii). Hence this gives an alternative definition of the cartesian product.

14 ORDERINGS

We first give a number of examples of ordering relations R on a given set X:

1. $X = \{\{a\},\{b\},\{a,b\}\}$ $(a \neq b)$ and $xRy := x \subseteq y$.

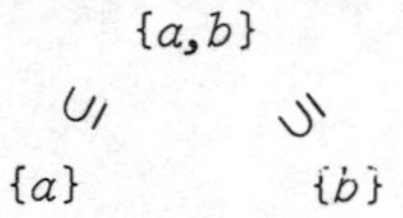

2. $X = \mathcal{P}(\{a,b\}) = \{\emptyset,\{a\},\{b\},\{a,b\}\}$ $(a \neq b)$ and $xRy := x \subseteq y$.

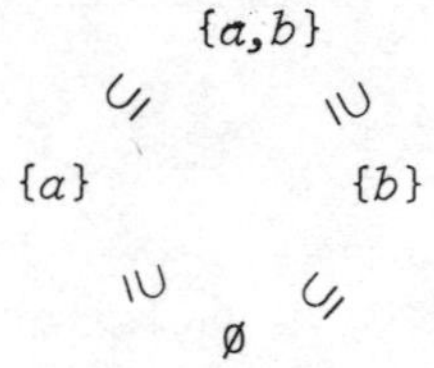

3. $X = \{2,3,4,6,8,12\}$ and $xRy := x$ is a divisor of y.

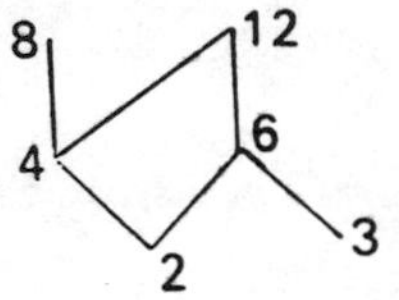

4. $X = \{1,2,3,4,6,8,12,24\}$ and $xRy := x$ is a divisor of y.

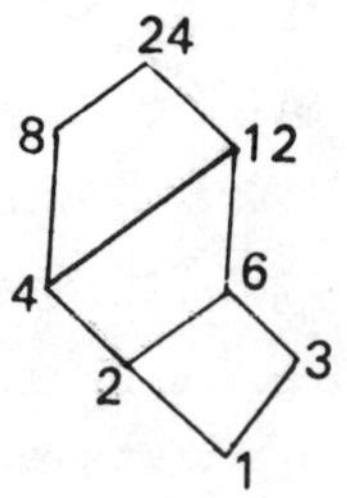

5. $X = \mathbb{Z}$ and $xRy := x \leqslant y$.

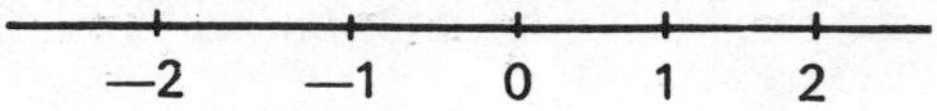

6. $X = \mathbb{Q}$ and $xRy := x \leqslant y$.

7. $X = \mathbb{N}$ and $xRy := x \leqslant y$.

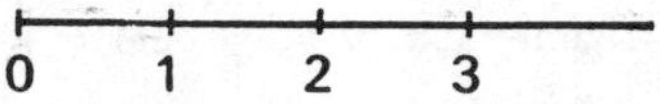

8. $X = \mathbb{N} \times \mathbb{N}$ and $\langle a,b\rangle \; R \; \langle x,y\rangle := a \leqslant x \vee (a{=}x \wedge b \leqslant y)$.

$\langle 0,0\rangle$, $\langle 0,1\rangle$, $\langle 0,2\rangle$, ..., $\langle 1,0\rangle$, $\langle 1,1\rangle$, $\langle 1,2\rangle$, ..., $\langle 2,0\rangle$, ...

The ordering in example 8 is similar to the well-known ordering of words in a dictionary. Therefore we call such an ordering the *lexicographic ordering* on $\mathbb{N} \times \mathbb{N}$.

The common features of the above examples are collected in the following

Definition 14.1: A relation R on a set X is a *partial ordering* on X iff

i) for all $x \in X$, xRx (R is reflexive), and

ii) for all $x,y \in X$, if xRy and yRx, then $x = y$ (R is anti-symmetric), and

iii) for all x,y and $z \in X$, if xRy and yRz, then xRz (R is transitive).

Check that in the examples 1,...,8 R is a partial ordering on X.

Instead of "R is a partial ordering on X" one sometimes says "X is a set, partially ordered by R", "R partially orders X", or "$\langle X,R\rangle$ is a partially ordered set".

If it is clear from the context what partial ordering relation is involved, we may write more briefly "X is a partially ordered set".

The *diagrams* drawn in examples 1 to 7 indicate the given partial orderings, the direction left-to-right or down-to-up corresponding to the given ordering relation.

Notation:
It is well-established custom to denote partial ordering relations by a suggestive symbol, such as $\leqslant$. Whenever convenient we will adopt this notation.

Let X be a set, partially ordered by $\leqslant$, and let $Y \subseteq X$.

Definition 14.2: y_0 is a *minimal element* of Y iff $y_0 \in Y$ and there is no element $y \in Y$ such that $y \leqslant y_0$ and $y \neq y_0$.
y_0 is a *least, (smallest, first) element* of Y iff $y_0 \in Y$ and for all $y \in Y$, $y_0 \leqslant y$.

In symbols:

y_0 is a minimal element of $Y := y_0 \in Y \wedge \forall y \in Y\, [y \leqslant y_0 \rightarrow y_0 = y]$.

y_0 is a least element of $Y := y_0 \in Y \wedge \forall y \in Y\, [y_0 \leqslant y]$.

Note that a least element of Y is necessarily a minimal element of Y. But, conversely, a minimal element of Y is not necessarily a least element of Y: in example 1 $\{a\}$ and $\{b\}$ are minimal elements of X, but X does not have a least element.

Note also that a set $Y \subseteq X$ has at most one least element (since if y_0 and y_1 are both least elements, then it follows from the definition that $y_0 \leqslant y_1$ and $y_1 \leqslant y_0$, and hence, by anti-symmetry, $y_0 = y_1$). Hence in the case that Y has a least element, we can speak of *the* least element of Y.

In example 2, $\emptyset$ is the least element of X and hence also a minimal element of X.

In example 3, 2 and 3 are minimal elements of X, but X does not have a least element.

In example 4, 1 is the least element of X and hence also a minimal element of X.

In examples 5 and 6, X has no minimal element and (hence) no least element either.

"*Maximal element* of Y" and "*greatest (largest, last) element* of Y" are defined analogously.

Let X be a set, partially ordered by $\leqslant$, and let $Y \subseteq X$.

Definition 14.3: x is a *lower bound* for Y in X iff ($x \in X$ and) for all $y \in Y$, $x \leqslant y$.

Note that if $x \leqslant y$, then $x \in X$ (since $\leqslant \subseteq X^2$). x is an *infimum* (greatest lower bound) for Y in X iff

i) x is a lower bound for Y in X, and

ii) for all $z \in X$, if z is a lower bound for Y in X, then $z \leqslant x$.

In symbols:

x is a lower bound for Y in X iff $\forall y \in Y\,[x \leqslant y]$.

x is an infimum for Y in X iff

$\forall y \in Y\,[x \leqslant y] \wedge \forall z \in X\,[\forall y \in Y\,[z \leqslant y] \rightarrow z \leqslant x]$.

Note that a subset Y of a partially ordered set X can have at most one infimum in X (again, by anti-symmetry). Hence, in the case that Y has an infimum in X, we can speak of *the* infimum of Y in X, which is then denoted by "inf(Y)".

If Y has a least element, then this is clearly the infimum for Y in X; but an infimum for Y in X need not belong to Y and thus need not be a least element of Y.

In example 1, X does not have an infimum in X. In example 2, $\emptyset$ is the least element of X, hence it is also the infimum for X in X. In example 3, X does not have an infimum in X. In example 3, 2 is the least element and hence also the infimum of $\{2,4,6,8,12\}$ in X. In example 4, 1 is the least element of X and hence also the infimum for X in X; further, the infimum for $\{2,3,4,6,8,12,24\}$ in X is 1. In example 5, X does not have an infimum in X.

In example 6, let $Y_1 := \{x \in \mathbb{Q} \mid 0 \leqslant x < 1\}$ and

$Y_2 := \{x \in \mathbb{Q} \mid 0 < x < 1\}$.

0 is the least element of Y_1 and hence also the infimum for Y_1 in $X = \mathbb{Q}$. 0 is the infimum for Y_2 in $X = \mathbb{Q}$, but because $0 \notin Y_2$, 0 is not a least element of Y_2. (In fact, Y_2 does not have a least element).

"*Upper bound* for Y in X" and "*supremum* (least upper bound) for Y in X" are defined analogously. Notation: sup(Y).

We will prove now that every partially ordered set $\langle X,\leqslant\rangle$ can be represented as a partially ordered set $\langle Y, \subseteq \rangle$, Y being a collection of subsets of X.

Theorem *14.4* *(Representation theorem* for partially ordered sets): If $\langle X,\leqslant \rangle$ is a partially ordered set, then there is a set Y of subsets of X, such that $\langle X, \leqslant \rangle$ is isomorphic to $\langle Y, \subseteq \rangle$.

Proof: For $x \in X$ let $X(x) := \{y \in X \mid y \leqslant x\}$. Now let $Y = \{X(x) \mid x \in X\}$ and define f: $X \rightarrow Y$ by $f(x) = X(x)$. One easily checks that f is a bijection from X to Y, such that $x_1 \leqslant x_2 \leftrightarrow f(x_1) \subseteq f(x_2)$. □

Example: In example 3 we see that $\langle X, \leqslant \rangle$ can be represented by $\langle Y, \subseteq \rangle$, with $Y = \{\{2\}$; $\{2,4\}$, $\{2,4,8\}$, $\{3\}$, $\{2,3,6\}$, $\{2,3,4,6,12\}\}$ (see Figure 30):

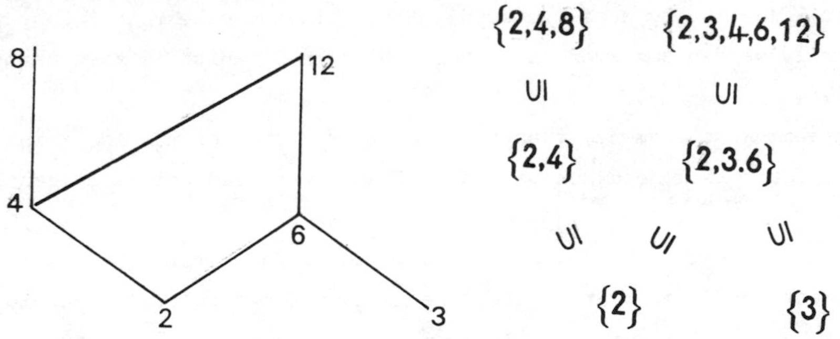

Fig. 30

There are actually two notions of partial ordering used in the literature; the one (which we have been using) satisfying reflexivity, and generally denoted by "$\leqslant$ " ; and the other satisfying irreflexivity (i.e. for all x, $\neg xRx$), and generally denoted by "$<$". (These two notions are connected in a simple way, as we will show below).

We can distinguish these two notions by speaking of "weak" and "strict" partial orderings, respectively. We give both definitions side by side:

<u>Definition</u>

R is a *weak partial ordering* on X iff

i) $\forall x \in X\,[xRx]$

ii) $\forall x, y \in X\,[xRy \wedge yRx \rightarrow x = y]$

iii) $\forall x, y, z \in X\,[xRy \wedge yRz \rightarrow xRz]$

<u>Definition</u>

R is a *strict partial ordering* on X iff

i) $\forall x \in X\,[\neg xRx]$ (irreflexivity)

ii) $\forall x, y \in X\,[xRy \rightarrow \neg yRx]$

iii) $\forall x, y, z \in X\,[xRy \wedge yRz \rightarrow xRz]$

Next we indicate how we can go from weak to strict partial orderings, and back again.

<u>Lemma 14.5</u>: Let R be a weak partial ordering on X. Define $x\hat{R}y :=$ $xRy \wedge x \neq y$. Then $\hat{R}$ is a strict partial ordering on X.

Proof: i) Suppose $x\hat{R}x$. Then $x \neq x$, which is impossible. Hence $\neg x\hat{R}x$.
ii) Suppose $x\hat{R}y$; then xRy and $x \neq y$. Suppose also $y\hat{R}x$; then yRx. So (by anti-symmetry of R) $x = y$. But this contradicts $x \neq y$. Hence $x\hat{R}y$ implies $\neg y\hat{R}x$.
iii) Trivial.

<u>Lemma 14.6</u>: If R is a strict partial ordering on X and $\check{R}$ is defined by $x\check{R}y := xRy \vee x = y$, then $\check{R}$ is a weak partial ordering on X.

The proof is left as an exercise to the reader.

Note that $\check{\hat{R}} = R$ and $\hat{\check{R}} = R$.

The above results show that it makes very little difference whether we take "partial ordering" as meaning weak or strict partial ordering. In fact we shall (following common mathematical practice) generally take "partial ordering" to mean "weak partial ordering" (as in fact we have been doing; cf. Definition 14.1).

We come to the concept of total or linear ordering (i.e. a partial ordering satisfying an extra condition: connectivity). Again, there are two definitions which can be given: "weak" and "strict" total orderings (corresponding to "weak" and "strict" partial orderings and generally denoted by "$\leqslant$" and "$<$" respectively). In this case (again following common mathematical practice) we will generally take "total ordering" to mean "strict total ordering" (as in the following definition). Further, we will often say simply "ordering" for "total ordering".

Definition 14.7: R is a (*total or linear*) *ordering* on X iff

i) for all $x \in X$, $\neg xRx$

ii) for all $x,y \in X$, $xRy \rightarrow \neg yRx$

iii) for all $x,y,z \in X$, $xRy \wedge yRz \rightarrow xRz$

iv) for all $x,y \in X$, $xRy \vee x = y \vee yRx$ (connectivity).

$\langle \underset{\sim}{Z}, R \rangle$ with $xRy := x < y$ and $\langle \underset{\sim}{Q}, R \rangle$ with $xRy := x < y$ (cf. examples 5 and 6) are (totally or linearly) ordered sets. (These examples suggest the term "linear ordering ".)

Whenever we refer to a subset Y of a partially or totally ordered set $\langle X,R \rangle$, we will usually think of this subset Y as being partially or totally ordered (respectively) by the restriction of R to Y, i.e. $R \cap (Y \times Y)$.

The ordered set $\langle \underset{\sim}{N}, < \rangle$ has the property that each non-empty subset of $\underset{\sim}{N}$ has a least element. The ordered sets $\langle \underset{\sim}{Z}, < \rangle$ and $\langle \underset{\sim}{Q}, < \rangle$ do not have this property.

Definition 14.8: A relation R on a set X is a *well-ordering* on X iff

i) R is an ordering on X, and

ii) each non-empty subset of X has a least element.

So in the cases $\langle \underset{\sim}{N}, R \rangle$ with $xRy := x < y$ and $\langle \underset{\sim}{N} \times \underset{\sim}{N}, R \rangle$ with $\langle a,b \rangle \; R \; \langle x,y \rangle := a < x \vee (a = x \wedge b < y)$ (cf. example 8), R is a well-ordering on $\underset{\sim}{N}$ and $\underset{\sim}{N} \times \underset{\sim}{N}$ respectively. But in the cases $\langle \underset{\sim}{Z},R \rangle$ and $\langle \underset{\sim}{Q},R \rangle$ with $xRy := x < y$, R is not a well-ordering on $\underset{\sim}{Z}$, $\underset{\sim}{Q}$ respectively.

Although it seems rather obvious, we shall show that $\underset{\sim}{N}$ is well-ordered by $<$. In the older literature the well-ordering on $\underset{\sim}{N}$ is sometimes called the least number principle.

It clearly suffices to show that each non-empty subset of $\{0,1,...,n\}$ has a least element for all n, for suppose $A \subseteq \underset{\sim}{N}$ and $n \in A$, then we have only to scan $\{0,1,...,n\}$ for a least element.

We now show that each set $\{0,1, ..., n\}$ is well-ordered, by induction on n.
$n = 0$, obvious.
Suppose that $\{0,1, ..., n\}$ is well-ordered and consider $A \subseteq \{0,1, ..., n+1\}$ with $A \neq \emptyset$. If $A \cap \{0,1, ..., n\} = \emptyset$, then $n+1$ is the least element of A. Otherwise $A \cap \{0,1, ..., n\} \neq \emptyset$ and, by the induction hypothesis, this set has a least element k. Clearly k is the required least element of A.

Conversely, the well-ordering on $\underset{\sim}{N}$ implies the principle of mathematical induction (cf. 14.10).

Definition 14.9: Let R be a partial ordering on X; $a,b \in X$.
b is an *R-successor* of a (or: a is an *R-predecessor* of b) iff aRb and $a \neq b$.
b is an *immediate successor* of a (or: a is an *immediate predecessor* of b) iff
1. aRb and $a \neq b$, and
2. $\neg\exists c \in X\,[c \neq a \wedge c \neq b \wedge aRc \wedge cRb]$.

In example 3, 4 and 6 are immediate successors of 2, while 2 and 3 are immediate predecessors of 6. In example 8, each pair $\langle a,b \rangle$ has an immediate successor, but the pairs $\langle 1,0 \rangle$, $\langle 2,0 \rangle$, $\langle 3,0 \rangle$, etc. do not have immediate predecessors, so not all elements of a well-ordered set need have immediate predecessors, as one might have supposed from the analogy of $\langle \underset{\sim}{N}, < \rangle$.

Other examples of well-ordered sets $\langle X,R \rangle$ are:

9. i) $X = \{0,1\}$, $xRy := x < y$. $0 < 1$
   ii) $X = \{0,1,2\}$, $xRy := x < y$. $0 < 1 < 2$

10. $X = \underset{\sim}{N} \cup \{\underset{\sim}{N}\}$, $xRy := (x \in \underset{\sim}{N} \wedge y \in \underset{\sim}{N} \wedge x < y) \vee (x \in \underset{\sim}{N} \wedge y = \underset{\sim}{N})$.

$$0,1,2,3, ..., \underset{\sim}{N}$$

Note that every element of X , except $\underset{\sim}{N}$, has an immediate predecessor.

11. $\underset{\sim}{N}^+ := \underset{\sim}{N} \cup \{\underset{\sim}{N}\}$. $X = \underset{\sim}{N}^+ \cup \{\underset{\sim}{N}^+\}$.

$xRy := (x \in \underset{\sim}{N} \wedge y \in \underset{\sim}{N} \wedge x < y) \vee (x \in \underset{\sim}{N} \wedge y = \underset{\sim}{N}) \vee$
$x \in \underset{\sim}{N} \wedge y = \underset{\sim}{N}^+) \vee (x = \underset{\sim}{N} \wedge y = \underset{\sim}{N}^+)$.

$0,1,2,3, \ldots, \underset{\sim}{N}, \underset{\sim}{N}^+$

Note that $\underset{\sim}{N}$ does not have an immediate predecessor, but $\underset{\sim}{N}^+$ does, namely $\underset{\sim}{N}$.

Consider now $\langle \underset{\sim}{N}, R \rangle$ where $xRy := x > y$. This is a totally ordered set, but not well-ordered:

$\ldots, 3,2,1,0$

On the other hand we have seen that $\langle \underset{\sim}{N}, < \rangle$ is a well-ordered set $(0,1,2,3, \ldots)$.

In general a set can be ordered in several ways:

12. $\langle \underset{\sim}{N}, R \rangle$ with $xRy :=$ (x is even $\wedge$ y is odd) $\vee$ (x is even $\wedge$ y is even $\wedge\ x < y$) $\vee$ (x is odd $\wedge$ y is odd $\wedge\ x < y$),
is also a well-ordered set.

$0,2,4, \ldots, 1,3,5, \ldots$

One of the most pleasing facts about well-ordered sets is that we can prove statements about their elements by a process similar to mathematical induction:

Theorem 14.10 *(Principle of transfinite induction):*
Let $\langle X, R \rangle$ be a well-ordered set, and let Φ be a property of elements of X.
$\forall x \in X\ [\forall y \in X\ [yRx \to \Phi(y)] \to \Phi(x)] \to \forall x \in X\ [\Phi(x)]$.

This can be described in words as follows.
Call a property Φ of elements of X *R-progressive* if for every $x \in X$, if all the R-predecessors of x have the property Φ, then so does x. Then the principle of transfinite induction states: if Φ is R-progressive, then every element of X has the property Φ.

Proof: Let $\langle X,R \rangle$ be a well-ordered set and suppose Φ is R-progressive, i.e.
$\forall x \in X\, [\forall y \in X\, [yRx \rightarrow \Phi(y)] \rightarrow \Phi(x)]$. (I)

Now suppose $\neg \forall x \in X\, [\Phi(x)]$, in other words
$\exists x \in X\, [\neg \Phi(x)]$. (II)

By II, $\{z \in X \mid \neg \Phi(z)\}$ is a non-empty subset of X.
Hence, because $\langle X,R \rangle$ is well-ordered, $\{z \in X \mid \neg \Phi(z)\}$ has a least element, say x_0. So $\neg \Phi(x_0) \wedge \forall z \in X\, [zRx_0 \rightarrow \Phi(z)]$. By I it follows from this that $\Phi(x_0)$. Contradiction with $\neg \Phi(x_0)$. □

Applying the principle of transfinite induction to $\langle \underset{\sim}{N}, < \rangle$, we get:
$\forall n \in \underset{\sim}{N}\, [\forall y \in \underset{\sim}{N}\, [y < n \rightarrow \Phi(y)] \rightarrow \Phi(n)] \rightarrow \forall n \in \underset{\sim}{N}\, [\Phi(n)]$.

Compare this with the principle of mathematical induction, which asserts:
$\Phi(0) \wedge \forall n \in \underset{\sim}{N}\, [\Phi(n) \rightarrow \Phi(n+1)] \rightarrow \forall n \in \underset{\sim}{N}\, [\Phi(n)]$.

Theorem 14.11: The principle of transfinite induction for $\langle \underset{\sim}{N}, < \rangle$ is equivalent to the principle of mathematical induction.

Proof: We have already verified that $\underset{\sim}{N}$ is well-ordered by $<$ and hence, by 14.10, the principle of transfinite induction holds for $\underset{\sim}{N}$. So we have to prove the principle of mathematical induction for $\underset{\sim}{N}$ assuming the principle of transfinite induction. We want to show that
$\Phi(0) \wedge \forall n\, [\Phi(n) \rightarrow \Phi(n+1)] \rightarrow \forall n\, [\Phi(n)]$.
Since we have the principle of transfinite induction at our disposal, we look for a suitable progressive predicate. As $\Phi(0) \wedge \forall n\, [\Phi(n) \rightarrow \Phi(n+1)]$ is given, the predicate $\Psi(n) := \forall y \leqslant n\, [\Phi(y)]$ seems a reasonable candidate.

If $\forall y < n+1 [\Psi(y)]$, then $\Phi(n)$ and hence, by assumption, $\Phi(n+1)$, so $\Psi(n+1)$ holds. Therefore $\forall y < n\, [\Psi(y)] \rightarrow \Psi(n)$ (note that the case $n=0$ is trivially covered).

Apply the principle of transfinite induction:
$\forall n\, [\Psi(n)]$, and hence $\forall n\, [\Phi(n)]$. □

The reader may wish to provide a direct proof of the principle of transfinite induction from the principle of mathematical induction.

It is an immediate consequence of the definition of well-ordering, that a subset Y of a set X, which is well-ordered by a relation R, is again itself well-ordered by the restriction of this relation to Y, i.e. by $R \cap (Y \times Y)$.

Next we repeat the definition of isomorphism for the case of ordered sets:

Definition 14.12: Suppose $\langle X,R \rangle$ and $\langle X',R' \rangle$ are (totally) ordered sets. $f\colon X \to X'$ is an isomorphism from $\langle X,R \rangle$ to $\langle X',R' \rangle$ iff
1. $f\colon X \to X'$ is bijective, and
2. for all $x,y \in X$, $xRy \leftrightarrow f(x)\, R'f(y)$.

Definition 14.13: The *initial segment* X_a of a well-ordered set $\langle X,< \rangle$ determined by an element a of X is the subset of X consisting of all $<$-predecessors of a.

$$X_a := \{y \in X \mid y < a\}.$$

Lemma 14.14: If f is an isomorphism from a well-ordered set $\langle X,< \rangle$ to a subset of itself, then for every $x \in X$, $x \leqslant f(x)$ (i.e. $x < f(x) \vee x = f(x)$).

Proof: By transfinite induction. Induction hypothesis: $\forall y < x\ [y \leqslant f(y)]$. Suppose $f(x) < x$; then (since f is an isomorphism) $f(f(x)) < f(x)$. But $f(x) < x$, so by the induction hypothesis $f(f(x)) \geqslant f(x)$. Contradiction. Hence $x \leqslant f(x)$. □

The totally ordered set $\langle \mathbb{Z},< \rangle$ is mapped isomorphically to itself by any one of the infinite set of isomorphisms $f_a\colon \mathbb{Z} \to \mathbb{Z}$, defined by $f_a(x) := x + a$, where a is any (fixed) element of $\mathbb{Z}$.

In the case of well-ordered sets, by contrast, if an isomorphism exists at all, it is unique. This is asserted by the next theorem.

Theorem 14.15: There is at most one isomorphism from one given well-ordered set to another.

Proof: Suppose f and g are two isomorphisms from a well-ordered set $\langle X,<_X \rangle$ to a well-ordered set $\langle Y,<_Y \rangle$. Then $f^{-1} \circ g$ is an isomorphism from X to itself, and hence, by Lemma 14.14, $x \leqslant_X (f^{-1} \circ g)(x)$ for every $x \in X$.

Hence for every $x \in X$, $f(x) \leqslant_Y g(x)$ (because f is an isomorphism).

In the same way we can prove: for every $x \in X$, $g(x) \leqslant_Y f(x)$.

Hence (because $<_Y$ is a total order on Y) for every $x \in X$, $f(x) = g(x)$, so $f = g$. □

N.B. There may be many isomorphisms of X to a subset of Y, so it is essential to recall that an isomorphism is surjective.

The proof that we have just given illustrates the usefulness of Lemma 14.14, and we now use this lemma again to show that it is impossible for a well-ordered set to be mapped isomorphically to any of its own initial segments.

Theorem 14.16: A well-ordered set $\langle X, < \rangle$ is not isomorphic to any of its initial segments.

Proof: Suppose $f: X \simeq X_a$ for some $a \in X$. Then, by Lemma 14.14, for every $x \in X$, $x \leqslant f(x)$. In particular $a \leqslant f(a)$. Thus $f(a) \notin X_a$, which is contrary to hypothesis. □

Theorem 14.17: Any well-ordered set $\langle X, < \rangle$ is isomorphic to the set of all its initial segments, ordered by strict inclusion ($\subset$).

Proof: Let $\langle X, < \rangle$ be a well-ordered set and let Y be the set of all initial segments X_z with $z \in X$. Then $f: X \to Y$, defined by $f(x) := X_x$, is a bijection. And, because $a < b \leftrightarrow X_a \subset X_b$ for all $a, b \in X$, this bijection is an isomorphism from $\langle X, < \rangle$ to $\langle Y, \subset \rangle$. (Cf. the proof of Theorem 14.4).

Exercises

1. Show that in example 12 the following holds
$\Phi(0) \wedge \Phi(1) \wedge \forall x \in \underset{\sim}{N}\, [\Phi(x) \to \Phi(x+2)] \to \forall x \in \underset{\sim}{N}\, [\Phi(x)]$.

2. For each natural number n let $F(n)$ be the number of non-isomorphic partial orderings on a set with n elements. Show that $F(2) = 2$, $F(3) = 5$, $F(4) = 16$ and $F(5) = 63$.

3. Draw diagrams for the following partially ordered sets:
   a) The set of all subsets of a set with 3 elements, partially ordered by $\subseteq$.
   b) The set of natural numbers 1, ..., 25, partially ordered by divisibility.

4. Prove that the set of all partitions of a given set is partially ordered by the relation R, defined by aRb iff the partition a is a refinement of the partition b.

5. Show that X is partially ordered by R in the cases a) and b), and find a pair of incomparable elements in each of the two sets:
   a) X is the set of all real-valued functions of a real variable, and fRg means: $f(x) \leqslant g(x)$ for all x.
   b) X is the set of all positive real-valued functions of a real variable, and fRg means: $f = g$ or $\lim_{x\to\infty} f(x)/\,g(x) = 0$.

6. A relation on a set X is said to be a *quasi-ordering* or *pre-ordering* if it is reflexive and transitive. Show that any quasi-ordering R can be reduced to a partial ordering by passing to the set of equivalence classes $[a] := \{b \in X \mid aRb \wedge bRa\}$.
   Show that the relations defined below are quasi-orderings, and reduce them to partial orderings in the manner just indicated:
   a) X is the set of all complex numbers and wRz means: the real part of w is less than or equal to that of z.
   b) X is the set of positive integers, and aRb means: the number of divisors of a is less than or equal to the number of divisors of b.

7. Give an example of two totally ordered sets, which are not isomorphic, although each of them is isomorphic to a subset of the other. (So the analogue of Theorem 15.16 of Cantor-Bernstein does not hold for isomorphisms between ordered sets.)

8. Prove that a totally ordered set is well-ordered if and only if it does not contain any infinite descending sequence (i.e. a subset isomorphic to the set of negative integers in their natural order). (This uses the axiom of choice; see the discussion after 16.3).

9. Show that if every subset of a totally ordered set has both a first and a last element, then the set is finite.

10. Show by examples that Theorems 14.14, 14.15 and 14.16 are all false if "well-ordered" is replaced everywhere by "totally ordered".

11. Find an example of a totally ordered set A and an isomorphism $f: A \to A$ such that $f(x) \neq x$ for all $x \in A$.

12. If for every $i \in I$, R_i is a partial ordering on A, then also $\bigcap_{i \in I} R_i$ is a partial ordering on A.

13. Prove that the converse of a partial ordering on A is again a partial ordering on A.

14. Let A and B be partially ordered sets and let $f: A \to B$ be an isomorphism. Prove each of the following.
   a) a is a maximal element of A if and only if $f(a)$ is a maximal element of B.
   b) a is the greatest element of A if and only if $f(a)$ is the greatest element of B.
   c) Suppose $C \subseteq A$; x is an upper bound of C if and only if $f(x)$ is an upper bound of $f(C)$.
   d) $b = \sup(C)$ if and only if $f(b) = \sup f(C)$.

15. Let R be an ordering on X. Show that X is finite if and only if R and $\breve{R}$ are well-orderings.

16. Let A and B be partially ordered sets. Prove the following.
   a) Suppose $A \times B$ is ordered lexicographically; if $\langle a,b \rangle$ is a maximal element of $A \times B$, then a is a maximal element of A.
   b) Suppose $A \times B$ is ordered anti-lexicographically ($\langle a_1, b_1 \rangle \; R \; \langle a_2, b_2 \rangle := a_2 \leqslant_A a_1 \vee (a_2 = a_1 \wedge b_2 \leqslant_B b_1)$, where $\leqslant_A$ is the partial ordering on A and $\leqslant_B$ is the partial ordering on B); if $\langle a,b \rangle$ is a maximal element of $A \times B$, then b is a maximal element of B.

17. Suppose $\langle B, \leqslant \rangle$ is a partially ordered set.
    i) If $A \subseteq B$ has a greatest element a, and B has a greatest element b, then $a \leqslant b$.
    ii) If $A \subseteq B$ and $A' \subseteq B$, such that A and A' each have a supremum in B, then $A \subseteq A' \rightarrow \sup(A) \leqslant \sup(A')$.

18. Let A be a totally ordered set. Prove that A is well-ordered if and only if each initial segment of A is well-ordered.

19. Let $\langle A, < \rangle$ be a well-ordered set. For $a \in A$, let a' be the immediate successor of a. An element q in A is called a limit-element of A if q is not the least element of A and q does not have an immediate predecessor. Prove:
    a) q is a limit-element of A if and only if $\forall a \in A\ [a < q \rightarrow a' < q]$.
    b) q is a limit-element of A if and only if $q = \sup \{x \in A | x < q\}$.

20. Let A and B be well-ordered sets. Prove: if $f\colon A \rightarrow B$ and $g\colon B \rightarrow A$ are isomorphisms, then $g = f^{-1}$.

21. Formulate an analogue of Definition 14.7 for weak total orderings (i.e., satisfying reflexivity).

15 EQUIVALENCE (CARDINALITY)

One of the favourite roles of mathematics in basic applications is to provide a convenient framework for measuring quantities. Usually one proceeds by simply counting the set one wants to measure, e.g. the set of guests at a party, the set of votes in favour of a payrise. This procedure, however, has the disadvantage that it cannot be generalized in a straightforward way to arbitrary sets (such as sets of reals, of triangles, of continuous functions).

Fortunately one can quite often restrict oneself to comparing sets (e.g. the set of guests and the set of wine glasses). In that way one avoids the process of counting. Cantor, Frege and Russell exploited this idea by a systematic study of the comparison of sets.

So instead of asking "how many elements has the set *A*?" we ask "has *A* just as many elements as *B*?". The proper formulation of this particular comparison makes use of "one-one-correspondence" or "matching". For example the set {Plato, Augustine, Wittgenstein} has just as many elements as the set $\{\sqrt{2}, \pi, ln3\}$, simply because we can match the sets in a suitable way:

Plato —— $\sqrt{2}$
Augustine —— π
Wittgenstein —— $ln3$

Early scientists were rather puzzled by the effects of the matching-concept (cf. Galileo's paradox [van Dalen - Monna, 1971, p. 4]).

The notion of "matching" two sets is made precise in the following definition.

Definition 15.1: V is *equivalent* to W iff there is a bijection $f: V \to W$ (i.e. $f: V \underset{I}{=} W$).

We write "$V \underset{I}{=} W$" for "there exists an f such that $f: V \underset{I}{=} W$".

In the literature one also meets the terms "equipollent", "similar" or "are of the same cardinality" (german: gleichmächtig).

By Theorem 13.13, $\underset{I}{=}$ is an equivalence relation (see remark), i.e.
i) for all sets V, $V \underset{I}{=} V$
ii) for all sets U and V, if $U \underset{I}{=} V$, then $V \underset{I}{=} U$
iii) for all sets U,V and W, if $U \underset{I}{=} V$ and $V \underset{I}{=} W$, then $U \underset{I}{=} W$.

Remark: Although $\underset{I}{=}$ has the properties of an equivalence relation, it is not a relation, for the simple reason that it is not a set. Recall that an equivalence relation on A is a subset of $A \times A$, so being a set is essential. At the moment it is not opportune to go into the matter, but we will return to it in the section on the paradoxes (cf. 19.3).

The pioneers of set theory did what every mathematician does when presented with an equivalence relation: they formed equivalence classes under $\underset{I}{=}$.

As a result they possessed a (suggestive but defective) analogue of "the number of elements of ...", called the "cardinality", or "cardinal of ...". This naive concept has turned out to be subject to the same paradoxes that worried Cantor and Frege, so it has been dropped from set theory (however, for a revival cf. II, 10), but the terminology has stuck to the informal language of set theory. So we will often say "V and W have the same cardinality" when we should say "V and W are equivalent". The reader has to keep in mind that the use of cardinal, or cardinality, at the present level is purely metaphorical.

Example 0: $\{1,2,3\} \underset{I}{=} \{7,8,9\}$ and not $\{1,2,3\} \underset{I}{=} \{2,3\}$.

Example 1: $\mathbb{N} \underset{I}{=} \mathbb{N}_{even}$, where $\mathbb{N}_{even} := \{x \in \mathbb{N} | x \text{ is even}\}$.
Namely $f\colon \mathbb{N} \to \mathbb{N}_{even}$, defined by $f(n) = 2n$, is a bijection from $\mathbb{N}$ to $\mathbb{N}_{even}$.

$\mathbb{N}$		$\mathbb{N}_{even}$
0	$\overset{f}{\to}$	0
1	$\to$	2
2	$\to$	4
3	$\to$	6
4	$\to$	8
⋮		⋮

Hence, the proposition "The whole is greater than its part" (Euclid) turns out to be false for infinite sets. $\mathbb{N}_{even}$ is a proper subset of $\mathbb{N}$, but $\mathbb{N}_{even}$ is still equivalent to $\mathbb{N}$. However, we shall presently see that "a proper part is smaller than the whole" is true for finite sets.

Example 2: $\mathbb{R}_+ \underset{I}{=} \mathbb{R}$, where $\mathbb{R}_+ := \{x \in \mathbb{R} | x > 0\}$. For $log\colon \mathbb{R}_+ \to \mathbb{R}$ is a bijection from $\mathbb{R}_+$ to $\mathbb{R}$ (see figure 31).

Note that $\mathbb{R}_+$ is a proper subset of $\mathbb{R}$, while $\mathbb{R}_+$ is equivalent to $\mathbb{R}$.

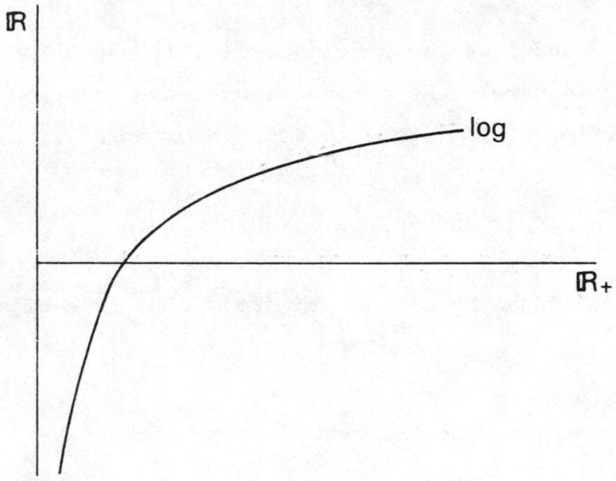

Fig. 31

Example 3: (-1, 1) $\underset{I}{=}$ $\underset{\sim}{R}$, where (-1, 1) := $\{x \in \underset{\sim}{R} | -1 < x < 1\}$.
For f: (-1, 1) → $\underset{\sim}{R}$, defined by $f(x) = tg\ (\frac{\pi}{2}\ x)$, is a bijection from (-1, 1) to $\underset{\sim}{R}$ (see figure 32).

(-1, 1) is again a proper subset of $\underset{\sim}{R}$, which is equivalent to $\underset{\sim}{R}$.

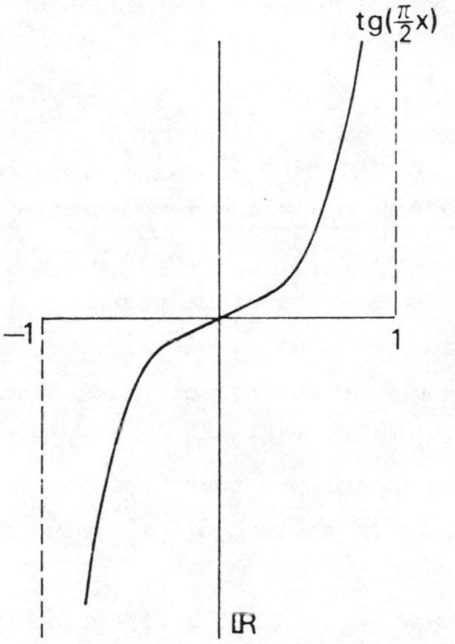

Fig. 32

From example 2 and 3 we can conclude: $(-1, 1) \underset{1}{=} \mathbb{R}_+$. The bijection, belonging to it, is namely the composition of the function f in example 3 and the inverse function of *log* (this is *exp*) in example 2.

$$(-1, 1) \underset{tg(\frac{\pi}{2} x)}{\rightarrow} \mathbb{R} \underset{exp(x)}{\rightarrow} \mathbb{R}_+$$

$g: (-1, 1) \rightarrow \mathbb{R}_+$, defined by $g(x) = exp\ (tg(\frac{\pi}{2}\ x))$, is a bijection from $(-1, 1)$ to $\mathbb{R}_+$.

Example 4: $\mathbb{N} \underset{1}{=} P$, where P is the set of prime numbers.

In book IX of Euclid's "Elements" (300 B.C.) it is shown that there are infinitely many prime numbers. Euclid proceeds by constructing for each finite set of primes a prime which does not belong to it.

Let $\{p_1, \ldots, p_n\}$ be a finite set of prime numbers. Consider the number $p = (\prod_{i=1}^{n} p_i) + 1$, i.e. $p_1 \cdot p_2 \ldots p_n + 1$. p is divisible by at least one prime number. Let p_{n+1} be the smallest prime number, which divides $(\prod_{i=1}^{n} p_i) + 1$. Hence for some $m \in \mathbb{N}$,

$$(\prod_{i=1}^{n} p_i) + 1 = m.p_{n+1}.$$

This implies that $p_{n+1} \neq p_i$ for each $i \in \{1, \ldots, n\}$. For suppose $p_{n+1} = p_i$ for some i, $1 \leqslant i \leqslant n$, then $(\prod_{i=1}^{n} p_i) + 1 = m\ p_i$, so $1 = p_i\ (m - p_1 \cdot p_2 \ldots p_{i-1} \cdot p_{i+1} \ldots p_n)$.

But $p_i \geqslant 2$ and $m - p_1 \cdot p_2 \ldots p_{i-1} \cdot p_{i+1} \ldots p_n \in \mathbb{N}$.

Contradiction. Hence, p_{n+1} is a new prime number.

Using this construction it is an elementary, but uninspiring, task to define an injection f from $\mathbb{N}$ to the set of prime numbers:

$f(1) = 2$, $f(2) = 3$, $f(3) = 7$, $f(4) = 43$, $f(5) = 13$, ...

Because f is an injection it follows that $\mathbb{N} \underset{1}{=} \text{Ran}\ (f)$, in other words, $\mathbb{N}$ is equivalent to a subset of the set of all prime numbers.

In a more pedestrian way we can find a bijection from $\mathbb{N}$ to P by running through $\mathbb{N}$ and checking for each number whether it is a prime. This is basically the method known as the sieve of Erathostenes. Here we use the fact that there are infinitely many primes.

Example 5: The reader may easily verify that $\underset{\sim}{N} \underset{1}{=} \underset{\sim}{N} - \{1\}$, $\underset{\sim}{N} \underset{1}{=} \underset{\sim}{N} - \{0,1,2\}$ and $\underset{\sim}{N} \underset{1}{=} \underset{\sim}{N}_s$, where $\underset{\sim}{N}_s := \{x \in \underset{\sim}{N} | x \text{ is a square}\}$.

In mathematics we now and then run into objects or notions, which are "essentially the same". An example of such a case is the field of complex numbers $a + i\,b$ and the matrix ring $\begin{pmatrix} a & b \\ -b & a \end{pmatrix}$. Although the objects look radically different, they behave identically with respect to the algebraic operations. The algebraist uses in this case the term "isomorphic". In topology the key notion is "homeomorphic".

What is the corresponding notion in naive set theory? That is, what sets do we consider to be essentially the same? Metaphorically speaking we are in a limit case of the isomorphism of structures. Sets have no structure, so all that counts is their size (or cardinality). Hence the proper notion in this case is *equivalence*. For set theoretical purposes we do not distinguish between equivalent sets (the proper framework for these considerations, of course, is the category of sets).

We will illustrate our viewpoint by having a closer look at the power set. For a start we consider the set $\{a,b\}$ with two elements. $\mathcal{P}(\{a,b\}) = \{\emptyset, \{a\}, \{b\}, \{a,b\}\}$. Now consider $F = \{0,1\}^{\{a,b\}}$. The elements of F are functions from $\{a,b\}$ to $\{0,1\}$. We have indicated them in the following table.

	a	b
f_1	0	0
f_2	0	1
f_3	1	0
f_4	1	1

The reader, who can count, observes that F and $\mathcal{P}(\{a,b\})$ are equivalent, e.g. $T : \mathcal{P}(\{a,b\}) \underset{1}{=} F$ with $T(\emptyset) = f_1$, $T(\{a\}) = f_4$, $T(\{b\}) = f_2$, $T(\{a,b\}) = f_3$.

This function, however, lacks method. We can do better; after all the functions f_i are characteristic functions on $\{a,b\}$. Let us assign to each subset of $\{a,b\}$ its characteristic function: $K(\emptyset) = f_1$, $K(\{a\}) = f_3$, $K(\{b\}) = f_2$, $K(\{a,b\}) = f_4$.

We can generalize this result to arbitrary sets V:

Theorem 15.2: For all sets V , $\mathcal{P}(V) \underset{I}{=} \{0,1\}^V$.

Proof: The function $K: \mathcal{P}(V) \to \{0,1\}^V$, defined by $K(U) = K_U$, K_U being the characteristic function of U, is a bijection from $\mathcal{P}(V)$ to $\{0,1\}^V$ (theorem 13.9). □

Corollary 15.3: $\mathcal{P}(\underset{\sim}{N}) \underset{I}{=} \{0,1\}^{\underset{\sim}{N}}$.

Now we have established

$$\mathcal{P}(\{1, \ldots, n\}) \underset{I}{=} \{0,1\}^{\{1,\ldots,n\}}$$

we can use Theorem 13.3 to determine the number of elements of $\mathcal{P}(\{1,\ldots,n\})$, namely 2^n, the number of elements of $\{0,1\}^{\{1,\ldots,n\}}$. So we know that for finite sets the power set is much larger than the set itself. For infinite sets a similar proposition holds, only we cannot expect to prove it by just counting. Cantor provided us for this purpose with a revolutionary technique: *diagonalisation*. We will demonstrate it below.

Recall that each $f \in \{0,1\}^{\underset{\sim}{N}}$ can be represented by an infinite sequence of zero's and one's, namely by the sequence $f(0)$, $f(1)$, $f(2)$, ...

We see that $\{0,1\}^{\underset{\sim}{N}}$ has at least as many elements as $\underset{\sim}{N}$ has by considering the following elements of $\{0,1\}^{\underset{\sim}{N}}$:

1	1	1	1	1	...
0	1	1	1	1	...
0	0	1	1	1	...
0	0	0	1	1	...

and so on.

More precisely: there is an injection $f: \underset{\sim}{N} \to \{0,1\}^{\underset{\sim}{N}}$, namely the function f, defined by $f(n) = f_n$, where $f_n\ (i) = 1$ if $i \geqslant n$
0 else.

Definition 15.4: V is *equivalent to a subset* of W iff there is an injection $f: V \to W$ (i.e. $f: V \underset{I}{\leqslant} W$).

Notation: "$V \underset{1}{\leqslant} W$" stands for "there is an f such that $f\colon V \underset{1}{\leqslant} W$".

For "$V \underset{1}{\leqslant} W$" one also uses "the cardinality of V is at most the cardinality of W".

Examples: $\{2,3\} \underset{1}{\leqslant} \{4,5\}$, $\{2,3\} \underset{1}{\leqslant} \{4,5,6\}$,
$\mathbb{N} \underset{1}{\leqslant} \mathbb{Q}$, $\mathbb{N} \underset{1}{\leqslant} \{0,1\}^{\mathbb{N}}$.

Theorem 15.5: $\mathbb{N}$ is not equivalent to $\{0,1\}^{\mathbb{N}}$.

We have to show that there is no bijection $F\colon \mathbb{N} \to \{0,1\}^{\mathbb{N}}$. So it is sufficient to prove: every $F\colon\mathbb{N} \to \{0,1\}^{\mathbb{N}}$ is not surjective, in other words, for each $F\colon \mathbb{N} \to \{0,1\}^{\mathbb{N}}$ there is an $h \in \{0,1\}^{\mathbb{N}}$, such that $h \neq F(i)$ for all $i \in \mathbb{N}$.

Proof: Let $F\colon \mathbb{N} \to \{0,1\}^{\mathbb{N}}$. Then for all $i \in \mathbb{N}$, $F(i) \in \{0,1\}^{\mathbb{N}}$. For convenience we write F_i instead of $F(i)$. Because $F_i \in \{0,1\}^{\mathbb{N}}$ we can represent F_i by an infinite sequence of zero's and one's, namely by the sequence $F_i(0)$, $F_i(1)$, $F_i(2)$, ... Now we list the values of the F_i's in the following diagram:

	0	1	2	3	
F_0	0	1	0	0	...
F_1	1	0	1	0	...
F_2	0	0	1	1	...
⋮					

Consider the sequence $F_0(0)$, $F_1(1)$, $F_2(2)$, ... at the diagonal and define $h(i) := 1 - F_i(i)$, i.e. we interchange the zero's and the one's at the diagonal. Then for all $i \in \mathbb{N}$, $h(i) \neq F_i(i)$; hence $h \neq F_i$. □

Remark: The method, used in the proof, mentioned above, is called *diagonalisation* or the *diagonal method* of Cantor.

In view of the preceding, it is natural to say that the cardinality (power) of $\mathbb{N}$ is less than the cardinality of $\{0,1\}^{\mathbb{N}}$ (or that $\mathbb{N}$ is "smaller" than $\{0,1\}^{\mathbb{N}}$).

Definition 15.6: $V \underset{1}{<} W := V \underset{1}{\leqslant} W \wedge \neg V \underset{1}{=} W$.

Intuitively this means that the set V is *strictly smaller* than W.

So we can reformulate the above result as

Theorem 15.7 : $\underset{\sim}{N} \underset{1}{<} \{0,1\}^{\underset{\sim}{N}}$.

Since $\underset{1}{=}$ is an equivalence relation, the following holds: if U is not equivalent to V and V is equivalent to W, then U is not equivalent to W. Taking $\underset{\sim}{N}$ for U, $\{0,1\}^{\underset{\sim}{N}}$ for V and $\mathcal{P}(\underset{\sim}{N})$ for W, we can conclude:

Theorem 15.8: $\underset{\sim}{N}$ is not equivalent to $\mathcal{P}(\underset{\sim}{N})$, in other words, there is no bijection $g: \underset{\sim}{N} \to \mathcal{P}(\underset{\sim}{N})$.

The function $f: \underset{\sim}{N} \to \mathcal{P}(\underset{\sim}{N})$, defined by $f(n) = \{n\}$, is an injection from $\underset{\sim}{N}$ to $\mathcal{P}(\underset{\sim}{N})$. Hence

Theorem 15.9 : $\underset{\sim}{N} \underset{1}{<} \mathcal{P}(\underset{\sim}{N})$.

We can generalize this result to arbitrary sets V:

Theorem 15.10 : For all sets V, $V \underset{1}{<} \mathcal{P}(V)$.

To show this, it suffices to prove:

1. there is an injection $f: V \to \mathcal{P}(V)$, and
2. for every $g: V \to \mathcal{P}(V)$ there is a $U \in \mathcal{P}(V)$, such that for every $v \in V$, $U \neq g(v)$; in other words: no $g: V \to \mathcal{P}(V)$ is surjective and hence there is certainly no bijection $g: V \to \mathcal{P}(V)$.

Proof of 1: The function $f: V \to \mathcal{P}(V)$, defined by $f(v) = \{v\}$, is an injection.

Proof of 2: Let $g: V \to \mathcal{P}(V)$. Take $U := \{v \in V | v \notin g(v)\}$.
Then $U \in \mathcal{P}(V)$ and for every $v \in V$, $U \neq g(v)$. For suppose $U = g(v)$; then for every u, $u \in U \leftrightarrow u \in g(v)$. In other words, for every u, $u \notin g(u) \leftrightarrow u \in g(v)$.

Hence, in particular, $v \notin g(v) \leftrightarrow v \in g(v)$. Contradiction. □

Taking for V respectively $\underset{\sim}{N}$, $\mathcal{P}(\underset{\sim}{N})$, $\mathcal{PP}(\underset{\sim}{N})$, ..., the preceding theorem yields:

$$\underset{\sim}{N} \underset{1}{<} \mathcal{P}(\underset{\sim}{N}) \underset{1}{<} \mathcal{PP}(\underset{\sim}{N}) \underset{1}{<} \mathcal{PPP}(\underset{\sim}{N}) \text{ , etc.}$$

This suggests that there are many "degrees of infinity" ; the "degree of infinity of $\mathcal{PP}(\underset{\sim}{N})$" is higher than the one of $\mathcal{P}(\underset{\sim}{N})$, etc.

Definition 15.11: Let $a,b \in \underset{\sim}{R}$.

$[a,b] := \{x \in \underset{\sim}{R} | a \leqslant x \leqslant b\}$.

$(a,b) := \{x \in \underset{\sim}{R} | a < x < b\}$.

We will prove now that not only $\mathcal{P}(\underset{\sim}{N})$, but also $[0,1]$ is not equivalent to $\underset{\sim}{N}$.

We will present a direct proof here for historical reasons. The proof below is Poincaré's proof. The first direct proof was presented by Cantor.

Theorem 15.12: $\underset{\sim}{N}$ is not equivalent to $[0,1]$.

We have to show that there does not exist a bijection $g: \underset{\sim}{N} \to [0,1]$. So it suffices to prove that no function $g: \underset{\sim}{N} \to [0,1]$ is surjective; in other words, it suffices to prove: for each function $g: \underset{\sim}{N} \to [0,1]$ we can construct a real number b, $b \in [0,1]$, such that for every $n \in \underset{\sim}{N}$, $g(n) \neq b$.

Proof: Let $g: \underset{\sim}{N} \to [0,1]$. Given this $g: \underset{\sim}{N} \to [0,1]$, we can construct a chain of segments (in $\underset{\sim}{Q}$) $S_0, S_1, S_2, \ldots$, such that for every $n \in \underset{\sim}{N}$, $g(n)$ is not an element of S_n.

Note that $[0,1] = [0,\frac{1}{3}] \cup [\frac{1}{3},\frac{2}{3}] \cup [\frac{2}{3},1]$. At least one of those three subsets does not contain $g(0)$, say S_0.

Suppose $S_0, \ldots, S_n$ have already been defined, such that

1. for all i, $0 \leqslant i \leqslant n$, $g(i)$ is not an element of S_i
2. for all i, $0 \leqslant i < n$, $S_{i+1} \subseteq S_i$, and
3. for all i, $0 \leqslant i \leqslant n$, length of S_i equals 3^{-i-1}.

Let $S_n = [p_n, q_n]$. Now S_n is the union of

$$\left[p_n, \frac{2p_n + q_n}{3}\right], \left[\frac{2p_n + q_n}{3}, \frac{p_n + 2q_n}{3}\right] \text{ and } \left[\frac{p_n + 2q_n}{3}, q_n\right].$$

At least one of those three subsets of S_n does not contain $g(n+1)$, say S_{n+1}.

This chain of segments $S_0, S_1, S_2, \ldots$ determines a real number b (see Section 12), such that for every $n \in \underset{\sim}{N}$, b occurs in S_n, and hence $b \in [0,1]$. Now for every $n \in \underset{\sim}{N}$, $g(n)$ does not occur in S_n, while b does occur in S_n. Hence for every $n \in \underset{\sim}{N}$, $b \neq g(n)$ (see Figure 33). □

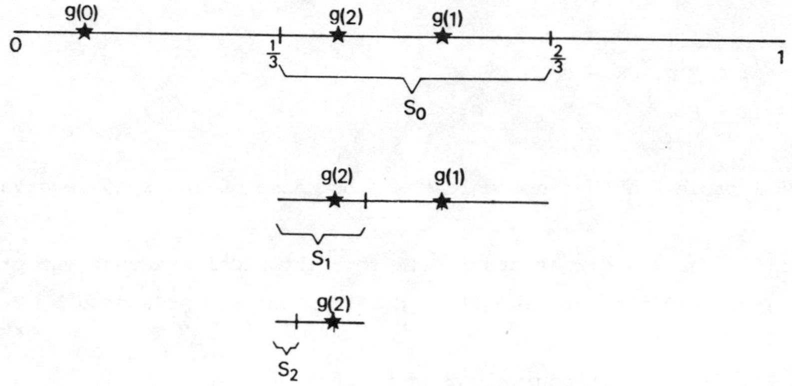

Fig. 33

The preceding theorem says that there does not exist a bijection from $\underset{\sim}{N}$ to $[0,1]$. (1)

On the other hand there does exist an injection from $\underset{\sim}{N}$ to $[0,1]$, for instance $f: \underset{\sim}{N} \to [0,1]$, defined by $f(n) = \frac{1}{n+1}$. (2)

From (1) and (2) we conclude

Theorem 15.13: $\underset{\sim}{N} \underset{I}{<} [0,1]$.

It is natural to call a set V denumerable if it is equivalent to $\underset{\sim}{N}$, since we can assign a number to each element of V.

Definition 15.14:

i) V is *denumerable* iff $\underset{\sim}{N} \underset{I}{=} V$; in other words, if there is a bijection $v: \underset{\sim}{N} \to V$.

ii) A bijection $v: \underset{\sim}{N} \to V$ is called a *denumeration* of V.

Given a bijection $v: \mathbb{N} \to V$, put $v_i := v(i)$. So, in the case that V is denumerable, V can be written as $\{v_i \mid i \in \mathbb{N}\}$ or as $\{v_0, v_1, v_2, \ldots\}$.

Theorem 15.12 tells us that $[0,1]$ is not denumerable, more precisely: for each denumeration $(g: \mathbb{N} \to [0,1])$ of elements of $[0,1]$ a real number b (between 0 and 1) can be constructed, such that b does not occur in that denumeration ($b \neq g(n)$ for all $n \in \mathbb{N}$).

On the other hand we can define only denumerably many individual real numbers (between 0 and 1). This restriction is inherent to our language.

Definition 15.15: V is *countable* iff V is finite or denumerable.

Remark: For the moment we tacitly use the term "finite". A definition is given in 16.1.

Warning: The usage of the terminology is not firmly established. Instead of "denumerable" some authors use "countably infinite".

The following theorem of *Cantor-Bernstein* is a convenient tool for establishing equivalence of sets.

Theorem 15.16: If $V \underset{1}{\leqslant} W$ and $W \underset{1}{\leqslant} V$, then $V \underset{1}{=} W$.

In other words: if there is an injection from V to W and an injection from W to V, then there is a bijection from V to W.

Proof: Suppose $V \underset{1}{\leqslant} W$ and $W \underset{1}{\leqslant} V$, i.e. there is an injection $f: V \to W$ and an injection $g: W \to V$.

$$V \xrightarrow{f} W \xrightarrow{g} V, \qquad \underset{g \circ f}{\longrightarrow}$$

Note that $W \underset{1}{=} \text{Ran}(g)$. So, in order to prove that $W \underset{1}{=} V$, it suffices to prove: $\text{Ran}(g) \underset{1}{=} V$. Now, $\text{Ran}(g) \subseteq V$ and $g \circ f$ is an injection from V to $\text{Ran}(g)$. Hence $V \underset{1}{\leqslant} \text{Ran}(g)$. So it suffices to prove the following lemma:

Lemma 15.17: If $C \subseteq V$ and $V \underset{1}{\leqslant} C$, then $C \underset{1}{=} V$.

Proof of Lemma 15.17: Suppose $C \subseteq V$ and $V \leqslant_1 C$. $V \leqslant_1 C$ means: there is an injection $h: V \to C$ (see Figure 34).

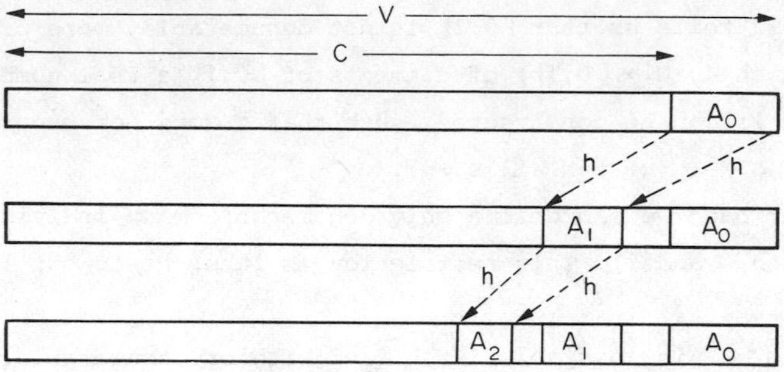

Fig. 34

Now let $A_0 := V - C$. V, together with its subset A_0, is reproduced in C by the function $h: V \to C$.

Let $A_1 := h(A_0)$, i.e. A_1 is the set of all h-images of elements of A_0. By applying h once more, C, together with its subset A_1, is reproduced in C.

Let $A_2 := h(A_1)$.

In general: $A_0 := V - C$
$A_{n+1} := h(A_n)$.

Next we define $F: V \to C$ as follows:
$F(a) := a$ if $a \notin \bigcup_n A_n$
$F(a) := h(a)$ if $a \in \bigcup_n A_n$.

We shall prove now that $F: V \to C$ is a bijection from V to C, in other words, that $V =_1 C$.

Claim 1: F is injective, i.e. if $a \neq b$, then $F(a) \neq F(b)$.
Proof: Suppose $a \neq b$.

i) $a \notin \bigcup_n A_n$ and $b \notin \bigcup_n A_n$. Then $F(a) = a$ and $F(b) = b$. Hence, because $a \neq b$, $F(a) \neq F(b)$.

ii) $a \notin \bigcup_n A_n$ and $b \in \bigcup_n A_n$. Then $F(a) = a$ and $F(b) = h(b)$. Hence $F(a) \notin \bigcup_n A_n$ and $F(b) \in \bigcup_n A_n$. Hence $F(a) \neq F(b)$.

iii) $a \in \bigcup_n A_n$ and $b \in \bigcup_n A_n$. Then $F(a) = h(a)$ and $F(b) = h(b)$. Hence, since $a \neq b$ and h is injective, $F(a) \neq F(b)$.

Claim 2: F is surjective, i.e. for every $c \in C$ there is an element $v \in V$, such that $c = F(v)$.

Proof: Suppose $c \in C$; so $c \notin A_0$. If $c \notin \bigcup_n A_n$, then $c = F(c)$. If $c \in \bigcup_n A_n$, say $c \in A_p$, then, since $c \notin A_0$, $p \geqslant 1$. So there is an element $c' \in A_{p-1}$, such that $c = F(c')$. □

Recall that $(0,1) := \{x \in \mathbb{R} | 0 < x < 1\}$ and $[0,1] := \{x \in \mathbb{R} | 0 \leqslant x \leqslant 1\}$ (15.11).

The theorem of Cantor-Bernstein now yields a simple proof of $(0,1) =_I [0,1]$.

<u>Theorem 15.18</u>: $(0,1) =_I [0,1]$.

Proof: $id_{(0,1)} : (0,1) \to [0,1]$ is an injection. Hence $(0,1) \leqslant_I [0,1]$. (1)

$f: [0,1] \to (0,1)$, defined by $f(x) = \frac{1}{4}(x+1)$, is an injection (see Figure 35). Hence $[0,1] \leqslant_I (0,1)$. (2)

From (1), (2) and the theorem of Cantor-Bernstein it follows that $(0,1) =_I [0,1]$. □

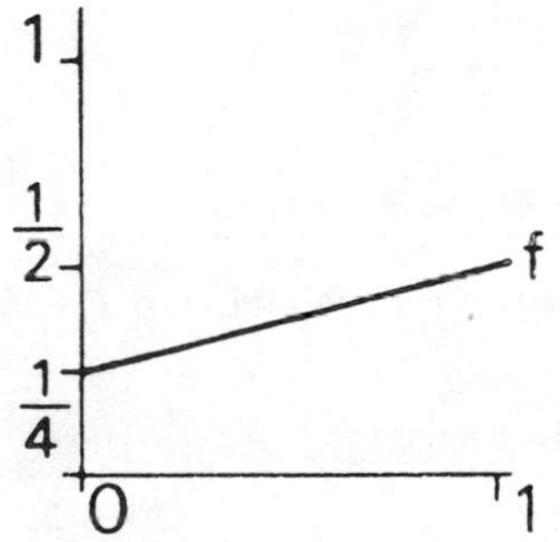

Fig. 35

The following theorems state that equivalence is compatible with the familiar set theoretic operations. The proofs are so simple that we leave them to the reader.

Theorem 15.19: Suppose $A \cap C = \emptyset$ and $B \cap D = \emptyset$.

i) If $A \underset{1}{\leqslant} B$ and $C \underset{1}{\leqslant} D$, then $A \cup C \underset{1}{\leqslant} B \cup D$.

ii) If $A \underset{1}{=} B$ and $C \underset{1}{=} D$, then $A \cup C \underset{1}{=} B \cup D$.

Theorem 15.20:

i) If $A \underset{1}{\leqslant} B$ and $C \underset{1}{\leqslant} D$, then $A \times C \underset{1}{\leqslant} B \times D$.

ii) If $A \underset{1}{=} B$ and $C \underset{1}{=} D$, then $A \times C \underset{1}{=} B \times D$.

iii) If $A \underset{1}{\leqslant} B$ and $C \underset{1}{\leqslant} D$, then $A^C \underset{1}{\leqslant} B^D$.

iv) If $A \underset{1}{=} B$ and $C \underset{1}{=} D$, then $A^C \underset{1}{=} B^D$.

Theorem 15.21:

i) If $A \underset{1}{\leqslant} B$, then $\mathcal{P}(A) \underset{1}{\leqslant} \mathcal{P}(B)$.

ii) If $A \underset{1}{=} B$, then $\mathcal{P}(A) \underset{1}{=} \mathcal{P}(B)$.

Theorem 15.22:

i) If $A \underset{1}{\leqslant} B$, $a \in A$ and $b \in B$, then $A - \{a\} \underset{1}{\leqslant} B - \{b\}$.

ii) If $A \underset{1}{=} B$, $a \in A$ and $b \in B$, then $A - \{a\} \underset{1}{=} B - \{b\}$.

Theorem 15.23: If $A \cap B = \emptyset$, then

i) $C^{A \cup B} \underset{1}{=} C^A \times C^B$

ii) $(C^A)^B \underset{1}{=} C^{A \times B}$

iii) $A \times (B \times C) \underset{1}{=} (A \times B) \times C$

iv) $A \times B \underset{1}{=} B \times A$

v) $A^C \times B^C \underset{1}{=} (A \times B)^C$

The Informal Theory of Cardinal Numbers.

We ordinarily associate with every finite set a certain abstract entity which we call the 'number of elements' in the set. The possible numbers form an unending sequence, 0, 1, 2, 3, ... , and we determine which one of them applies to any particular set A by counting the members of A. It may readily be seen that two finite sets A and B have the same number, in this elementary sense, if and only if they can be paired off against each other, more precisely if they are equivalent.

For infinite sets, common sense does not so easily envisage a corresponding use of numbers. But every set, finite or infinite, may reasonably be thought of as having a magnitude, and we may attempt to generalize the elementary notion of number so that it will apply to all sets, without restriction.

Our problem is to define formally some entity that can serve to represent the common magnitude of all those sets which are equivalent to any given set. For the present we will use brute force:

Postulate for Cardinal Numbers: With every set A is associated a well defined set $|A|$, called the cardinal number of A, in such a way that the condition $A \underset{I}{=} B \leftrightarrow |A| = |B|$ is satisfied.

We shall not be in a position to define $|A|$ for an arbitrary set A until a much later stage in the development of set theory; see chapter II, section 10. In order to help the intuition, the reader may think of all subsets of a given set S, and define the cardinality, $|A|$, of a set A as the equivalence class of A under the relation '$\underset{I}{=}$' . This is a "safe" imitation of Frege's cardinal definition. $|\emptyset| = 0$, $|\{\emptyset\}| = 1$, $|\{\emptyset,\{\emptyset\}\}| = 2$, ...
Aleph-zero, $\aleph_0$, is by definition the cardinal number of $\underset{\sim}{N}$, i.e. $\aleph_0 := |\underset{\sim}{N}|$.
$\mathfrak{c}$ is by definition the cardinal number of $\underset{\sim}{R}$, i.e. $\mathfrak{c} := |\underset{\sim}{R}|$.

We introduce arithmetical operations on arbitrary cardinal numbers. The means for doing this lie ready to hand, in the set theoretic operations which yield the union of two sets, the cartesian product of two sets, and the set of functions from one given set to another. Our definition, moreover, will in the special case of finite cardinals conform to the well-known operations on natural numbers.

Definition: Let $\mathfrak{n}$ and $\mathfrak{m}$ be any two given cardinal numbers, and let A and B

be disjoint sets such that $\mathrm{n} = |A|$ and $\mathrm{m} = |B|$. Then

$$\mathrm{n} + \mathrm{m} := |A \cup B|$$
$$\mathrm{n.m}\ (= \mathrm{n\,m}) := |A \times B|$$
$$\mathrm{n}^{\mathrm{m}} := |A^B| \,.$$

Note that the operations are well-defined, since $\underset{1}{=}$ is a congruence relation with respect to disjoint union, product and exponent (15.9 (*ii*), 15.20 (*ii*), (*iv*)).

We can now show, by elementary set-theoretic argument, that most of the formal rules of the arithmetic of the natural numbers remain valid in the wider system of cardinal arithmetic. In the table which now follows, we give some of the more important of these rules, accompanied by set-theoretic equivalents. It is to be understood that, wherever unions are involved, the sets in question are disjoint.

(*i*)	$(\mathrm{n} + \mathrm{m}) + \mathrm{p} = \mathrm{n} + (\mathrm{m} + \mathrm{p})$	, $(A \cup B) \cup C = A \cup (B \cup C)$	(5.2)
(*ii*)	$\mathrm{n} + \mathrm{m} = \mathrm{m} + \mathrm{n}$	, $A \cup B = B \cup A$	(5.2)
(*iii*)	$(\mathrm{n\,m})\mathrm{p} = \mathrm{n}(\mathrm{m\,p})$	, $(A \times B) \times C \underset{1}{=} A \times (B \times C)$	(15.23)
(*iv*)	$\mathrm{n\,m} = \mathrm{m\,n}$	, $A \times B \underset{1}{=} B \times A$	(15.23)
(*v*)	$\mathrm{n}(\mathrm{m} + \mathrm{p}) = \mathrm{n\,m} + \mathrm{n\,p}$	, $A \times (B \cup C) = (A \times B) \cup (A \times C)$	(9.3)
(*vi*)	$\mathrm{n}^{\mathrm{m}}\,\mathrm{n}^{\mathrm{p}} = \mathrm{n}^{\mathrm{m}+\mathrm{p}}$	, $A^B \times A^C \underset{1}{=} A^{B \cup C}$	(15.23)
(*vii*)	$\mathrm{n}^{\mathrm{p}}\,\mathrm{m}^{\mathrm{p}} = (\mathrm{n\,m})^{\mathrm{p}}$	, $A^C \times B^C \underset{1}{=} (A \times B)^C$	(15.23)
(*viii*)	$(\mathrm{n}^{\mathrm{m}})^{\mathrm{p}} = \mathrm{n}^{\mathrm{mp}}$	, $(A^B)^C \underset{1}{=} A^{B \times C}$	(15.23)

Many of the results concerning cardinal numbers that we have obtained in the course of this chapter assume a very simple form when expressed in arithmetical language, as for instance the following:

(1) If $\mathrm{n} = |A|$, then $|\mathcal{P}(A)| = 2^{\mathrm{n}}$. (Theorem 15.2.)

(2) $2^{\aleph_0} = \mathfrak{c}$, where $\mathfrak{c}$ is the cardinal number of the continuum. (Theorem 18.5.)

(3) For any infinite m, $\mathrm{m} + \aleph_0 = \mathrm{m}$. (Corollary 17.6.)

(4) For any finite n, $\mathrm{n}.\aleph_0 = \aleph_0$. (Corollary 17.8.)

Furthermore, results of this kind can often be proved easily by use of the arithmetical formalism. Thus, for instance, we can obtain

(5) For any finite n, $\mathfrak{c}^{\mathrm{n}} = \mathfrak{c}$

as a corollary to (2) and (4), by applying the principles (*i*) - (*viii*) in the following way:

$$\mathfrak{c}^{\mathrm{n}} = (2^{\aleph_0})^{\mathrm{n}} = 2^{\aleph_0 . \mathrm{n}} = 2^{\aleph_0} = \mathfrak{c}\,.$$

Exercises

1. Prove: if $A \subseteq B$ and $A \underset{1}{=} A \cup C$, then $B \underset{1}{=} A \cup B \cup C$.

2. Prove: if $A - B \underset{1}{=} B - A$, then $A \underset{1}{=} B$.

3. If $B \subseteq D$ and $C \subseteq A$ and $A \cup B \underset{1}{=} B$, then $C \cup D \underset{1}{=} D$.

4. Suppose $A \cap C = \emptyset$, $B \cap D = \emptyset$, $A \underset{1}{=} C$ and $B \underset{1}{=} D$.
If $A \cup C \underset{1}{=} B \cup D$, then $A \underset{1}{=} B$.

5. Let f be a function. Prove: $f \underset{1}{=} \text{Dom}(f)$.
Does $f \underset{1}{=} \text{Ran}(f)$ hold?

6. Suppose $a \notin A$, $b \notin B$. If $A \cup \{a\} \underset{1}{=} B \cup \{b\}$, then $A \underset{1}{=} B$.

7. If Z is a partition of A, then the following holds:
i) $\mathcal{P}(Z) \underset{1}{\leq} \mathcal{P}(A)$
ii) $Z \underset{1}{<} \mathcal{P}(A)$.

8. Let $B = \{f \in \mathbb{N}^{\mathbb{N}} | f \text{ is bijective}\}$ (B is the set of permutations of $\mathbb{N}$).
Prove that $B \underset{1}{=} \mathbb{N}^{\mathbb{N}}$.

16 FINITE AND INFINITE

Definition 16.1: V is *finite* iff there is some natural number n ($n \in \mathbb{N}$) such that $V \underset{1}{=} \{x \in \mathbb{N} | x < n\}$.

Examples: $\emptyset$ is finite, for $\emptyset \underset{1}{=} \{x \in \mathbb{N} | x < 0\}$.
$\{-7, 8\frac{1}{2}, 90\}$ is finite, for $\{-7, 8\frac{1}{2}, 90\} \underset{1}{=} \{x \in \mathbb{N} | x < 3\}$.

Definition 16.2: V is *infinite* iff V is not finite.

Examples of infinite sets: $\mathbb{N}$, $\mathbb{Z}$, $\mathbb{Q}$, $\mathbb{R}$, $\mathbb{N}^2$, $\mathbb{Z}^2$, $\mathbb{Q}^2$, $\mathbb{R}^2$, $\{x \in \mathbb{Q} | 0 \leqslant x \leqslant 1\}$, $\{x \in \mathbb{R} | 0 \leqslant x \leqslant 1\}$, $\{0,1\}^{\mathbb{N}}$, $\mathbb{N}^{\mathbb{N}}$, $\mathbb{N}^{\mathbb{R}}$, $\mathcal{P}(\mathbb{N})$, $\mathcal{P}(\mathbb{R})$, $\mathcal{P}\mathcal{P}(\mathbb{N})$, ...

<u>Theorem 16.3</u> *(AC)*: An infinite set contains a denumerable subset (i.e. if V is infinite, then $\mathbb{N} \leqslant_1 V$).

Proof: Suppose V is infinite.

Because V is non-empty, V contains at least one element; we choose one: x_0.

Because V is infinite, $V - \{x_0\} \neq \emptyset$; we choose an element of $V - \{x_0\}$: x_1.

Suppose the elements $x_0, \ldots, x_n$ have been chosen. Because V is infinite, $V - \{x_0, \ldots, x_n\} \neq \emptyset$; we choose an element of $V - \{x_0, \ldots, x_n\}$: x_{n+1}.

Thus we produce by repeating the process a denumerable subset of V, namely $\{x_0, x_1, x_2, \ldots\}$. □

In the proof above, we have done something that needs further justification: we have created a set by making infinitely many choices. It is clear that we can always choose an element from a non-empty set; this is exactly what "non-empty" means: $A \neq \emptyset$ iff $\exists x\, [x \in A]$. And "we can choose an element from A" also means: $\exists x\, [x \in A]$. By an easy induction argument one also shows that one can make a finite number of choices from non-empty sets:
$A_1 \neq \emptyset \wedge \ldots \wedge A_n \neq \emptyset \rightarrow \exists x_1 [x_1 \in A_1] \wedge \ldots \wedge \exists x_n [x_n \in A_n]$.

However, making infinitely many choices is quite a different matter. We present a suggestive example as an illustration. Let $\forall n \in \mathbb{N}\, [A_n \neq \emptyset]$; then we want to formulate "we can choose from each A_n an element". A naive solution would be: $\exists x_1 [x_1 \in A_1] \wedge \exists x_2 [x_2 \in A_2] \wedge \exists x_3 [x_3 \in A_3] \wedge \ldots$.
However, the three dots indicate our failure; an infinite sentence does not clarify our intentions. A better attempt would be "we have a method which allows us to pick from each A_n an element". Now we can translate this in the language of set theory: there is a function f, defined on $\mathbb{N}$, such that for each n, $f(n) \in A_n$. In symbols: $\exists f\, [\mathrm{Dom}(f) = \mathbb{N} \wedge \forall n \in \mathbb{N}\, [f(n) \in A_n]]$.
The infinite sequence of choices is embodied in a function f, which "chooses" elements for us. This function is called a *choice function*. It is by no means clear, although plausible on the ground of experience with finite sets, that such a choice function *exists*. It is, so to speak, a condition on the

richness of the universe to assume that each family of non-empty sets has a choice function.

The latter statement, "*each family of non-empty sets has a choice function*", is called the *axiom of choice* (*AC*) (see Figure 36).

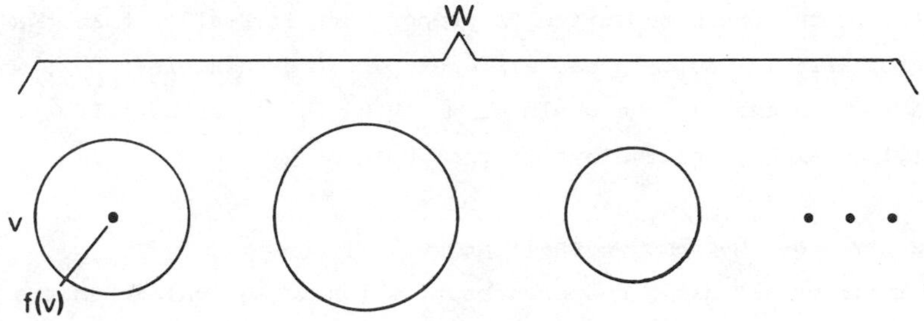

Fig. 36

The choice function of the family W has domain W, i.e. f chooses from $v \in W$ the element $f(v)$.

We will apply the axiom of choice rather loosely. In chapter II and III we return to the matter of AC (see Chapter II, Section 9).

For the concrete examples of infinite sets, we have introduced so far, Theorem 16.3 is no surprise. It is a trifle to exhibit denumerable subsets. Some examples are shown below.

Set	*Denumerable subset*
$\underset{\sim}{R}$	$\underset{\sim}{N}$
$[0,1]$	$\{\frac{1}{n+1} \mid n \in \underset{\sim}{N}\}$
$\underset{\sim}{Q}$	$\{2^{-n} \mid n \in \underset{\sim}{N}\}$
$\mathcal{P}(\underset{\sim}{N})$	$\{\{n\} \mid n \in \underset{\sim}{N}\}$
$\{0,1\}^{\underset{\sim}{N}}$	$\{K_{\{n\}} \mid n \in \underset{\sim}{N}\}$
$\underset{\sim}{N}^{\underset{\sim}{N}}$	$\{f_n \mid n \in \underset{\sim}{N}\}$ with $f_n(x) = n$ for all x.

No recourse to AC was necessary, the required subsets could quite simply be constructed (or defined). As happens to be the case, one can almost always in the familiar concrete examples exhibit denumerable subsets, without the use of *AC*. Facts of this sort may provide a historical explanation of why the observation of *AC* came so late in the development of mathematics.

In the examples 1, 2 and 3 of Section 15, sets are mentioned, which are equivalent to a proper subset of themselves, namely $\underset{\sim}{N} \underset{1}{=} \underset{\sim}{N}_{even}$, $\underset{\sim}{R} \underset{1}{=} \underset{\sim}{R}_{+}$ and $\underset{\sim}{R} \underset{1}{=} (-1,1)$. Hence, the proposition "a proper part is smaller than the whole" is false for infinite sets: $\underset{\sim}{N}$ has a proper part, for example $\underset{\sim}{N}_{even}$, which is as big as (equivalent to) the whole of $\underset{\sim}{N}$; $\underset{\sim}{R}$ has a proper part, for instance $\underset{\sim}{R}_{+}$ or $(-1,1)$, which is equivalent to the whole of $\underset{\sim}{R}$.

In the next two theorems we shall prove:

I Every infinite set has a proper subset, which is equivalent to the whole.

II If V is finite, then indeed a proper subset of V is smaller than the whole of V.

Let $\Phi(V)$:= V has a proper subset, which is equivalent to the whole of V.

Then I and II say that the property Φ is characteristic for infinite sets, i.e. every infinite set has this property Φ and every set with the property Φ is infinite.

Theorem 16.4 (*AC*): If V is infinite, then there is a proper subset of V, which is equivalent to V.

Proof: Let V be an infinite set.

By Theorem 16.3 there is a subset $W = \{x_0, x_1, x_2, \ldots\}$ of V, which is denumerable. $V - \{x_0\}$ is a proper subset of V, which is equivalent to V. For the function $g\colon V \to V - \{x_0\}$, defined by

$g(x) = x$ if $x \notin \{x_0, x_1, x_2, \ldots\}$

$g(x_i) = x_{i+1}$ for all $i \in \underset{\sim}{N}$,

is a bijection from V to $V - \{x_0\}$ (see Figure 37). □

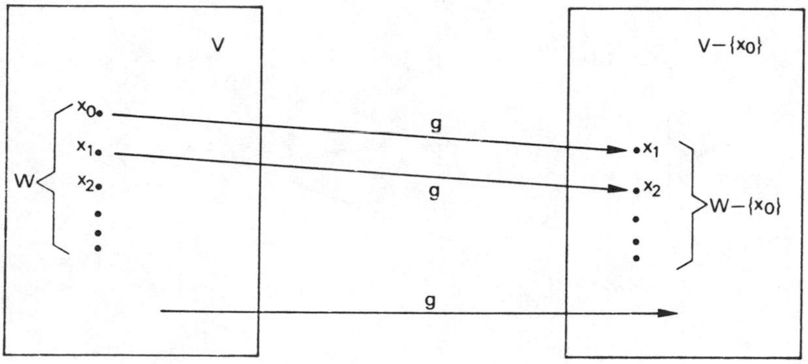

Fig. 37

In the proof of Theorem 16.4 the axiom of choice is used only (via Theorem 16.3) in order to construct a denumerable subset W of V.

For those sets V, for which we can indicate a concrete, denumerable, subset W without using the axiom of choice, for instance for $[0,1]$, $\mathbb{Q}$, $\mathcal{P}(\mathbb{N})$ and $\{0,1\}^{\mathbb{N}}$, we can thus prove Theorem 16.4 without making use of the axiom of choice.

The converse of the last theorem follows from the following

<u>Theorem 16.5</u>: A finite set V is not equivalent to a proper subset.

Proof: We prove this theorem by means of mathematical induction on the number of elements of V.

If the number of elements of V is 0, then $V = \emptyset$ and because $\emptyset$ does not have a proper subset, the theorem is trivially true.

Now suppose that the theorem has been proved for all sets with at most n elements (induction hypothesis), and that V is a set with n+1 elements, say $V = \{a_1, \ldots, a_n, a_{n+1}\}$. We have to show now that V does not have a proper subset, which is equivalent to V. So suppose $W \subseteq V$, $W \neq V$ and $V =_1 W$, in other words:

i) $W \subseteq V$,

ii) there is an element of V, which is not in W; it is no restriction to take a_{n+1},

iii) there is a bijection $f: V \to W$.

iv) Let $a_k := f(a_{n+1})$. Hence $a_k \in W$ (see Figure 38).

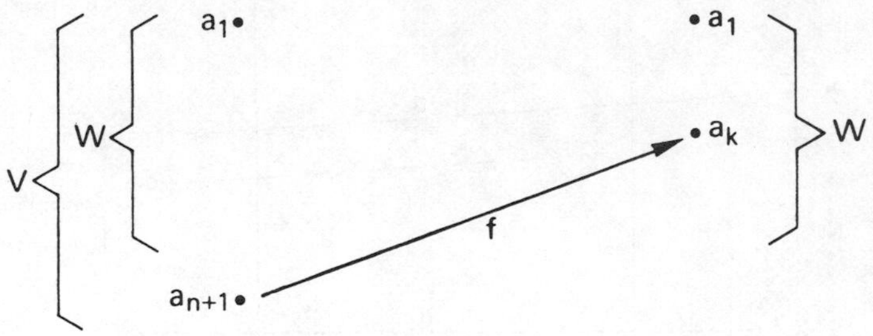

Fig. 38

Now the restriction $f \restriction W$ of f to W is an injection from W to W. (1)

The f-image of W, this is $\{f(w) | w \in W\}$, is a subset of W. (2)

And by induction hypothesis W does not have a proper subset, which is equivalent to W. (3)

From (1), (2) and (3) it follows that $\{f(w) | w \in W\} = W$; so, in particular, that $a_k = f(a_j)$ for some $a_j \in W$. Contradiction with iii) and iv). □

Combining the last two theorems, we get

Corollary 16.6 (*AC*) : V is infinite if and only if it is equivalent to a proper subset.

We have defined "V is infinite" as "V is not finite". Corollary 16.6 provides another characterization of infinity, known under the name "Dedekind infinite".

Definition 16.7: V is *Dedekind infinite* iff V has a proper subset, which is equivalent to V.

Using this terminology, our corollary says: V is infinite if and only if V is Dedekind infinite.

Theorem 16.8: V is Dedekind infinite if and only if $\underset{\sim}{N} \leqslant_1 V$.

Proof of $\leftarrow$: Suppose f: $\underset{\sim}{N} \underset{I}{\preccurlyeq} V$. Define $W := V - \{f(0)\}$. Then $W \underset{\neq}{\subset} V$ and $V \underset{I}{=} W$, for g: $V \rightarrow W$, defined by $g(f(i)) = f(i+1)$

$$g(x) = x \text{ if } x \neq f(i) \ , \ i = 0,1,2,\ldots \ ,$$

is a bijection from V to W.

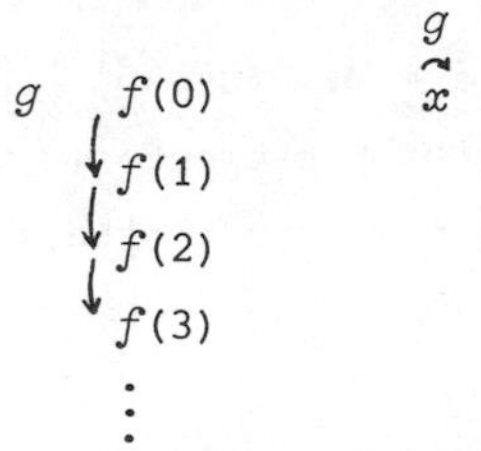

Proof of $\rightarrow$: Let $W \underset{\neq}{\subset} V$, $a \in V$, $a \notin W$ and f: $V \underset{I}{=} W$.
Define g:$\underset{\sim}{N} \rightarrow V$ by $g(0) = a$

$$g(k) = \underbrace{f(\ldots(f(a)))}_{k \text{ times}}.$$

More precisely: $g(0) := a$

$$g(k+1) := f(g(k)).$$

We have to prove now that g is an injection, i.e. $i \neq j \rightarrow g(i) \neq g(j)$. We prove this by means of mathematical induction on i.

For i=0 and $0 \neq j$, $g(0) = a \notin W$ and $g(j) \in W$, hence $g(0) \neq g(j)$.

For $i=k+1$ and $k+1 \neq j$, we prove that $g(k+1) \neq g(j)$ by means of mathematical induction on j.

For j=0, $g(0) = a \notin W$ and $g(k+1) \in W$, hence $g(k+1) \neq g(0)$.

For $j=l+1$ it follows because of $k+1 \neq j$, that $k \neq l$; hence, by induction hypothesis, $g(k) \neq g(l)$. Because f is an injection, $f(g(k)) \neq f(g(l))$, in other words, $g(k+1) \neq g(l+1)$, or $g(i) \neq g(j)$. □

Theorem 16.8 may seem a bit pedantic after the preceding results. The point, however, is that *AC* is not used in the proof, so that 16.8 gives a "clean" equivalent of Dedekind infinite.

An alternative formulation of 16.8 is

Corollary 16.9: V is Dedekind infinite iff it contains a denumerable subset.

Corollary 16.9 gives a positive meaning to the concept of infinity. It is therefore preferred in certain constructive approaches to mathematics.

Exercises

1. Suppose that A and B are finite sets and that $A \cap B = \emptyset$. Prove by mathematical induction: if A has m elements and B has n elements ($m,n \in \underset{\sim}{N}$), then $A \cup B$ has $m+n$ elements. Conclude that the union of two finite sets is finite.

2. Prove, by making use of exercise 1, that if A is infinite and B is finite, $B \subseteq A$, then $A-B$ is infinite.

3. Show that A is infinite if and only if for all $n \in \underset{\sim}{N}$, A has a subset with n elements.

4. Prove by mathematical induction: if A has m elements and B has n elements (m, $n \in \underset{\sim}{N}$), then $A \times B$ has $m.n$ elements. Conclude that the cartesian product of two finite sets is finite.

17 DENUMERABLE SETS

There is a curious application of set theory to the field of hotel-management, which is attributed to David Hilbert. We will present here a free version.

In a town, which we shall refer to as town X, for reasons not to be specified, there is a remarkable hotel, the *Hilbert Hotel*, which is distinguished from the average hotel by its size. The Hilbert Hotel is widely known for the fact that it contains denumerably many rooms, numbered 0,1,2,3, ...

At the day of a big congress the Hilbert Hotel was fully booked, when a late guest wished to register. Of course, he was kindly, but firmly, shown the door, but because of his persistence (the Hilbert Hotel was the only hotel in X !) the desk clerk called for the manager. The manager apologized profusely, quoting the hotel-axiom: full is full. Fortunately the daughter of the manager, who could not sleep, appeared. The clever girl considered the problem for a moment and then gave a solution, which was both brilliant and simple: "Dad, ask each guest to move to the room with the next number, then this gentleman can take room number 0". This solved the whole problem; the manager, immensely relieved, noticed that in this way he could accomodate another hundred guests (Theorem 17.4).

However, just when everybody was about to retire cheerfully for the night, a bus with the complete delegation of Smoryn drove up. Now, if this delegation had been finite, the manager could have accommodated it easily. Unfortunately, the delegation of Smoryn was denumerable! It was again the daughter of the manager, who provided a solution: "It is quite simple, Dad, this time you ask every guest to move to the room of which the number is twice the number of his present room. Then all rooms with an odd number will be vacant, and the representatives of Smoryn can be put in these rooms" (Theorem 17.4). The reader can easily verify that her proposal was correct; everything worked out perfectly.

So far the Hilbert Hotel had overcome all difficulties. The real problems started only, when the next day each guest wanted to accommodate denumerably many friends. How was the manager to provide everybody with a room? Even the daughter found that problem not totally trivial (as she said herself). She retreated in the scullery, from which she emerged after a quarter of an hour, with the words: "Dad, it is trivial after all". Indeed, her solution turned out to be not so complicated. We invite the reader to give a solution himself (see Theorem 17.7).

The Hilbert Hotel is still flourishing. What has become of the clever daughter is not known to us; some say she took up the study of mathematics, others say she has married and lives comfortably and contently in a modest cottage at the edge of the forest.

Now back to real life.

Theorem 17.1: A subset of a denumerable set is finite or denumerable (countable).

Proof: Let A be denumerable via f, i.e. $f\colon \underset{\sim}{N} \underset{I}{=} A$. If the subset B of A is finite, the theorem has been proved; so suppose that B is not finite.

Define a denumerating function (denumeration) g by $g(0) = f(k)$, where k is the least number such that $f(k) \in B$. Suppose that $g(0), \ldots, g(n)$ have been defined, then put $g(n+1) = f(l)$, where l is the least number such that $f(l) \in B - \{g(0), \ldots, g(n)\}$.

Note that $\mathrm{Dom}(g) = \underset{\sim}{N}$, since B is not finite, and that g is injective. So $g: \underset{\sim}{N} \underset{1}{\sim} B$, hence B is denumerable. □

Theorem 17.2: $\underset{\sim}{Z}$ is denumerable.

Proof: $h: \underset{\sim}{N} \to \underset{\sim}{Z}$, defined by $h(x) = y+1$ if $x = 2y+1$
$= -y$ if $x = 2y$, is a bijection from $\underset{\sim}{N}$ to $\underset{\sim}{Z}$ (see the diagram below). □

$\underset{\sim}{N}$	0	1	2	3	4	5	6	...
	↓	↓	↓	↓	↓	↓	↓	
$\underset{\sim}{Z}$	0	1	−1	2	−2	3	−3	...

The following lemma is trivial, but handy for applications.

Lemma 17.3: If $f: \underset{\sim}{N} \to A$ is surjective, then A is countable.

Proof: Let $f: \underset{\sim}{N} \to A$ be surjective.

We define a function g as follows:

$g(0) = f(0)$

$g(n+1) = f(k)$ if $A \neq \{g(0), \ldots, g(n)\}$, where k is the smallest number such that $f(k) \notin \{g(0), \ldots, g(n)\}$. Note that the domain of g is either a finite set $\{0, 1, \ldots, m\}$ or $\underset{\sim}{N}$. Since g clearly is a bijection, we have shown that A is finite or denumerable. □

If $A = \mathrm{Ran}(f)$ and $\mathrm{Dom}(f) = \underset{\sim}{N}$, then we say that A is *enumerated* by f. One can view f as a machine producing elements of A, but f may make repetitions. The function g, exhibited above, is obtained by running through $\underset{\sim}{N}$ and picking only "new" values of f. We have shown now that for infinite sets "enumerable" and "denumerable" amounts to the same. This phenomenon is exploited in recursion theory.

Theorem 17.4: If A and B are denumerable and if C is finite, then $A \cup B$ and $A \cup C$ are denumerable.

Proof:

i) Let $f\colon \underset{\sim}{N} \underset{1}{=} A$ and $g\colon \underset{\sim}{N} \underset{1}{=} B$. Define $h\colon \underset{\sim}{N} \to A \cup B$ by $h(2n) = f(n)$
$h(2n+1) = g(n)$.
Then h is a surjection from $\underset{\sim}{N}$ to $A \cup B$. Hence, by lemma 17.3, $A \cup B$ is denumerable.

ii) Let $f\colon \underset{\sim}{N} \underset{1}{=} A$ and $g\colon \{0,\ldots,n\} \underset{1}{=} C$. Define $h\colon \underset{\sim}{N} \to A \cup C$ by
$h(i) = g(i)$ for $i \leqslant n$
$h(n+i+1) = f(i)$.
Then h again is a surjection from $\underset{\sim}{N}$ to $A \cup C$. Hence, by lemma 17.3, $A \cup C$ is denumerable. The case $C = \emptyset$ is trivial.

ad i) A: $f(0)$ $f(1)$ $f(2)$ ...
B: $g(0)$ $g(1)$ $g(2)$ ...
$A \cup B$: $f(0), g(0), f(1), g(1), f(2), g(2), \ldots$

ad ii) C: $g(0), \ldots, g(n)$
A: $f(0), f(1), f(2), \ldots$
$A \cup C$: $g(0), \ldots, g(n), f(0), f(1), f(2), \ldots$ □

Summary: The union of two denumerable sets is denumerable; the union of a denumerable set and a finite set is denumerable.

From Theorem 17.4 we easily deduce the following

Corollary 17.5: If A is Dedekind infinite and B is countable, then $A \cup B \underset{1}{=} A$.

Proof: Suppose A is Dedekind infinite and B is countable. By corollary 16.9, A can be written as $A = A_0 \cup A_1$, where $A_0 \cap A_1 = \emptyset$ and A_0 is denumerable. By Theorem 17.4, $A_0 \cup B \underset{1}{=} A_0$. From this it immediately follows that $A \cup B \underset{1}{=} A$; for $A \cup B = (A_0 \cup B) \cup A_1$ and $A_0 \cup B \underset{1}{=} A_0$, hence $A \cup B \underset{1}{=} A_0 \cup A_1$ and $A_0 \cup A_1 = A$. □

<u>Corollary 17.6</u> (*AC*): If A is infinite and B is countable, then $A \cup B =_I A$.

Proof: Use Theorem 16.4 (*AC*) and Corollary 17.5. □

<u>Theorem 17.7</u>: i) $\underset{\sim}{N}^2 =_I \underset{\sim}{N}$

ii) The cartesian product of two denumerable sets is denumerable.

Proof: i) In the picture below, we have shown a small part of $\underset{\sim}{N}^2$, geometrically speaking the lattice points with non-negative integer coordinates. We want to make a "walk" through the plane passing once and only once through each lattice point. A systematic walk is indicated by the arrows: follow a line from bottom to top and then jump to the next one to the right. In this way we get a bijection from $\underset{\sim}{N}^2$ to $\underset{\sim}{N}$:

$\langle 0,0\rangle \to 0$

$\langle 1,0\rangle \to 1$

$\langle 0,1\rangle \to 2$

$\langle 2,0\rangle \to 3$

and so on.

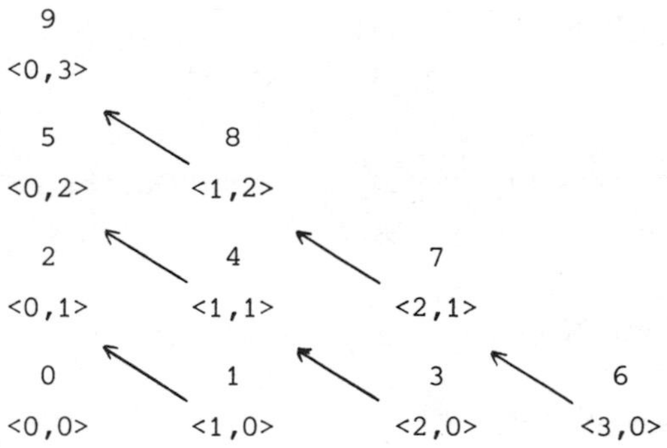

The point $\langle m,n\rangle$ is on the $(m+n+1)^{th}$ anti-diagonal (starting with the 0th anti-diagonal). Hence there are $(m+n)$ complete anti-diagonals or $1+2+\ldots+(m+n)$ points before $\langle m,n\rangle$. Recall that $1+2+\ldots+(m+n) = \frac{1}{2}(m+n)(m+n+1)$. In addition there are n points before $\langle m,n\rangle$ on its own anti-diagonal.

Hence, $<m,n>$ is the $[\frac{1}{2}(m+n)(m+n+1) + n]^{th}$ point (we start with the 0th point).

Define $J(<m,n>) := \frac{1}{2}(m+n)(m+n+1) + n$.

From the construction of J it follows that J is a bijection from $\underset{\sim}{N}^2$ to $\underset{\sim}{N}$ (a reader, who is not convinced may give a proof by mathematical induction on $m+n$ and n). Now we have shown that $\underset{\sim}{N}^2 =_1 \underset{\sim}{N}$.

Since the function J occurs rather frequently, we will simplify the notation a bit: write $J(m,n)$ instead of $J(<m,n>)$.

Because J is a bijection, we have $\forall z \exists! x \exists! y\ [J(x,y) = z]$. ("$\exists! x$" means "there is exactly one x"). Hence there are functions K and L such that $J(K(z),\ L(z)) = z$. Substituting $J(x,y)$ for z, we get $J(K(J(x,y)),\ L(J(x,y))) = J(x,y)$. Because J is a bijection, it follows that
$K(J(x,y)) = x$ and
$L(J(x,y)) = y$.

J is called a *coding* from $\underset{\sim}{N}^2$ onto $\underset{\sim}{N}$ and K and L are called its *inverses* ("projections" would be more appropriate).

ii) Let $f: \underset{\sim}{N} =_1 A$ and $g: \underset{\sim}{N} =_1 B$. Then $F: \underset{\sim}{N} \to A \times B$, defined by $F(x) = <f(K(x)),\ g(L(x))>$, is a bijection and so $F: \underset{\sim}{N} =_1 A \times B$.

<u>Corollary 17.8</u>: The cartesian product of a finite collection of denumerable sets is denumerable.

Proof: Apply mathematical induction on the number of factors of the product.

Remark: The basic trick in proving that the cartesian product of two denumerable sets is denumerable, is the reduction to a "coding" of $\underset{\sim}{N}^2$ on $\underset{\sim}{N}$ (i.e. $\underset{\sim}{N}^2 =_1 \underset{\sim}{N}$).

There are many such codings (take one such coding and apply permutations to $\underset{\sim}{N}$), some of which have nice properties. We give one more example. Consider the following array:

1	3	5	...	(2 does not divide x)
2	6	10	...	(4 does not divide x, but 2 does)
4	12	20	...	(8 does not divide x, but 4 does)
.				
.				
.				

Obviously, each positive natural number appears once and only once.

Now, put $f(m,n) = 2^m + n.2^{m+1}$. Then $f(m,n)$ is the coding of the pair $\langle m,n \rangle$. Thus $\underset{\sim}{N}^2 \underset{1}{=} \underset{\sim}{N} - \{0\}$ and hence $\underset{\sim}{N}^2 \underset{1}{=} \underset{\sim}{N}$.

The following theorem is a generalization of 17.4. The formulation i) is necessarily somewhat awkward, ii) is more elegant, but it requires the axiom of choice.

Theorem 17.9:

i) Let denumerably many denumerable sets V_i ($i \in \underset{\sim}{N}$) and functions $f_i : \underset{\sim}{N} \underset{1}{=} V_i$ be given, then $\bigcup_{i \in \underset{\sim}{N}} V_i$ is denumerable.

ii) (AC) The union of denumerably many denumerable sets is denumerable.

Proof: i) Define $F: \underset{\sim}{N} \to \bigcup_{i \in \underset{\sim}{N}} V_i$ by $F(n) = f_{K(n)}(L(n))$. Then F is a surjection from $\underset{\sim}{N}$ to $\bigcup_{i \in \underset{\sim}{N}} V_i$, because, if $a \in V_i$, then $a = f_i(j)$ for some j and hence $a = F(J(i,j))$ (see the proof of 17.7). Now, by Lemma 17.3, $\bigcup_{i \in \underset{\sim}{N}} V_i$ is finite or denumerable. Since $V_0 \subseteq \bigcup_{i \in \underset{\sim}{N}} V_i$, it follows that $\bigcup_{i \in \underset{\sim}{N}} V_i$ is denumerable.
ii) Let $\{V_i \mid i \in \underset{\sim}{N}\}$ be a family of denumerable sets. From the axiom of choice it follows that there are functions $f_i : \underset{\sim}{N} \underset{1}{=} V_i$. Now apply i). □

Corollary 17.10:

i) Let denumerably many countable sets V_i ($i \in \underset{\sim}{N}$) and functions $f_i : \underset{\sim}{N}_i \underset{1}{=} V_i$ be given (where $\underset{\sim}{N}_i = \underset{\sim}{N}$ or $\underset{\sim}{N}_i = \{0,\ldots,n_i\}$ for some $n_i \in \underset{\sim}{N}$), then $\bigcup_{i \in \underset{\sim}{N}} V_i$ is countable.

ii) (AC) The union of denumerably many countable sets is countable.

Proof: i) Consider first the union of all finite V_i's and prove that their union is finite or denumerable. Then apply Theorem 17.4 or 17.9 i).
ii) Analogous to the proof of 17.9 ii).

□

Let A be a set and k a natural number.

Note that $A^k = \{<a_1,\ldots,a_k> | a_1 \in A \wedge \ldots \wedge a_k \in A\}$, in other words, A^k is the set of all finite sequences of elements of A, of length k. In particular, $A^0 = A^\phi = \{\emptyset\}$.

By Corollary 17.8 A^k is denumerable, if A is denumerable. It is an elementary exercise to show that, for each $k > 0$, the bijection $f_k : \underset{\sim}{N} \underset{1}{=} A^k$ can be explicitly exhibited. So the following theorem is an immediate consequence of Theorem 17.9 i):

<u>Theorem 17.11</u>: If A is denumerable, then $\bigcup_{k \in \underset{\sim}{N}} A^k$ is also denumerable.

In particular, $\bigcup_{k \in \underset{\sim}{N}} \underset{\sim}{N}^k$, this is the set of *all* finite sequences of natural numbers (with arbitrary length), is denumerable.

<u>Corollary 17.12</u>: The set of all finite subsets of a denumerable set is denumerable.

Proof: Let $A^* = \bigcup_{k \in \underset{\sim}{N}} A^k$ and $\mathcal{P}_\omega(A) = \{B \subseteq A | B \text{ is finite}\}$; then $f\colon A^* \to \mathcal{P}_\omega(A)$, defined by $f(<a_1,\ldots,a_k>) = \{a_1,\ldots,a_k\}$, is surjective. Since A^* is denumerable (17.11), there is a surjection of $\underset{\sim}{N}$ onto $\mathcal{P}_\omega(A)$. Clearly $\mathcal{P}_\omega(A)$ is not finite, therefore (17.3) it is denumerable. □

Remark: as in the case of $\underset{\sim}{N}^2$, we can set up an effective coding of $\mathcal{P}_\omega(\underset{\sim}{N})$ in $\underset{\sim}{N}$: define $F(\{n_0,\ldots,n_k\}) = 2^{n_k} + 2^{n_{k-1}} + \ldots + 2^{n_0}$ (where $n_0 < n_1 < \ldots < n_k$), $F(\emptyset) = 0$. One easily checks that $F\colon \mathcal{P}_\omega(\underset{\sim}{N}) \underset{1}{=} \underset{\sim}{N}$. Note that in the binary system the coding is extremely simple: $F(\{n_0,\ldots,n_k\}) = a_p \ldots a_0$ where $a_i = 1$ if $i \in \{n_0,\ldots,n_k\}$, $a_i = 0$ else. E.g. $F(\{0,2,5\}) = 100101$, $F(\{3\}) = 1000$. $F(B)$ is called the canonical index of B. We can now prove 17.12, using the canonical index.

<u>Theorem 17.13</u>: $\underset{\sim}{Q}$ is denumerable.

Proof: Observe that $\underset{\sim}{Q} \underset{1}{=} \{<x,y> | x \in \underset{\sim}{Z} \wedge y \in \underset{\sim}{N} - \{0\} \wedge g\ c\ d\ (x,y) = 1\}$, where $g\ c\ d\ (x,y)$ is the greatest common divisor of x and y. Hence $\underset{\sim}{Q} \underset{1}{\leqslant} \underset{\sim}{Z} \times \underset{\sim}{N}$. So $\underset{\sim}{Q} \underset{1}{\leqslant} \underset{\sim}{N}$. Also $\underset{\sim}{N} \underset{1}{\leqslant} \underset{\sim}{Q}$. By the Cantor-Bernstein theorem $\underset{\sim}{Q} \underset{1}{=} \underset{\sim}{N}$. □

We now consider a famous theorem of F.P. Ramsey, which in a rather unexpected way produces denumerable sets. In a plausible way the theorem is a generalization of the familiar pigeon hole principle.

The (finite) pigeon hole principle states that if a set of n elements is partitioned into less than n sets, at least one set contains more than one element. A simple infinite version is: if a denumerable set is partitioned into finitely many sets, at least one of those is denumerable.

Ramsey generalized this to two-element sets, three-element sets, etc.

First some notations.

Definition 17.14: $[A]^n := \{X \subseteq A \mid X \text{ has } n \text{ elements}\}$.

It is a familiar fact that each partitioning of a set in m parts determines a function f onto $\{0,\ldots,m-1\}$, and vice versa. So we can formulate facts about partitions in terms of functions.

Theorem 17.15 *(Ramsey)*: If A is denumerable and $f\colon [A]^n \to \{0,1,\ldots,m-1\}$ $(n,m \in \mathbb{N};\ n,m \geq 1)$, then there is a set X, such that $X \subseteq A$, X is denumerable and such that f is constant on $[X]^n$.

Proof: Let A be denumerable and let $f\colon [A]^n \to \{0,1,\ldots,m-1\}$ (it is no restriction to assume f to be surjective). We use mathematical induction on n.

(1) $[A]^1$ is the set of all singletons $\{a\}$ for $a \in A$.

$[A]^1 = B_0 \cup \ldots \cup B_{m-1}$, with $B_i = f^{-1}(\{i\})$, $B_i \cap B_j = \emptyset$ $(i \neq j)$. Now, by a kind of converse of 17.4 at least one of the B_i must be denumerable (exercise 1). Let X be the set of all a such that $\{a\}$ belongs to that particular B_i; then f is constant on $[X]^1$. Note that we have applied here a straightforward generalization of the pigeon hole principle.

(2) Suppose that Ramsey's theorem has been proved for n.

Let $f\colon [A]^{n+1} \to \{0,1,\ldots,m-1\}$.

We introduce an auxiliary sequence of sets A_i and elements a_i with the properties

i) $A \supset A_0 \supset A_1 \supset A_2 \supset \ldots$
ii) A_i is denumerable $(i \in \mathbb{N})$
iii) for all $i \in \mathbb{N}$, $a_i \in A_i - A_{i+1}$.

Construction of the sequences:

For $i < n$, A_i and a_i are chosen arbitrarily such that $i)$, $ii)$ and $iii)$ hold. Now suppose that $A_0,\dots,A_p$ and $a_0,\dots,a_p$ have been defined ($p \geqslant n-1$). For $a \in A_p - \{a_p\}$ let the function $f_a^p : [\{0,\dots,p\}]^n \to \{0,1,\dots,m-1\}$ be defined by

$$f_a^p(\{j_0,\dots,j_{n-1}\}) = f(\{a_{j_0},\dots, a_{j_{n-1}}, a\}). \tag{1}$$

Because $[\{0,\dots,p\}]^n$ and $\{0,1,\dots,m-1\}$ are finite there are only finitely many different f_a^p. Hence at least one of the functions f_a^p belongs to denumerably many indices a (another application of the pigeon hole principle). Let a_{p+1} be the first element of $A_p - \{a_p\}$, such that $\{a \in A_p - \{a_p\} \mid f_a^p = f_{a_{p+1}}^p\}$ is denumerable, and

let $A_{p+1} := \{a \in A_p - \{a_p\} \mid f_a^p = f_{a_{p+1}}^p\}$.

Then $A_p \supseteq A_{p+1}$, A_{p+1} is denumerable, $a_p \in A_p - A_{p+1}$ and $a_{p+1} \in A_{p+1}$.

Claim: If $j_0 <\dots< j_n < j_{n+1}$, then $f(\{a_{j_0},\dots,a_{j_{n-1}}, a_{j_n}\}) =$

$f(\{a_{j_0},\dots,a_{j_{n-1}}, a_{j_{n+1}}\})$.

Proof: Suppose $j_0 <\dots< j_n < j_{n+1}$. Then $A_{j_0} \supseteq \dots \supseteq A_{j_{n-1}} \supseteq A_{j_{n-1}+1} \supseteq A_{j_n} \supseteq A_{j_{n+1}}$. Hence

$a_{j_n}, a_{j_{n+1}} \in A_{j_{n-1}+1}$, since $a_{j_n} \in A_{j_n}$ and $a_{j_{n+1}} \in A_{j_{n+1}}$. So $f_{a_{j_n}}^{j_{n-1}} = f_{a_{j_{n+1}}}^{j_{n-1}}$. By (1) it follows that $f(\{a_{j_0},\dots,a_{j_{n-1}}, a_{j_n}\}) = f(\{a_{j_0},\dots,a_{j_{n-1}}, a_{j_{n+1}}\})$.

From the claim it follows immediately that for $j > j_0,\dots,j_{n-1}$, $f(\{a_{j_0},\dots,a_{j_{n-1}}, a_j\}) = f(\{a_{j_0},\dots,a_{j_{n-1}}, a_{j+1}\}) = f(\{a_{j_0},\dots,a_{j_{n-1}}, a_{j+2}\}) =$

$\dots$.

Now let $B := \{a_i \mid i \in \mathbb{N}\}$ and define $g: [B]^n \to \{0,1,\dots,m-1\}$ by $g(\{a_{j_0},\dots,a_{j_{n-1}}\}) = i$ if for all $j > j_0,\dots, j_{n-1}$, $f(\{a_{j_0},\dots, a_{j_{n-1}}, a_j\}) = i$.

By the induction hypothesis there is a set X, such that $X \subseteq B$, X is denumerable and such that g is constant on $[X]^n$. Then clearly f is constant on $[X]^{n+1}$. □

Because every infinite set has a denumerable subset (Theorem 16.3 (*AC*)), Ramsey's theorem can be easily generalized.

Corollary 17.16 (*AC*): If A is infinite and $f\colon [A]^n \to \{0,1,\ldots,m-1\}$ $(n,m \in \mathbb{N};\ n,m \geqslant 1)$, then there is an infinite subset X of A, such that f is constant on $[X]^n$.

Ramsey's theorem was published in 1929; in recent years considerable interest has been focussed on it, since it turned out to be a starting point for infinite combinatorics.

Exercises

1. If $A \cup B$ is denumerable, then A or B is denumerable.

2. Let A be infinite. Use Theorem 16.3 (*AC*) to show that A is denumerable if and only if for every infinite subset $B \subseteq A$, $A =_1 B$.

3. Let L be the set of all polynomials $a_0 + a_1 x + \ldots + a_n x^n$ over the integers (i.e. $a_i \in \mathbb{Z}$ for all i, $0 \leqslant i \leqslant n$). Prove that L is denumerable.

4. An algebraic real is a real root of an equation of the form $a_0 + a_1 x + \ldots + a_n x^n = 0$, where the coefficients a_i are integers. Prove that the set of all algebraic reals is denumerable.

5. A real number is transcendental if it is not algebraic. Prove that the set of all transcendental numbers is not denumerable.

6. Prove that the set of all points with rational co-ordinates in the Euclidean plane is denumerable.

7. Prove that every pairwise disjoint collection of intervals of reals is finite or denumerable.

8. Let $f\colon \mathbb{R} \to \mathbb{R}$ be strictly monotonic (i.e. $x < y \to f(x) < f(y)$). Prove that the set of all points x, such that f is discontinuous in x, is finite or denumerable.

9. Let $D \subseteq \mathbb{R}^2$ be denumerable. Prove that $D = X \cup Y$, such that $X \cap Y = \emptyset$ and such that every line, parallel to the x-axis, has at most finitely many points with X in common and such that every line, parallel to the y-axis, has at most finitely many points with Y in common.

10. Let $V \subseteq \mathbb{R}^2$. Prove that the set of isolated points of V is countable.

11. Prove that every denumerable, totally ordered, set is isomorphic to a subset of the set $\mathbb{Q}$ of rational numbers in their natural order.

12. Let $V \subseteq [0,1] \subseteq \mathbb{R}$.
    i) If for every $a \in V$, $a < 1$, the set $\{x \in V \mid x < a\}$ is countable, then V is countable.
    ii) If V, in the natural order, is well-ordered, then V is countable.
    iii) Prove ii) for $V \subseteq \mathbb{R}$.

18 UNCOUNTABLE SETS

So far we have, apart from denumerable sets, only considered a few examples of infinite sets, e.g. $\{0,1\}^{\mathbb{N}}$, $\mathbb{R}$ (15.5, 15.12). Theorem 15.10 told us that there are many uncountable sets, consider for example $\mathcal{P}(\mathbb{N})$, $\mathcal{P}\mathcal{P}(\mathbb{N})$, etc.

In this section we will restrict ourselves to some familiar uncountable sets, and determine the relation between them with respect to $\leqslant_1$.

For a start we consider intervals on $\mathbb{R}$.

Definition 18.1: For $a,b \in \mathbb{R}$, with $a < b$, we define

$[a,b] := \{x \in \mathbb{R} \mid a \leqslant x \leqslant b\}$ $\quad$ $[a,b) := \{x \in \mathbb{R} \mid a \leqslant x < b\}$

$(a,b] := \{x \in \mathbb{R} \mid a < x \leqslant b\}$ $\quad$ $(a,b) := \{x \in \mathbb{R} \mid a < x < b\}$.

The various intervals are very much alike as the following theorem shows:

Theorem 18.2: $[a,b] =_1 (a,b] =_1 [a,b) =_1 (a,b)$ and $(a,b) =_1 (c,d)$.

Proof: Find suitable injections from intervals in intervals and apply the theorem of Cantor-Bernstein (15.16).

We will only sketch a proof of $(a,b) =_1 (c,d)$.
First we translate a and c to 0.
$f_1 : (a,b) =_1 (0,b-a)$, with $f_1(x) = x-a$.
$f_2 : (c,d) =_1 (0,d-c)$, with $f_2(x) = x-c$.
Now we stretch (or shrink) $(0,b-a)$ a bit.
$f_3 : (0,b-a) =_1 (0,d-c)$, with $f_3(x) = \frac{d-c}{b-a}x$.
Then $f_2^{-1} \circ f_3 \circ f_1 : (a,b) =_1 (c,d)$.

In section 15, example 3 we showed

Lemma 18.3: $\mathbb{R} =_1 (-1,1)$

As a result we have

Theorem 18.4: $\mathbb{R} =_1 (a,b)$

We will now investigate the relation between $\{0,1\}^{\mathbb{N}}$ and $\mathbb{R}$. Obviously there is some relation, just think of the reals in $(0,1)$, written in a binary expansion. Unfortunately the matter is not settled so easily, since distinct expansions may represent the same number. Example: $\frac{1}{2} = 0.10000\ldots = 0.01111\ldots$ (where $0.a_1a_2a_3 \ldots$ stands for $\sum_{i=1}^{\infty} a_i 2^{-i}$). We use the fact that irrational numbers will never have the luxury of two binary (or decimal for that matter) expansions. Some elementary number theory is sufficient to prove that the rationals are exactly those reals with periodic expansions (use the sum formula for geometric series).

Theorem 18.5: $\mathbb{R} =_1 \{0,1\}^{\mathbb{N}}$.

Proof: We will show that $(0,1) =_1 \{0,1\}^{\mathbb{N}}$.

Let A be the set of irrationals in $(0,1)$ and B the set of rationals in

(0,1). Evidently A is Dedekind infinite (exhibit a denumerable subset of A); therefore, by 17.5, $A =_1 A \cup B = (0,1)$. So it suffices to consider A. $f\colon A \leqslant_1 \{0,1\}^{\underset{\sim}{N}}$, where f assigns to each irrational number the sequence of 0's and 1's given by its representing binary expansion. By the remark above, f is an injection. Next we inject $\{0,1\}^{\underset{\sim}{N}}$ into A by forcefully destroying any periodicity: let g assign to any sequence $a_0 a_1 a_2 a_3 a_4 \ldots$ the binary expansion $0.a_0 0a_1 01a_2 011a_3 0111a_4 01111\ldots$. To be precise: insert after each a_n one 0 and n 1's. By definition $g\colon \{0,1\}^{\underset{\sim}{N}} \leqslant_1 A$.

By the Cantor-Bernstein theorem we have therefore $A =_1 \{0,1\}^{\underset{\sim}{N}}$. Combined with the preceding remarks, this yields $\underset{\sim}{R} =_1 \{0,1\}^{\underset{\sim}{N}}$. □

The main feature of Theorem 18.5 is that we have eliminated the topological (or continuity) aspects of $\underset{\sim}{R}$ in determining its size (cardinality). Traditionally $\underset{\sim}{R}$ is called the *continuum*; the term is, however, also metaphorically used for $\{0,1\}^{\underset{\sim}{N}}$ (which, anticipating a set theoretical definition of 2 (cf. 20.4), is usually written as $2^{\underset{\sim}{N}}$).

The choice of the set $\{0,1\}$, in the above theorem, is convenient, since it yields

<u>Corollary 18.6</u>: $\underset{\sim}{R} =_1 \mathcal{P}(\underset{\sim}{N})$.

Proof: By 15.3 and 18.5. □

We could, however, have chosen any countable set instead of $\{0,1\}$.

<u>Theorem 18.7</u>: If A is countable, then $\{0,1\}^{\underset{\sim}{N}} =_1 A^{\underset{\sim}{N}}$.

Proof: We will show the special case $\underset{\sim}{N}^{\underset{\sim}{N}} =_1 \{0,1\}^{\underset{\sim}{N}}$. Note that each function from $\underset{\sim}{N}$ to $\underset{\sim}{N}$ is a subset of $\underset{\sim}{N}^2$, so $\underset{\sim}{N}^{\underset{\sim}{N}} \subseteq \mathcal{P}(\underset{\sim}{N}^2)$. Therefore $\underset{\sim}{N}^{\underset{\sim}{N}} \leqslant_1 \mathcal{P}(\underset{\sim}{N}^2)$. From $\underset{\sim}{N} =_1 \underset{\sim}{N}^2$ one easily concludes that $\mathcal{P}(\underset{\sim}{N}) =_1 \mathcal{P}(\underset{\sim}{N}^2)$, so $\underset{\sim}{N}^{\underset{\sim}{N}} \leqslant_1 \mathcal{P}(\underset{\sim}{N}^2) =_1 \mathcal{P}(\underset{\sim}{N}) =_1 \{0,1\}^{\underset{\sim}{N}}$. On the other hand $\{0,1\}^{\underset{\sim}{N}} \subseteq \underset{\sim}{N}^{\underset{\sim}{N}}$, so $\{0,1\}^{\underset{\sim}{N}} \leqslant_1 \underset{\sim}{N}^{\underset{\sim}{N}}$. Now apply the Cantor-Bernstein theorem.

For finite A, observe that $\{0,1\}^{\underset{\sim}{N}} \subseteq A^{\underset{\sim}{N}} \subseteq \underset{\sim}{N}^{\underset{\sim}{N}}$.
Also compare exercise 6. □

As far as our limited experience goes, we have shown that (leaving aside larger sets, such as $\mathcal{P}\mathcal{P}(\underset{\sim}{N})$, $\mathcal{P}\mathcal{P}\mathcal{P}(\underset{\sim}{N})$) most familiar infinite sets are either denumerable or equivalent to the continuum.

Cantor, on the basis of similar evidence, conjectured in 1878 that each infinite subset of $\underset{\sim}{R}$, either is denumerable, or is equivalent to the continuum. This conjecture is known as the *Continuum Hypothesis* (*CH*). A precise formulation reads:

$$\text{if } \underset{\sim}{N} \leqslant_I A \leqslant_I \underset{\sim}{R}\text{, then } A =_I \underset{\sim}{N} \text{ or } A =_I \underset{\sim}{R}.$$

So far the continuum hypothesis has withstood all attempts to settle it. By the results of Gödel (1938) and Cohen (1963) we know that it is independent of the basic axioms of set theory (such as given by Zermelo and Fraenkel). The matter of its truth or falsity in the intended universe of set theory however remains unsettled. Gödel, in his paper "What is Cantor's Continuum Problem?" (in [Benacerraf and Putnam] and in Am. Math. Monthly 54 (1947)) has analysed the evidence, which turns out to be rather in favour of a rejection.

We present one more remarkable theorem by Cantor, which at the time (1877) greatly surprised the mathematical world. Cantor showed that the real plane ($\underset{\sim}{R}^2$) is equivalent to the continuum ($\underset{\sim}{R}$). The result seemed, at first sight, counter-intuitive - it apparently conflicts with the invariance of dimension. As a matter of fact the mapping, involved in the proof, is not topological, hence the change in dimension.

The proof below is from first principles. A simple proof is indicated in exercise 5.

Theorem 18.8: $\underset{\sim}{R}^2 =_I \underset{\sim}{R}$.

Proof: By 18.2 and 18.4 it suffices to show that $(0,1] \times (0,1] =_I (0,1]$. We assign to each number $x \in (0,1]$ a unique decimal expansion $0.x_1x_2x_3 \ldots$, by suppressing the terminating expansions. We now divide the sequence of digits in this expansion into "blocks" in the following way: each non-zero digit that is not preceded by zero is to form a block by itself, but any string of consecutive zero's is to form a single block with the non-zero digit immediately following it. For example, $x = 0.47019002\ldots$ and $y = 0.0323801\ldots$ are divided into "blocks" as follows:
$0.|4|7|01|9|002|\ldots$ and $0.|03|2|3|8|01|\ldots$. Now if $\langle x,y\rangle \in (0,1] \times (0,1]$,

we form a real number $z = 0.z_1 z_2 \ldots$ by interlacing the decimal expansions for x and y, i.e. by stringing together the consecutive blocks of x and y, taken alternating. For example, if $x = 0.47019002\ldots$ and $y = 0.0323801\ldots$, then $z = 0.403720139800201\ldots$. Observe that, conversely, from z we can uniquely reconstruct x and y, hence we have the required bijection from $(0,1]^2$ to $(0,1]$. $\square$

Query: Why couldn't we just place the digits of x (resp. y) on the even (resp. odd) places of z?

Exercises

1. Prove that the following sets are equivalent to $\{0,1\}^{\underset{\sim}{N}}$.

   a) $\{f \in \{0,1,2\}^{\underset{\sim}{N}} \mid \forall n \in \underset{\sim}{N} \, [f(n) \neq f(n+1)]\}$

   b) $\{A \in \wp(\underset{\sim}{N}) \mid \sum_{n=1}^{\infty} K_A(n) \frac{1}{n} \text{ is convergent}\}$

   c) $\{f \in \underset{\sim}{N}^{\underset{\sim}{N}} \mid \forall n \in \underset{\sim}{N} \, [f^{-1}(\{n\}) \underset{1}{=} \underset{\sim}{N}]\}$

   d) $\{f \in \underset{\sim}{Q}^{\underset{\sim}{N}} \mid \lim_{n \to \infty} f(n) = 0\}$

2. Let V be an uncountable subset of $\underset{\sim}{R}$. Prove that the set of all accumulation points of V is uncountable.

3. (Cantor, 1874). Let $\langle a_i \rangle_{i \in \underset{\sim}{N}}$ be a sequence of real numbers and let $[p,q]$ be any (non-empty) interval.

   i) Define the following sequences:

   $x_0 := p \qquad ; \; y_0 := q$

   $x_{n+1} :=$ the first number in the sequence $\langle a_i \rangle$, which is between x_n and y_n, if it exists.

   $x_{n+1} := x_n$ else.

   $y_{n+1} :=$ the first number in the sequence $\langle a_i \rangle$, which is between x_{n+1} and y_n, if it exists.

   $y_{n+1} := y_n$ else.

   ii) The sequences $\langle x_i \rangle$ and $\langle y_i \rangle$ have a supremum and infimum respectively (the sequences may be both finite and infinite).

iii) ½ (sup (x_i) + inf (y_i)) does not belong to the sequence $\langle a_i \rangle$.

(iv) $\underset{\sim}{R}$ is non-denumerable.

4. Prove that the set of all partitions of $\underset{\sim}{N}$ is equivalent to $\{0,1\}^{\underset{\sim}{N}}$.

5. Prove: i) $\{0,1\}^{\underset{\sim}{N}} \times \{0,1\}^{\underset{\sim}{N}} =_1 \{0,1\}^{\underset{\sim}{N}}$.

   ii) $\underset{\sim}{R} \times \underset{\sim}{R} =_1 \underset{\sim}{R}$ (via i)).

6. Prove: i) If A is finite, then $A^{\underset{\sim}{N}} =_1 \{0,1\}^{\underset{\sim}{N}}$.

   Hint: encode each infinite sequence of elements of A into a zero-one-sequence.

   ii) If A and B are finite, then $A^{\underset{\sim}{N}} =_1 B^{\underset{\sim}{N}}$. $(A \neq \emptyset,\ B \neq \emptyset)$.

7. Let $R_1, R_2, R_3, \ldots$ be a sequence of equivalence relations on $\underset{\sim}{N}$. Construct by means of diagonalisation an equivalence relation R on $\underset{\sim}{N}$, such that for all $i \in \underset{\sim}{N}$, $\langle \underset{\sim}{N}, R_i \rangle$ is not isomorphic to $\langle \underset{\sim}{N}, R \rangle$. Hint: look at the partitions, belonging to the R_i (see also exercise 4).

8. Give a concrete bijection from $[0,1)$ to $(0,1)$.

9. Let $V \subseteq \mathcal{P}(\underset{\sim}{N})$ be such that for each $A \subseteq \underset{\sim}{N}$ either $A \in V$ or $A^c \in V$ (but not both). Show that $V =_1 \{0,1\}^{\underset{\sim}{N}}$.

19 THE PARADOXES

Till now we have, following Cantor, uncritically used the following naive comprehension principle: if $\Phi(x)$ is a property, then $\{x|\Phi(x)\}$ is a set. In other words: each property determines a set of all objects that have the property. The principle thus asserts the *existence* of certain sets. The use of the abstraction term $\{x|\Phi(x)\}$ is determined by the rule
$y \in \{x|\Phi(x)\} \leftrightarrow \Phi(y)$.

In 1902 Bertrand Russell showed in a letter to Frege ([Van Heyenoort], p. 124) that the principle leads to a paradox. The argument is extremely simple and uses no technical machinery at all.

Apply the comprehension principle to the property (described by) $x \notin x$. Let $y = \{x | x \notin x\}$. Let us ask whether $y \in y$? By our convention,

$$\begin{aligned} y \in y &\leftrightarrow y \in \{x | x \notin x\} \\ &\leftrightarrow y \notin y. \end{aligned}$$

So, by reductio ad absurdum, we have $y \notin y$. And likewise $y \in y$. This is the promised paradox.

In technical terms, Russell's argument shows that set theory with the naive comprehension principle is *inconsistent*.

Cantor, himself, had already previously (1895) discovered various paradoxes, that is to say he showed that the existence of certain sets entailed contradictions. (cf. [Van Heyenoort], p. 113; [van Dalen - Monna], Ch. 1, §2 .)

One of Cantor's "inconsistent sets" - the set of all sets (the basic ingredient of Cantor's paradox) - will be dealt with below. The reactions to the paradoxes ranged from mild amusement to despair. In set theory a number of proposals were made to salvage as much as possible. The various proposals lead to various theories. We will treat only the theory of Zermelo - Fraenkel, because of its practical success, and mention the other theories briefly.

A first reaction is "$x \notin x$ produced a contradiction, negation is impeccable, so we must suppress $x \in x$". Russell, in his theory of types, has chosen this approach - assign types to variables (sets) and allow expressions, such as $x \in y$, only if the type of x is one less than the type of y. A modification is presented by Quine in his "New Foundations" (connected to Russell's Zig Zag theory).

Another plausible observation is the following: the sets involved in the derivation turn out to be very large - for example the set of all sets, the set of all well-orderings (ordinals), the set of all sets not being an element of themselves - so allow only "small" sets. This "limitation of size" attitude was adopted by Zermelo. He noted that the full force of the comprehension principle was hardly ever used, one mostly uses it to create subsets of a given set. That is, one introduces smaller sets and not larger ones.

So instead of the comprehension principle he put forward his *separation axiom* (Aussonderungs Axiom): If A is a set and $\Phi(x)$ a property, then there is a set of elements of A satisfying $\Phi(x)$.

The set specified above is denoted by $\{x \in A | \Phi(x)\}$ and we put by convention,

$$y \in \{x \in A | \Phi(x)\} \leftrightarrow y \in A \wedge \Phi(y).$$

If we abandon the naive comprehension principle and adopt the separation axiom instead, we can no longer accept the proof of Russell's paradox. This is a step forward, but it, of course, does not exclude the possibility of another proof of it. Specifically, if the revised system of set theory were inconsistent, we could derive every statement (as every student of logic knows), hence, also Russell's paradox. However, this is the only possibility to regain Russell's paradox; with some material from the next chapter one can prove that the paradoxical set does not exist in Zermelo's system, provided it is consistent.

At the moment we can use the idea of the proof of Russell's paradox to obtain, with the help of the separation axiom, a positive result.

Theorem 19.1: From the separation axiom it follows that no set contains all sets.

Proof: For any set A we will exhibit a set B such that $B \notin A$.

Define (by virtue of the separation axiom) the set $B = \{x \in A | x \notin x\}$. By the definition of B we have $B \in B \leftrightarrow B \in A \wedge B \notin B$.
Suppose $B \in A$, then $B \in B \leftrightarrow B \notin B$.
Contradiction. Hence $B \notin A$. □

Corollary 19.2: There is no set of all sets. (i.e. Cantor's paradox is avoided).

A reformulation of 19.2 is: the *universe* (or totality) of all sets is not a set.

What are we giving up if we drop the naive comprehension principle in favour of the axiom of separation (apart from a contradiction which we do not want anyway)? Evidently we cannot make larger sets, such as pairs, unions, powersets, etc. Therefore one has to add a number of axioms, allowing the construction of specific sets, necessary for the development of set theory. This project will be carried out in the next section.

We shall now illustrate one more pitfall in naive set theory. We already mentioned that the obvious definition of cardinal number, as an equivalence class under $=_I$, was incorrect. The reason will now become clear.

Theorem 19.3: The existence of the set $\{x \mid x =_I \{\emptyset\}\}$ implies the existence of the set of all sets.

Proof: Suppose $A = \{x \mid x =_I \{\emptyset\}\}$ exists. By definition A is the set of *all* singletons. So, for each z, $z \in \{z\}$ and $\{z\} \in A$, and hence $z \in \bigcup A$. Therefore $\bigcup A$ is the set of all sets. □

Remark 1: We used the sum axiom: if A is a set, so is $\bigcup A$; but this is a harmless axiom.
Remark 2: $\bigcup A$ could contain more than just sets, namely objects which are not sets (so called ur-elements). In that case one would apply the separation axiom to obtain the set of all sets (where $\Phi(x) := \exists y\ [y \in x]$). However, we will, unless specifically introduced, not consider ur-elements.

A more practical formulation of 19.3 is: there is no set containing all x equivalent to $\{\emptyset\}$. Or, the Frege-Russell cardinals do not exist.

Yet another approach to the foundations of set theory was proposed by Von Neumann (unaware of similar ideas of Cantor, [Van Heyenoort] p.113). His idea (in a more convenient terminology) was to consider *classes* as basic objects and to award some classes the status of set. In Von Neumann's system (as modified by Bernays and Gödel) there is a comprehension principle:
If $A(x)$ is a formula in which the bound variables range over sets and x is a set variable, then there is a class X such that $\forall x\ [x \in X \leftrightarrow A(x)]$.

This comprehension principle does not suffer from the same deficiencies as the naive comprehension principle. Mainly because it asserts the existence of a *class* of *sets*. Note that, because the bound variables range over sets, no reference is made to the totality of *all* classes.

Von Neumann's system is slightly more flexible, since it allows one to introduce more objects than Zermelo's system does. For example, there is a class of all sets. By mimicking Theorem 19.1, one easily shows that the class of all sets is not a set; we call it a *proper class*.

The choice between the systems of Zermelo-Fraenkel and Von Neumann-Bernays-Gödel is largely a matter of taste, since both theories have the same strength, in the following, precise form: let A be a statement, containing only set variables, then A is provable in the Zermelo-Fraenkel system if and only if it is provable in the Von Neumann - Gödel - Bernays system (the latter theory is *conservative* over the former). For a proof see [Cohen, p. 77].

For a discussion of classes, properties, etc. see the Appendix.

Finally a word about the axiomatization of set theory. There is a long tradition of axiomatics in mathematics; in particular geometry has been handed down to us as an axiomatic system. The details of the axiomatization of set theory turn out to contain several fine distinctions (cf. Appendix), therefore the use of the language of first order logic will be adopted, after Thoralf Skolem. However, no formal proofs will be given and no sophisticated knowledge of logic will be required. A nodding acquaintance with logical usage will be sufficient.

20 THE SET THEORY OF ZERMELO - FRAENKEL (**ZF**).

We first present a *formal language* for Zermelo-Fraenkel set theory (**ZF** for short), that is to say, a collection of basic symbols together with instructions how to combine these to get formulas:

Basic symbols:

i) a countably infinite collection of *variables;*

ii) the *epsilon-symbol* and the *identity-symbol*: $\in$,= ;

iii) *connectives* $\wedge$, $\vee$, $\rightarrow$,$\neg$, $\leftrightarrow$;

iv) *quantifiers* $\forall$ and $\exists$; v) *parentheses* (,)

Formulas:

1. *atomic formulas* $x \in y$ and $x = y$, where x and y are any variables;
2. whenever Φ and Ψ are formulas, so are $(\Phi \wedge \Psi)$, $(\Phi \vee \Psi)$, $(\Phi \to \Psi)$, $(\neg \Phi)$ and $(\Phi \leftrightarrow \Psi)$;
3. whenever Φ is a formula then so are $\forall x\, [\Phi]$ and $\exists x\, [\Phi]$ (where x is any variable);
4. every formula is built in finitely many steps from atomic formulas using instructions 2 and 3.

Notation: Usually we omit the outermost parentheses in a formula. Furthermore we agree that $\wedge$ and $\vee$ bind stronger than $\to$ and $\leftrightarrow$. Finally we write $x \notin y$ and $x \neq y$ for $\neg\, x \in y$ and $\neg\, x = y$ respectively. For example: $x \in y \wedge x \neq z \to z \in y$ is an imprecise, but convenient, notation for $((x \in y \wedge \neg\, x = z) \to z \in y)$.

If, between brackets immediately after a letter denoting some formula, one or more variables occur, this indicates the possibility of free occurrences of those variables in the formula; for instance, x may occur free in $\Phi(x)$. In that case $\Phi(y)$ is the formula obtained from Φ by substituting y for the free occurrences of x.

Examples:

1. x occurs *free* in $\Phi(x) = \exists z[\, x \in z]$, but z does not occur free in $\exists z[\, x \in z]$ (we say that z has a *bound* occurrence in that formula). $\Phi(y)$ is the formula $\exists z[\, y \in z]$. In $\forall x\, [\Phi(x)]$, which is $\forall x\, \exists z[\, x \in z]$, x, as well as z, occurs bound.
2. In $\Psi(x,y) = \exists y\, [y \in x] \to \exists x\, [x \in y]$ both x and y occur free as well as bound. This is why we only substitute in places where a variable occurs free.
   $\Psi(y,y) = \exists y\, [y \in y] \to \exists x\, [x \in y]$.

The language defined above is taken from formal logic where such languages are used in formulating so called first-order theories. The theory of Zermelo-Fraenkel itself is such a first-order theory and can be treated completely formally in logic. This means that the theorems of **ZF** can be derived by means of predicate logic only. Such a treatment can be found in [Gödel (1940)] and [Shoenfield (1967)].

We shall use the formal language only to formulate assertions in a precise way, but not to give formal proofs in predicate logic. As a rule we shall interpret our formulas and give ordinary mathematical proofs. A reader familiar with formal logic can easily formalize our informal proofs.

It is surprising how little logic is needed to handle a sizeable part of mathematics!

As a matter of fact, we have in this section started a new approach, that is to say, we are dealing now with an *axiomatic theory*, in which we may use only that which is explicitly allowed by the axioms, rules and definitions. The so called *axiomatic method* will be familiar to most readers, for instance from geometry or algebra.

In geometry one talks about points, lines and planes without knowing (or wanting to know) what these are. It is exclusively on the basis of axioms and logical reasoning that theorems are being derived. Widely known expositions are Euclid's (± 300 BC) and Hilbert's (1899).

Formal logic provides an excellent framework in which the axiomatic method may be applied: in order to derive theorems by way of logic no conceptual insight concerning the objects mentioned in the axioms is required. (The fundamentals of formal logic are exposed in Shoenfield [1967] and Enderton [1972]).

One could leave the work of giving the proofs to a machine, at least in principle. As has been said we shall not take this extreme view. But it should be stressed that everything done in Chapter I should be considered anew. Fortunately we have been careful there and most of the proofs can be copied, by just formalizing the informal reasoning.

Remark: Zermelo laid down his system of axioms in 1908; the extensions of Fraenkel date from 1922. The use of a formal language in formulating the theory was proposed by Skolem in 1922.

We now present the axioms; first in words and then in formal language. To be able to refer to axioms in an easy way we provide them with mnemonic labels.

The axioms of Zermelo and Fraenkel

I EXT *Extensionality axiom*: Sets having the same elements are equal.
Formally· $\forall z\ [z \in x \leftrightarrow z \in y] \rightarrow x = y$.

II EMP *Axiom of the empty set*: There is a set which has no elements (which is empty).
Formally: $\exists\, x\ \forall\, y\ [y \notin x]$.

Since, by EXT, a set is determined by its elements, there is at most one empty (elementless) set; hence there is exactly one, which we denote in the sequel by $\emptyset$.

III SEP *Separation axiom*: For every set a and every property Φ of sets there exists a set whose elements are exactly those of a having the property Φ.
Formally: $\exists b\ \forall x\ [x \in b \leftrightarrow x \in a \wedge \Phi(x)]$.
(b is not allowed to occur in Φ.)

Formally this is not just one axiom but an infinite series of axioms: one corresponding to every formula Φ (and variables a,b and x); usually one expresses this by saying that SEP is an *axiomschema* which generates axioms by specifying the parameter Φ. We cannot express SEP as one formula because in the formal language we cannot talk about properties in general - we can talk about sets only. And in the formal language we can only use in SEP those properties expressable by means of a formula of the language (which may contain parameters besides the variable x).

The extensionality axiom implies that there is exactly one set, the members of which are those of a, satisfying Φ; we denote this set by $\{x \in a \,|\, \Phi\ (x)\}$.

Viewing SEP as a principle for constructing sets, the restriction on Φ, that it does not talk of b, is intuitively clear.

Once we have SEP, we may drop EMP if only we are willing to admit at least one set. (One can consider this supposition to be a logical one: otherwise the theory would be devoid of any sense whatsoever; also EMP follows from the axiom INF, to be discussed later.) For suppose that we have a set a, then $\{x \in a | x \neq x\}$ is empty, since no object is distinct from itself.

Finally we may introduce, now we have SEP, some important set-theoretic operations:

$a \cap b := \{x \in a | x \in b\}$

$b - a := \{x \in b | x \notin a\}$.

In case a is non-empty, for instance $c \in a$, we define $\bigcap a := \{x \in c | \forall y\, [y \in a \rightarrow x \in y]\}$. (One easily shows that $\bigcap a$ does not depend on the choice of c.)

IV PAIR *Pairing axiom*: Given any sets a and b, there exists a set c whose elements are exactly a and b.

Formally: $\exists c\, \forall x\, [x \in c \leftrightarrow x = a \vee x = b]$.

Extensionality, once again, guarantees the uniqueness of the set whose existence is required by PAIR; we denote it by $\{a,b\}$ (the *unordered pair* of a and b).

$\{a\} := \{a,a\}$ is the *singleton* of a and $<a,b> := \{\{a\}, \{a,b\}\}$ is called the *ordered pair* of a and b, since we have the

Lemma 20.1: $<a,b> = <c,d> \rightarrow a = c \wedge b = d$.

Proof: compare Theorem 8.2. □

V SUM *Sum-set axiom*: For every set a there exists a set b, whose elements are exactly those objects occurring in at least one element of a.

Formally: $\exists b\, \forall x\, [x \in b \leftrightarrow \exists y\, [y \in a \wedge x \in y]]$.

Again, EXT guarantees the uniqueness of the set b, required by SUM. This unique set is called the *sum-set* of a, and is denoted by $\bigcup a$, $\bigcup_{y \in a} y$ or $\bigcup\{y | y \in a\}$.

We may introduce now the *union* of two sets a and b by putting $a \cup b := \bigcup\{a,b\}$; note that this set contains exactly the elements of both a and b.

Finally we may define

$\text{Dom}(a) := \{x \in \bigcup\bigcup a \mid \exists y\ [\langle x,y\rangle \in a]\}$, and

$\text{Ran}(a) := \{y \in \bigcup\bigcup a \mid \exists x\ [\langle x,y\rangle \in a]\}$. (Compare Corollary 10.8).

VI POW *Powerset axiom*: For every set a there exists a set b, the elements of which are exactly the subsets of a.

Formally: $\exists b\ \forall x\ [x \in b \leftrightarrow \forall y\ [y \in x \rightarrow y \in a]]$, or using inclusion:

$$\exists b\ \forall x\ [x \in b \leftrightarrow x \subseteq a].$$

x is *subset* of a, in symbols: $x \subseteq a$, if every element of x also is element of a. Again, the required b is unique; it is called the *powerset* of a and denoted by $\mathcal{P}(a)$.

Whenever t is an operation and E a property, $\{t(x,y) \mid E(x,y)\}$ stands for the collection of objects $t(x,y)$ for which $E(x,y)$. This need not be a set at all; in such a case we consider $\{t(x,y) \mid E(x,y)\}$ merely as a notational device (compare the appendix).

We can now introduce the (*cartesian*) *product* of a and b, by defining $a \times b := \{\langle x,y\rangle \mid x \in a \wedge y \in b\}$.
If a and b are sets, then so is $a \times b$, due to PAIR, SUM, POW and SEP. For we have: $z \in a \times b \leftrightarrow z \in \mathcal{P}\mathcal{P}(a \cup b) \wedge \exists x\ \exists y[x \in a \wedge y \in b \wedge z = \langle x,y\rangle]$
(cf. corollary 9.2).

f is called a *function* if f is a set of ordered pairs such that for every $x \in \text{Dom}(f)$ there is exactly one $y \in \text{Ran}(f)$ with $\langle x,y\rangle \in f$ (this y is denoted by $f(x)$).

Notation: $Fnc\ (f)$.

f is a *function from* a *to* b, if $\text{Dom}(f) = a$ and $\text{Ran}(f) \subseteq b$.

b^a is the set of all functions from a to b.

Note that $b^a \subseteq \mathcal{P}(a \times b)$, hence the foregoing axioms imply that b^a exists as a set, if a and b are sets.

PAIR, EXT, SUM and EMP alone entail the existence of infinitely many sets, for instance $\emptyset$, $\{\emptyset\}$, $\{\emptyset,\{\emptyset\}\}$, $\{\emptyset, \{\emptyset\}, \{\emptyset, \{\emptyset\}\}\}$, ... , as do EMP and POW: $\emptyset$, $\mathcal{P}(\emptyset)$, $\mathcal{P}\mathcal{P}(\emptyset)$, ...

Each of these sets has finitely many elements though. The axiom of infinity , which we now introduce, demands the existence of an infinite set.

Definition 20.2: $x^+ := x \cup \{x\}$.

Definition 20.3: $0 := \emptyset$

$1 := 0^+$

$2 := 1^+$

in general, $n+1 := n^+$.

In mathematics, the set $\underset{\sim}{N}$ of natural numbers probably is the best known example of an infinite set.

It is on purpose that we use the well-known symbols 0,1,2, ... in 20.3 : the sets 0,1,2,3, ... are going to play the role of natural numbers.

It may appear slightly odd to just single out certain sets and declare them to be natural numbers. What should be kept in mind however is that all we need is a well-behaved set together with a successor operation. We shall show presently that 0,1,2,3, ... together with the operation $^+$ satisfy our requirements.

Lemma 20.4: 1. $n = \{0, \ldots, n-1\}$

2. the sets 0,1,2,3, ... are pairwise distinct.

Proof: 1. Induction. $1 = 0^+ = 0 \cup \{0\} = \emptyset \cup \{\emptyset\} = \{0\}$, and if $n = \{0,\ldots n-1\}$, then $n^+ = n \cup \{n\} = \{0,\ldots,n\}$.

2. Induction. If we know already that $0,\ldots,n-1$ are pairwise distinct, then n differs from $0, \ldots, n-1$ too, since according to the induction hypothesis and 20.4 (1) n has more elements than each of $0, \ldots, n-1$. □

Definition 20.5: a is a *successor set* if $0 \in a$ and $\forall x\ [x \in a \to x^+ \in a]$.

In words: a is a successor set if a is closed under the operation $^+$ and contains 0.

We use the abbreviation $SUC\ (a)$ for $0 \in a \wedge \forall x\ [x \in a \to x^+ \in a]$.

If a is any set for which $SUC\ (a)$ we see that $0,1,2,3, \ldots \in a$, this means, informally spoken, that a is infinite.

Note that "finite" and "infinite" are used rather loosely here. In our newly begun axiomatic life we have not as yet defined these terms. Informally, we may see that for instance the set 3 is finite, but up to now we do not have a general criterion for finiteness. Nevertheless we have, intuitively, a reasonable conviction that every successor set is infinite. Therefore we use successor sets in the next axiom.

VII INF *Axiom of infinity* : There exists a successor set.
Formally: $\exists a\, [\,SUC\ (a)]$.

The axioms preceding INF happened to be such that exactly one set fulfilled the requirements. Unfortunately this is not true of the axiom of infinity. The existence of at least one successor set entails the existence of very many others. With some difficulty, however, we can determine a specific one, viz. the least such.

Lemma 20.6: $\exists a\, [\,SUC\ (a) \wedge \forall x\, [\,SUC\ (x) \rightarrow a \subseteq x\,]]$.

Proof: According to INF some successor set b exists. Take a to be the intersection of all successor sets, which are subsets of b:
$a := \bigcap \{x \in \mathcal{P}(b) \mid SUC\ (x)\}$. Then a is a successor set:
i) If $SUC\ (x)$, then $0 \in x$; hence, since a is an intersection of successor sets, $0 \in a$.
ii) If $u \in a$, then $u \in x$ for every successor set $x \in \mathcal{P}(b)$. Hence $u^{+} \in x$ for all such x; which implies $u^{+} \in a$.

Also, a is the least successor set. For suppose that $SUC\ (c)$. It can be easily verified that $SUC\ (b \cap c)$. Thus, $a \subseteq b \cap c \subseteq c$. □

On the basis of EXT it now is clear that exactly one smallest successor set exists; we call this one ω.

Definition 20.7: $\omega := \{x \mid \forall a\, [\,SUC\ (a) \rightarrow x \in a]\}$.

Definition 20.8: The members of ω are called *natural numbers*.

In defining ω we have made some progress again: the individual natural numbers had been defined already one by one, now we have introduced the set of all natural numbers. The importance of this lies in the fact that we now may quantify over natural numbers:
$\forall x \in \omega\,[\Phi(x)]$ means: for all natural numbers x, $\Phi(x)$ holds.
Moreover we are now in a position to do anything in which natural numbers are needed, such as forming sequences, real numbers, etc.

On the basis of our axioms we can prove now the principle of mathematical induction.

Theorem 20.9: $\Phi(0) \wedge \forall x \in \omega\,[\Phi(x) \rightarrow \Phi(x^+)] \rightarrow \forall x \in \omega\,[\Phi(x)]$

Proof: Suppose that $\Phi(0) \wedge \forall x \in \omega\,[\Phi(x) \rightarrow \Phi(x^+)]$.
According to SEP, $\{x \in \omega \mid \Phi(x)\}$ is a set and by hypothesis it is a successor set. Since ω is the smallest such, $\omega \subseteq \{x \in \omega \mid \Phi(x)\}$, hence $\forall x \in \omega\,[\Phi(x)]$. □

VIII REP *Axiom of replacement*: The image of a set under an operation (functional property) is again a set.

More precisely: whenever E is a property of pairs of sets, such that to every x there is exactly one y for which $E(x,y)$, and a is any set, then there exists a set, the elements of which are exactly those y for which an $x \in a$ exists such that $E(x,y)$. Formally:

$\forall x\, \exists! y\,[\Phi(x,y)] \rightarrow \exists b\, \forall y\,[y \in b \leftrightarrow \exists x\,[x \in a \wedge \Phi(x,y)]]$, where $\Phi(x,y)$ is any formula in which b does not occur.

Here $\exists! y\, \Phi(x,y)$ abbreviates $\exists y\, \forall z\,[\Phi(x,z) \leftrightarrow y = z]$, z being any variable not occurring in Φ (i.e. "there exists exactly one ...").

Some comments are now in order.

1. Just as with SEP we have an axiomschema here; formally REP is an infinite collection of axioms: one corresponding to every choice of the formula Φ. Again, this is due to the fact that, in our formal language, we cannot talk about operations (or functional properties) other than those describable by means of formulas. The restriction Φ not containing

b has the same justification as in the case of SEP.

2. In all practical applications of the replacement axiom, Φ will be a formula in which the variables x and y really appear; hence Φ will be understood as a condition on x and y.

   The left-hand side, $\forall x\ \exists! y\ [\Phi(x,y)]$, of the axiom states that Φ describes an operation: to every set x exists exactly one y for which Φ holds.

   The right-hand side concludes the existence of the image b of a under the operation described by Φ.

Example: Take Φ to be $\forall z\ [z \in y \leftrightarrow z=x]$. Note that Φ states that $y = \{x\}$; hence extensionality implies $\forall x\ \exists! y\ \Phi$. Therefore the replacement axiom (with this Φ) implies the existence, for any set a, of the set b of singletons of elements of a.

In this case we could have derived this conclusion already from powerset and separation axiom, since $b = \{y \in \mathcal{P}(a) \mid \exists x \in a\ \forall z\ [z \in y \leftrightarrow z = x]\}$. But later on we shall meet cases in which the use of REP is essential.

3. As a second example we present the derivation of the separation axiom from the axiom of the empty set and REP. Suppose that Ψ is a formula in which b does not occur; we shall establish on the basis of these axioms that for any set a, $\exists b\ \forall x\ [x \in b \leftrightarrow x \in a \wedge \Psi]$.
   To this end we distinguish two cases:
   i) $\neg\ \exists x\ [x \in a \wedge \Psi]$. Then $b := \emptyset$ satisfies the condition.
   ii) $\exists x\ [x \in a \wedge \Psi]$, for instance $y_o \in a$ satisfies Ψ. Take Φ to be the formula $[(x \in a \wedge \Psi) \wedge y=x] \vee [\neg\ (x \in a \wedge \Psi) \wedge y=y_o]$.
   Note that b does not occur in Φ and $\forall x\ \exists!\ y\ \Phi$. The conclusion of REP provides for the set b as desired.

4. Among the Zermelo-Fraenkel axioms the replacement axiom forms the only contribution of Fraenkel. It probably is historically correct to say that Zermelo left it out by mistake; the first purpose of his axiomatization was to provide for a precise context for proving the well ordering theorem (which we treat later on), and due to his ingenious construction in this proof he simply did not need the replacement axiom at all.

5. The replacement axiom has a conceptually simpler formulation which is also closer to practice. It is a good mathematical habit to associate *operations* with functional conditions of some importance. Whenever Φ describes such a functional correspondence, (i.e. if $\forall x\ \exists! y\ \Phi$) an *operation* t is introduced and $t(x)=y$ is written just in case y corresponds to x according to Φ (that is, if x and y satisfy Φ).

As a matter of fact, we already have repeatedly applied this way of introducing operations. For instance, with the formula $\forall z\ [z \in y \leftrightarrow z=x]$ (compare the example in comment 2) we have associated the operation of singleton formation $\{.\}$ and the writing of $y = \{x\}$. Analogous remarks can be made concerning the operations of powerset $\mathcal{P}$ and sum-set $\bigcup$, etc. We even have introduced operations of two and more arguments such as union and pair-formation. Normally this is done only for correspondences of some importance; but in theory we may as well assume this to be done for every correspondence.

The replacement axiom now states that the image $\{t(x) \mid x \in a\}$ of any set a under an operation t is again a set. This also explains its name: we obtain the image of a by *replacing* every element x of a by its image $t(x)$.

6. We may extend an operation t, defined on a set a, to the universe of all sets to the operation t' as follows: put $t'(x) := t(x)$ whenever $x \in a$ and $t'(x) := x$ otherwise. It is clear then that $\{t'(x) \mid x \in a\} = \{t(x) \mid x \in a\}$; therefore we may weaken the assumption of REP to $\forall x \in a\ \exists!\ y\ \Phi$, thereby strengthening REP itself.

7. In REP we may admit also operations of two (and more) arguments: for $\{t(x,y) \mid x \in a \wedge y \in b\} = \{t'(z) \mid z \in a \times b\}$, where t' is defined on ordered pairs by putting $t'(\langle x,y\rangle) := t(x,y)$.

IX REG *Axiom of regularity* (or *foundation*): Every non-empty set is disjoint from at least one of its elements.

Formally: $\exists x\ [x \in a] \rightarrow \exists b\ [b \in a \wedge \forall x[x \in a \rightarrow x \notin b]]$, or equivalently,

$$a \neq \emptyset \rightarrow \exists b\ [b \in a \wedge b \cap a = \emptyset].$$

An element b of a is called *epsilon-minimal* (or *minimal* for short) in a if $b \cap a = \emptyset$. REG therefore states that non-empty sets have minimal elements. In chapter II, section 1, we discuss REG and its role more thoroughly. For the moment we only remark that it excludes some pathological situations.

Theorem 20.10: i) $a \notin a$

ii) $\neg(a_1 \in a_2 \wedge a_2 \in a_3 \wedge \ldots \wedge a_{n-1} \in a_n \wedge a_n \in a_1)$

Proof: i) Suppose that $a \in a$. Look at $\{a\}$. This is a non-empty set, hence it has, according to REG, a minimal element. This must be a, since this is the only element of $\{a\}$. Therefore $a \cap \{a\} = \emptyset$. But $a \in a \cap \{a\}$ according to the assumption that $a \in a$, which is therefore contradictory.

ii) generalizes i) and states that there are no n sets $a_1, \ldots, a_n$ for which $a_1 \in a_2 \in a_3 \in \ldots \in a_{n-1} \in a_n \in a_1$. For otherwise $\{a_1, \ldots, a_n\}$ would not have a minimal element. □

Theorem 20.10 admits a further generalization:

Theorem 20.11: There is no function f defined on ω such that $f(i+1) \in f(i)$ for all $i \in \omega$.

Proof: Take a minimal element in $\mathrm{Ran}(f)$. □

Hence there are no infinite descending epsilon-sequences.

It is a consequence of the axiom of dependent choices that, given the other **ZF** -axioms, REG is equivalent with Theorem 20.11 (compare II, 9.8).

The axioms given above are basic and completely accepted. The resulting theory has practical value and is elegant. The axiomatization however is in no way complete, there are many set theoretic principles (statements) which are independent of **ZF** . The axiom of choice (*AC*) and the generalized continuum hypothesis (*GCH*) are not mentioned here just because of their more dubious status. A separate discussion follows in II.9 and II.10.

21 PEANO'S ARITHMETIC

One of the strongest arguments in favour of set theory is derived from its almost unlimited capacity for accommodating mathematical theories. As a matter of fact, with the machinery of **ZF** one can easily define groups, fields, modules, etc. The very least, however, we should expect from set theory, is to provide us with a theory of numbers. We already have (20.3, 20.7) a set ω of natural numbers and a successor function (20.2), so our task is to show that this structure $\langle \omega, ^{+}, 0 \rangle$ satisfies the well-known axioms of Peano.

Strictly speaking we are interpreting Peano's arithmetic in **ZF**, i.e. we have a translation of the language of arithmetic into the language of **ZF** with the property that the translation of an arithmetical statement A is provable in **ZF**, if A is provable in arithmetic. We will, however, avoid the logic involved, by simply proving the axioms of Peano in **ZF** for the structure $\langle \omega, ^{+}, 0 \rangle$.

The *axioms of Peano* are listed below.

1. $\forall x \in \omega\, [x^{+} \neq 0]$

2. $\forall x, y \in \omega\, [x^{+} = y^{+} \to x = y]$

3. $(A(0) \wedge \forall x \in \omega\, [A(x) \to A(x^{+})]) \to \forall x \in \omega\, [A(x)]$ (mathematical induction)

In words:

1. Each successor of a natural number is distinct from 0.
2. Natural numbers with identical successors are identical.
3. If 0 has the property A and if n^{+} has the property A, if n has it, then each natural number has the property A.

Theorem 21.1: The axioms of Peano are provable in **ZF**.

Proof: 1. $x^{+} = x \cup \{x\} \neq \emptyset$, since $x \in \{x\}$.

2. If $x \cup \{x\} = y \cup \{y\}$, then $x \in y \cup \{y\}$ and $y \in x \cup \{x\}$, or $(x \in y \vee x = y) \wedge (y \in x \vee y = x)$. Now suppose $x \neq y$, then $x \in y \wedge y \in x$. This contradicts 20.10. Hence $x = y$.

Note that this proof uses the axiom of regularity; we can avoid it at the cost of some labour.

3. has already been proved in 20.9. □

Having established the validity of Peano's axioms, we can, following in the footsteps of Dedekind (Was sind und sollen die Zahlen?), introduce addition, multiplication, exponentiation, etc. That is to say, we can define in **ZF** operations $+$, $\cdot$, exp ,etc. and add them to $\langle \omega, ^+, 0 \rangle$. Instead of introducing these functions separately we formulate a theorem which allows the definition of functions by recursion.

Theorem 21.2 *(Recursion theorem)*: Let $a \in X$ and $f: X \to X$. Then there exists a function $u: \omega \to X$, such that $u(0) = a$ and $u(y^+) = f(u(y))$ for all $y \in \omega$.

Proof: Let $a \in X$ and $f: X \to X$. Note that a function from ω to X is a subset of $\omega \times X$. We will construct u explicitly as a set of ordered pairs. u must be a function from ω to X, such that $u(0)= a$, in other words, $<0,a> \in u$, and such that for all $y \in \omega$, $u(y^+) = f(u(y))$, in other words, if $u(y)= x$, then $u(y^+) = f(x)$, or $<y,x> \in u \to <y^+, f(x)> \in u$.

In view of this end we define:

A is an R-set $:= A \subseteq \omega \times X \wedge <0,a> \in A \wedge \forall y \in \omega\ \forall x \in X\ [\ <y,x> \in A \to <y^+,f(x)> \in A]$.

Note that $\omega \times X$ is an R-set, so that the collection C of all R-sets is non-empty. Hence we can construct the intersection u of all R-sets (of the collection C). By the axioms of Zermelo-Fraenkel u is a set and $u \subseteq \omega \times X$. It is easy to see that u itself is an R-set, in other words, $u \subseteq \omega \times X$ and

$<0,a> \in u$ (1)

$<y,x> \in u \to <y^+, f(x)> \in u$ (2)

It now only remains to prove that u is a function from ω to X, in other words that $\forall y \in \omega\ \exists!\ x \in X\ [\ <y,x> \in u]$.

Thus let $\Phi(y) := \exists! x \in X\ [\ <y,x> \in u]$, then we will show: $\forall y \in \omega\ [\Phi(y)]$.

For an application of mathematical induction it is sufficient to prove:

i) $\Phi(0)$, in other words, $\exists! x \in X\ [\ <0,x> \in u]$, and

ii) $\forall y \in \omega\ [\Phi(y) \to \Phi(y^+)]$, in other words,

$\forall y \in \omega\ [\ \exists! x \in X\ [\ <y,x> \in u] \to \exists! x \in X\ [\ <y^+, x> \in u]]$.

Proof of $\Phi(0)$: From (1) it follows that $\langle 0,a\rangle \in u$. In order to prove that a is the only element with $\langle 0,a\rangle \in u$, suppose that $\langle 0,b\rangle \in u$ and $b \neq a$. Then consider $u - \{\langle 0,b\rangle\}$. Now it is easy to see that $u - \{\langle 0,b\rangle\}$ is an R-set, which is contrary to the fact that u is the smallest R-set.

Proof of $\forall y \in \omega\, [\Phi(y) \to \Phi(y^+)]$: Suppose that $y \in \omega$ and $\Phi(y)$, in other words that $y \in \omega$ and there exists exactly one $x \in X$ such that $\langle y,x\rangle \in u$. Because $\langle y,x\rangle \in u$, it follows from (2) that $\langle y^+,f(x)\rangle \in u$. In order to prove that $f(x)$ too is the only element with $\langle y^+,f(x)\rangle \in u$, suppose that $\langle y^+,b\rangle \in u$ and $b \neq f(x)$. Then consider $u - \{\langle y^+,b\rangle\}$. Again it is easy to see that $u - \{\langle y^+,b\rangle\}$ is an R-set, which is contrary to the fact that u is the smallest R-set.

This finishes the proof of the recursion theorem. □

An application of the recursion theorem is called a *definition by recursion* or a *recursive definition*.

The introduction of addition for natural numbers is a typical application of definition by recursion:

Let $X = \omega$, $f = {}^+$ (i.e. $f(x) = x^+$, $f\colon \omega \to \omega$) and $a = x$ ($x \in \omega$), then it follows from the recursion theorem that there exists a (unique) function $s_x : \omega \to \omega$ (as a set in **ZF**) such that
$s_x(0) = x$, and
$s_x(y^+) = (s_x(y))^+$ for all $y \in \omega$.

For $x,y \in \omega$ we define $x+y := s_x(y)$. Thus $+$ is a function from ω^2 to ω.

Similar techniques are applied in the definition of multiplication. For $x \in \omega$ the recursion theorem yields the existence of a function p_x such that $p_x(0) = 0$ and such that for all $y \in \omega$, $p_x(y^+) = p_x(y) + x$ $(= s_x(p_x(y)))$.

Define $x.y = p_x(y)$ for $x,y \in \omega$.

Having defined + and ., we should convince ourselves that the structure $\langle \omega, {}^{+}, +, ., 0 \rangle$ has the right properties. To be precise, we have to check the axioms for + and . :

$$\begin{cases} x + 0 = x \\ x + y^{+} = (x+y)^{+} \end{cases} \quad \text{and} \quad \begin{cases} x.0 = 0 \\ x.y^{+} = (x.y) + x. \end{cases}$$

The validity follows immediately from the definitions, e.g. $x+0 = s_x(0) = x$ and $x + y^{+} = s_x(y^{+}) = (s_x(y))^{+} = (x+y)^{+}$.

The reader is asked to give a definition of $\exp(x,y)$ $(=x^y)$.

It is an experimental fact that the greater part of number theory can be derived in Peano's system. As a consequence set theory incorporates number theory and, as we have seen, the theory of the rationals and the reals.

Set theory, however, yields a more powerful arithmetic than just Peano's arithmetic. To be precise: in **ZF** more (true) arithmetical statements are provable than in Peano's system. This follows from Gödel's theorems, which state (among other things) that the statement, expressing the consistency of Peano's arithmetic, is not provable in Peano's arithmetic. This particular statement <u>is</u> provable in **ZF** , thus **ZF** -arithmetic is stronger than Peano's arithmetic.

CHAPTER II
AXIOMATIC SET THEORY

1 THE AXIOM OF REGULARITY

In this section we consider more extensively the regularity-axiom. The reasons to accept it are both technical and intuitive in nature and we shall discuss them later on. As examples of technical reasons we mention the validity of the principle of *epsilon-induction* (§2) and the simple definition of the concept of an *ordinal number* (§3); an intuitive one is the fact that under assumption of this axiom the universe of all sets coincides with the *cumulative hierarchy* (§4), which gives a reasonably clear picture of this universe. In this section we establish the axiom's innocence by showing that its acceptance amounts to neglecting certain 'non-regular' sets (which are not accessible from regular ones by means of set-theoretic operations, given by the other axioms, anyway). For the axiom of regularity simply states the non-existence of non-regular sets: all sets are regular.

Suppose that $\mathbf{R}$ is some relation. We say that a is $\mathbf{R}$-*minimal* in a set A if $a \in A$ and there does not exist a $b \in A$ for which $b \mathbf{R} a$. (Notice the slight departure from definition I.14.2. We shall use the one given above in connection with irreflexive relations.)

A relation which is often considered in connection with a set A is ϵ_A, the *epsilon-relation* on A, defined by $\epsilon_A := \{\langle a,b\rangle \epsilon A^2 \mid a \in b\}$. An element of A which is ϵ_A-minimal in A is also called *epsilon-minimal* or just *minimal*. A set a is called *regular* if any set containing a as an element has a minimal element; symbolically, $\forall A[a \in A \rightarrow \exists b \in A\ \forall c \in A(c \notin b)]$.
A bit different, but equivalent: $\forall A[a \in A \rightarrow \exists b \in A\ \ (b \cap A = \emptyset)]$. It will be shown that lots of sets are regular. The main rôle in the argument is played by the

Lemma 1.1: A set is regular iff its elements are regular.
Proof: Suppose that a consists of regular sets and let $a \in A$. If a happens to

be minimal in A we are done. Otherwise some b is an element of $a \cap A$. As $b \in a$, b is regular. But then, since $b \notin A$, A has a minimal element. Therefore, a is regular.

Conversely, suppose that a is regular and let $b \in a$. To show that b is regular, assume that $b \in A$; we have to find a minimal element in A. Since $a \in A \cup \{a\}$ and a is regular, $A \cup \{a\}$ will have a minimal element c. c is distinct from a for $b \in a \cap (A \cup \{a\})$, therefore $c \in A$.

But then c is minimal in A also. □

Since the elements of regular sets are regular, the axiom of extensionality holds within the domain of regular sets (provided it holds in the domain of *all* sets): regular sets with the same regular elements are the same. Reason: if regular sets have the same regular elements then they have the same elements throughout, hence they are identical according to the axiom of extensionality.

It turns out that the other Zermelo-Fraenkel-axioms hold as well in the domain of regular sets.

The empty set is regular (either by 1.1 or by definition), and it satisfies the axiom of the empty set which therefore holds in the domain of regular sets.

If $b = \{x \in a | \Phi(x)\}$ and a is regular then all elements of b are regular and thereby b itself, hence the separation-axiom produces regular sets from regular sets. Thus, it also is valid in the domain of those sets.

Pairs are regular as soon as their elements are; hence the pairing-axiom remains valid if we throw out the non-regular sets.

If a is regular so is $\bigcup a$ by a few applications of 1.1; since $\bigcup a$ satisfies the sum-set axiom considered in the domain of regular sets, this one remains valid too.

The same thing can be said of the power set and the corresponding axiom.

If x is regular, so is $x^+ := x \cup \{x\}$; hence: if a is a successor so is a^1, the set of regular elements of a (separation), and a^1 is itself regular (in particular ω is regular). Also a^1 satisfies the axiom of infinity considered in the domain of regular sets, therefore INF is valid in this domain.

Finally, if $t(x)$ is regular, whenever $x \in a$ then $\{t(x) \mid x \in a\}$ is also regular so the replacement-axiom remains valid.

Now look at the statement, saying that all sets are regular, in symbols: $\forall a\ \forall A\ [a \in A \rightarrow \exists b \in A\ \forall c \in A\ (c \notin b)]$. This is trivially equivalent to $\forall A\ \forall a\ [a \in A \rightarrow \exists b \in A\ \forall c \in A\ (c \notin b)]$ which in turn amounts to $\forall A\ [\exists a(a \in A) \rightarrow \exists b \in A\ \forall c \in A\ (c \notin b)]$ which is the regularity-axiom IX! Suppose now that we are given a universe of sets in which the Zermelo-Fraenkel axioms I-VIII hold. By throwing out the non-regular sets (hence keeping the regular ones, which we have to study anyway) we may as well add to our axioms the regularity-axiom IX. For we have shown already that I-VIII hold in our new, restricted universe if they hold in the old one, and we presently show that IX holds there as well (irrespective of whether it holds or not in the old universe!): suppose that A is a regular set which has regular elements: Then by definition A has (by 1.1) a regular element b which is minimal among all elements of A. But then b is also minimal among the regular elements of A.

To sum up the discussion: the formal acceptance of the regularity axiom amounts to restricting our attention to regular sets only.

We close this paragraph with three remarks.

As was to be expected, there is no set $\mathbf{R}$ the elements of which are exactly the regular sets, since otherwise by 1.1 $\mathbf{R}$ had to be regular itself, therefore $\mathbf{R} \in \mathbf{R}$, but then $\{\mathbf{R}\}$ would not have a minimal element while it should according to the definition of regularity. In other words: Cantor's naive theory admits the paradox of the set of all regular sets.

To motivate the regularity-axiom we have shown that the universe $\mathbf{R}$ is closed under the set theoretic operations hitherto introduced and also that the Zermelo-Fraenkel axioms maintain their validity if we restrict the universe to $\mathbf{R}$. This means that the introduction of new axioms or operations requires additional motivation of this axiom, at least in principle. To give an example: one formulation of the *axiom of choice* which we shall discuss in detail in a later paragraph demands the existence, for every set A, of a function $f: \mathcal{P}(A) \rightarrow A$, for which $f(B) \in B$ whenever $\emptyset \neq B \subseteq A$. However it is trivially verified that *every* function $f: \mathcal{P}(A) \rightarrow A$ is regular whenever A is;

so are in particular the functions required by the axiom of choice for regular sets (cf. exercise 2).

Finally one can construe the property $\forall a(a \subseteq \mathbf{R} \to a \in \mathbf{R})$ of $\mathbf{R}$ given by 1.1 as an *inductive definition* of $\mathbf{R}$, meaning that $\mathbf{R} = \bigcap \{K|\forall a(a \subseteq K \to a \in K)\}$ (for *classes* K) were it not for the fact that the introduction of "superclasses" such as $\{K|\forall a(a \subseteq K \to a \in K)\}$ (the elements of which are classes) is prohibited in our theory. Forgetting for a minute the illegal character of the superclasses, it is possible to make this definition of $\mathbf{R}$ more meaningful by giving the following "proof": put $\mathbf{A} := \{K|\forall a(a \subseteq k \to a \in K)\}$. Then

1. $\mathbf{R} \supseteq \bigcap \mathbf{A}$ since $\mathbf{R} \in \mathbf{A}$ by 1.1; and
2. $\bigcap \mathbf{A} \supseteq \mathbf{R}$ since $\mathbf{R} \subseteq K$ for every $K \in \mathbf{A}$.

To prove the second of these two assertions we use the fact that to every regular set a there exists a regular $b \supseteq a$ with the property that $b \subseteq \mathcal{P}\, b$ (this is proved in detail in §2; $b = \mathrm{TC}(a)$ fulfills the requirements). Thus suppose that $K \in \mathbf{A}$ and $a \in \mathbf{R}$. we show that $a \in K$. Let the regular $b \supseteq a$ be such that $b \subseteq \mathcal{P}b$ and consider $b-K$. We first show that this set is empty. Otherwise, $b-K$, being regular, contains a minimal element c, which means that $c \cap b \subseteq K$. But $c \in b \subseteq \mathcal{P}b$, therefore $c \subseteq b$ and $c \cap b = c$; hence $c \subseteq K$ and $c \in K$ since $K \in \mathbf{A}$. This contradicts $c \in b - K$. Thus, $b - K = \emptyset$, in other words, $b \subseteq K$, hence $a \subseteq K$ and therefore $a \in K$ since $K \in \mathbf{A}$, which was to be proved. (Cf. the proof of Theorem 2.6.)

<u>Exercises</u>

1. The product of regular sets is regular.

2. If $f\colon \mathcal{P}(A) \to A$ and A is regular then f is regular.

3. In §1 it was shown that, under the assumption of all **ZF**-axioms minus the the regularity axiom, these axioms plus the regularity axiom are valid under the re-interpretation of the concept of a 'set' as: 'regular set'. We may also, instead of the set-concept, re-interpret the 'elements-of'-concept symbolized by $\in$ for instance by $a\ R\ b := a=b=\emptyset \vee (b\neq\{\emptyset\} \wedge a \in b)$. Show that all **ZF**-axioms are valid under this interpretation but the regularity-axiom is <u>*not*</u>.

4. Given the remaining **ZF**-axioms, the regularity axiom is equivalent with the statement $a \neq \emptyset \rightarrow \exists x \in a(a \cap \bigcup x = \emptyset)$.

5. The regularity-axiom implies the statement $a \neq \emptyset \rightarrow a \neq a \times a$. Does the converse hold as well?

2 INDUCTION AND RECURSION

We shall now have a closer look at two properties of $\in := \{<a,b> | a \in b\}$. Let R be any collection of pairs of sets. We write '$a\ R\ b$' for '$<a,b> \in R$'. a is called *R-predecessor* of b in case aRb.

Definition 2.1: (i) R is called *well-founded* if every non-empty set has an R-minimal element; in symbols: $\forall A[A \neq \emptyset \rightarrow \exists a \in A\ \forall b \in A \neg bRa]$.

(ii) R is called *local* if the R-predecessors of an object form a set; symbolically if $\forall a\ \exists b\ \forall c\ [c \in b \leftrightarrow cRa]$.

We can reformulate a bit the condition (ii) as follows:

In the sequal we take V to be the collection of *all* sets. (ii) then says that $\{c | c\ R\ a\} \in V$ for every a.

Notice that $\in$ is well-founded as well as local: the first property follows by the axiom of regularity and the second one on purely logical grounds.

If R is a well-founded collection including S as a subclass then S is well-founded also, and the same holds for locality.

Relations, being sets by definition, always are local because $\{c | c\ R\ a\} \subseteq \bigcup\bigcup R$.

In what follows, the concept of *transitive closure* of a relation or a class of pairs will play a central rôle. To illustrate it we give an informal example: let R be the relation of parenthood; its transitive closure is the relation of ancestorship — this is the smallest relation containing R as a subclass which is transitive.

In general, if R is a relation on a set A, we define its *transitive closure* to be $\bar{R} := \bigcap \{S \subseteq A^2 | R \cup (S \circ S) \subseteq S\}$. Notice that the condition $S \circ S \subseteq S$ expresses transitivity for S since $S \circ S = \{<a,c> | \exists b(aSb \wedge bSc)\}$. Thus, $\bar{R}$ is the intersection of all transitive relations containing R as a subset. For evident reasons, $\bar{R}$ is sometimes called the ancestor relation belonging to R.

Lemma 2.2: 1. $R \subseteq \bar{R}$;

2. $\bar{R} \circ \bar{R} \subseteq \bar{R}$;

3. $R \cup (S \circ S) \subseteq S \to \bar{R} \subseteq S$;

4. 1-3 are characteristic for $\bar{R}$.

Proof: 1 and 3 are immediate from the definition of $\bar{R}$.

2. Suppose that $a\bar{R}b$ and $b\bar{R}c$; we show that aSc for any S such that $R \cup (S \circ S) \subseteq S$ (this suffices according to the definition of $\bar{R}$). Take such an S. Then by 3. $\bar{R} \subseteq S$ hence aSb and bSc; therefore aSc since S is transitive.

4. Suppose that R' is such that 1'. $R \subseteq R'$; 2'. $R' \circ R' \subseteq R'$ and 3'. $R \cup (S \circ S) \subseteq S \to R' \subseteq S$. Then from 1', 2' and 3 it follows that $\bar{R} \subseteq R'$ and from 1, 2 and 3' that $R' \subseteq \bar{R}$, i.e., $R' = \bar{R}$. □

Generalizing this procedure to classes of pairs one immediately gets into a difficulty: if R is such a class $\bar{R}$ has to be the intersection of a 'super-collection' of classes - but such things are not allowed in our axiomatic treatment. Fortunately there is a way out:

Definition 2.3: If R is a collection of ordered pairs, then $\bar{R}$ is the collection of those ordered pairs $\langle a,b\rangle$ for which a natural number n and a function f exist such that Dom $f = \{m \mid m \leq n+1\}$ $(=n+2)$, $f(0)=a$, $f(n+1)=b$ and $\forall i \leq n\, [f(i)\ R\ f(i+1)]$.

One may simply check (see exercise 2) that parts 1,2 and 3 of lemma 2.2 hold for this new definition (for 3 mathematical induction is needed).

A property which often is used without reference is given by the

Lemma 2.4: $\bar{R} = R \cup (\bar{R} \circ R)$.

Proof: cf. exercise 4 . □

The behaviour of $\bar{R}$ with respect to well-foundedness and locality is given by the next lemma:

Lemma 2.5: (i) if R is local so is $\bar{R}$;

(ii) if R is local and well-founded so is $\bar{R}$.

Remark: Using the axiom of regularity one may show that $\bar{R}$ is well-founded if R is, without R being local; cf. exercise 9 of §4. We shall not need this result , however.

Proof: (i) Take any b; we show that $\{a|a\bar{R}b\} \in V$ whenever R is local. Let $A_n (n \in \omega)$ consist of those a for which a function f defined on $n+2$ exists with the property that $f(0)=a$, $f(n+1)=b$ and $\forall i \leqslant n[f(i)R\, f(i+1)]$. Since $A_o = \{a|aRb\}$, A_o is a set. Since $A_{n+1} = \bigcup\{\{a|aRc\}|c \in A_n\}$ this is a set if A_n is by replacement -and sum set- axioms; therefore, applying mathematical induction, all A_n are sets. To complete the proof, notice that $\{a|a\bar{R}b\} = \bigcup\{A_n|n \in \omega\}$ and again apply the replacement - and sum set- axioms.

(ii) Suppose that R is local and well-founded; let A be a non-empty set. We look for an $\bar{R}$-minimal element in A. Take any $a_o \in A$. If a_o happens to be $\bar{R}$-minimal we are through; so suppose not. Then $B := \{b|b\bar{R}a_o \wedge [b \in A \vee \exists a \in A(a\bar{R}b)]\}$ certainly is a non-empty set (apply part (i)). Let b be R-minimal in B. If $a\bar{R}b$ for a certain $a \in A$, then either aRb, which contradicts b being minimal, or for some c, $a\bar{R}c$ and cRb (use 2.4), again contradicting the choice of b. Therefore, $\neg\exists a \in A[a\bar{R}b]$. Hence, $b \in A$ and b is $\bar{R}$-minimal in A. □

In order to apply this to $\in$ we introduce the following important property. A set is called *transitive* if it contains each of its elements as a subset. In a formula: a is transitive iff $\forall b(b \in a \rightarrow b \subseteq a)$; equivalently: $\forall b \forall c(c \in b \wedge b \in a \rightarrow c \in a)$. (Other ways of expressing this property are: $\bigcup a \subseteq a$ and $a \subseteq \wp a$.) Confusion with the transitivity -concept for relations is highly unlikely since the only relation which is transitive as a set in the present sense is the empty set. Anyway -it will always be clear from the context which transitivity- notion is meant.

$TC(a) := \{b|b \bar{\in} a\}$ is called the *transitive closure* of a; 2.5 (i) implies that this is a set. The name is motivated by the fact that $TC(a)$ is transitive, contains a as a subset and is the smallest set of the kind. This is immediate from the definition of $\bar{\in}$ and lemma 2.2. Moreover 2.4 implies that $TC(a) = a \cup \bigcup_{b \in a} TC(a)$.

We give yet another application of 2.5. Suppose that R is some well-founded collection of pairs. I.e., every non-empty set has an R-minimal element. We may ask whether every non-empty *class* has an R-minimal element. The answer is yes, but we prove this for local R only (compare the remark just after 2.5; also, see exercise 8 of §4). Suppose that K is a non-empty collection, say, $a \in K$. Put $K' := \{b \in K | b \bar{R} a\}$. If $K' = \emptyset$ than a clearly is R-minimal in K. Therefore, suppose that $K' \neq \emptyset$. Since R is local we know by 2.5(i) that $\bar{R}$ is local also; hence K' being a set has an R-minimal element b. But then b is R-minimal in K too. We formulate this result in a slightly different but logically equivalent way as the

Theorem 2.6: If R is well-founded and local then the following *principle of R-induction* holds:

$\forall a[\forall b(bRa \to P(b)) \to P(a)] \to \forall a\, P(a)$ (where P is any property of sets).

Proof: Suppose that P applies to any a for which P applies to all its R-predecessors, but nevertheless that P does not apply to all sets. Take K to be the collection of those sets which do not have the property P; hence $K \neq \emptyset$. We have just shown above that K has an R-minimal element, say, a. By definition, every R-predecessor of a has P. By assumption, $P(a)$ holds. Thus, $a \notin K$ and a contradiction is arrived at. Therefore P *does* apply to all sets. □

Applying 2.6 to the well-founded local collection $\in$ we obtain the often-used principle of *$\in$-induction*:

$\forall a[\forall b(b \in a \to P(b)) \to P(a)] \to \forall a\, P(a)$. Considering membership in a collection as a property we get $\forall a[a \subseteq P \to a \in P] \to \forall a[a \in P]$. (Compare the so-called inductive definition of the collection of regular sets, end of §1. It is fair to say that the axiom of regularity restricts the universe only to the extend of making the principle of $\in$-induction valid). The name of this induction principle derives from the formal similarity with one form of the principle of mathematical induction, namely: restrict the values of a and b to natural numbers and read 'smaller than' for R.

We are now ready for the recursion theorem. Remember that $f \upharpoonright A = \{\langle a,f(a)\rangle | a \in A \cap \mathrm{Dom} f\}$.

Recursion Theorem 2.7: If R is well-founded and local and t is a (ternary) operation then there is exactly one (binary) operation σ which satisfies the *recursion equation*:

$\sigma(a,b) = t(a,b,\{\langle x,\sigma(a,x)\rangle | xRb\})$.

Remark: One can conceive of the recursion equation as a prescription how to compute $\sigma(a,b)$ from the values $\sigma(a,x)$ where x is an R-predecessor of b.

2.7 tells us that in some cases we may define an operation by giving only a recursion-equation for it since there is exactly one solution for such an equation.

Proof: There is at most one solution to the recursion equation, for suppose σ and σ' both satisfy it. To prove that $\sigma(a,b) = \sigma'(a,b)$ for all a and b we suppose that $\sigma(a,x) = \sigma'(a,x)$ for all R-predecessors x of b. Then

$$\begin{aligned}\sigma(a,b) &= t(a,b, \{<x,\sigma(a,x)>|xRb\})\\ &= t(a,b, \{<x,\sigma'(a,x)>|xRb\})\\ &= \sigma'(a,b),\end{aligned}$$

and the result follows from the principle of R-induction. To prove that at least one solution exists, fix the parameter a and call a function f *correct* (w.r.t. b) if (1) Dom $f = \{x|x\bar{R}b\}$ and (2) $f(x) = t(a,x,f\restriction\{y|yRx\})$ for all $x \in$ Dom f.

(i) If f and g both are correct they agree on arguments for which they are both defined. For suppose that $f(y)$ and $g(y)$ are equal when defined whenever yRx and that $x \in$ Dom $f \cap$ Dom g. Then since f and g are correct, $\{y|yRx\} \subseteq$ Dom $f \cap$ Dom g and

$$\begin{aligned}f(x) &= t(a,x,f\restriction\{y|yRx\})\\ &= t(a,x,g\restriction\{y|yRx\}) = g(x).\end{aligned}$$

Therefore, (i) follows by one application of R-induction.

(ii) With respect to every set there exists a (and hence exactly one) correct function.

To prove that we now apply $\bar{R}$-induction (using both 2.5 and 2.6). Suppose that f_x is the correct function on $\{y|y\bar{R}x\}$ when $x\bar{R}b$. Define f on $\{x|x\bar{R}b\}$ by $f(x) = t(a,x,f_x\restriction\{y|yRx\})$. We show that f is correct w.r.t. b. Let $x\bar{R}b$. If $y\bar{R}x$ then

$$\begin{aligned}f(y) &= t(a,y,f_y\restriction\{z|zRy\})\\ &= t(a,y,f_x\restriction\{z|zRy\}) \text{ by (i)}\\ &= f_x(y) \text{ since } f_x \text{ is correct.}\end{aligned}$$

Therefore, $f_x \subseteq f$. Hence

$$\begin{aligned}f(x) &= t(a,x,f_x\restriction\{y|yRx\})\\ &= t(a,x,f\restriction\{y|yRx\})\end{aligned}$$

which means that f is correct.

(iii) We finally define the solution required by the equation $\sigma(a,b) = t(a,b,f_b|\{x|xRb\})$, where f_b is the unique function correct w.r.t. b. Since σ is defined in the same way as the function f under (ii) we may check, copying the proof given there, that σ satisfies the recursion equation. □

Remark. (*i*) We shall trace the points at which the replacement-axiom is needed in the proof of 2.7. (Notice that it is needed already for the recursion-equation to make sense!). A minor point is that the above proof leans heavily on 2.5 for part (*ii*) which in turn is based on the replacement-axiom. The important point is that the definition of f under (*ii*) really defines a *function* (and not merely an *operation*) since the replacement-axiom tells us that the restriction of an operation to a set is a function.

(*ii*) The well-foundedness of R is essential. To give just one example: both functions $\mathbb{Z}\times\{0\}$ and $\mathbb{Z}\times\{1\}$ on $\mathbb{Z}$ satisfy the equation $f(x) = \max\{f(y)\,|\,y<x\}$; indeed, $<$ is not well-founded on $\mathbb{Z}$.

(*iii*) In 2.8, a functions as a parameter. Of course formulations with more parameters exist, but they are an immediate consequence of the theorem as given here: just replace 'a' by '$\langle a_1,\ldots,a_n\rangle$'.

Example: Define t by $t(a,b,s) := b \cup \bigcup \mathrm{Ran}(s)$. According to 2.7 there is exactly one σ such that

$$\begin{aligned}\sigma(a,b) &= t(a,b,\{\langle x,\sigma(a,x)\rangle\,|\,x \in b\})\\ &= b \cup \bigcup \mathrm{Ran}(\{\langle x,\sigma(a,x)\rangle\,|\,x\in b\})\\ &= b \cup \bigcup_{x\in b} \sigma(a,x).\end{aligned}$$

Since we know from 2.4 that $TC(b) = b \cup \bigcup_{x\in b} TC(x)$ we see that $\sigma(a,b) = TC(b)$. Also, this equation uniquely characterizes the transitive closure-operation.

Before we give an important application of the recursion theorem we need some definitions.

Definition 2.8 : The relation R on the set A is called

(*i*) *connected* on A if $\forall a,b \in A\,[a \neq b \rightarrow aRb \vee bRa]$;

(*ii*) a *well-ordering* on A if R is well-founded and connected on A;

(*iii*) *extensional* on A if for all $a,b \in A$: $\forall x(xRa \leftrightarrow xRb) \rightarrow a=b$.

Every well-founded relation is irreflexive and asymmetric and every well-founded connected relation is transitive (why?). Therefore well-orderings are also (irreflexive, total) orderings. Total orderings always are extensional. Examples of well-founded extensional relations are the epsilon-relations on transitive sets; we shall see presently that these examples are characteristic.

In the sequel, a *structure* will mostly be a pair $\langle A,R\rangle$ for which $R \subseteq A^2$. We say that it is connected (a well-ordering, extensional etc.) if R has that property with respect to A. The concept of *isomorphism* already has been introduced in chapter I; h is called an *isomorphism* between $\langle A,R\rangle$ and $\langle B,S\rangle$ if $h\colon A \to B$ is a bijection for which $S = \{\langle h(a),h(b)\rangle \mid aRb\}$. An *automorphism* of $\langle A,R\rangle$ is an isomorphism between $\langle A,R\rangle$ and $\langle A,R\rangle$. The *trivial automorphism* of $\langle A,R\rangle$ is the identity $\{\langle a,a\rangle \mid a \in A\}$. An *epsilon-structure* is a structure of the form $\langle A,\in_A\rangle$; it is called *transitive* if A is transitive (this does *not* imply that $\in_A$ is transitive!).

The next representation theorem for well-founded extensional structures, also called Mostowski's *collapsing lemma*, is one of the most important applications of the recursion theorem.

Theorem 2.9: Every well-founded extensional structure $\mathbf{A}$ is isomorphic with a transitive epsilonstructure $\mathbf{B}$; both the isomorphism and $\mathbf{B}$ are completely determined by $\mathbf{A}$.

Proof: Suppose that $\mathbf{A} = \langle A,R\rangle$ is well-founded and extensional. Suppose also that h is an isomorphism meeting the requirements, then if $a,b \in A$ we shall have $bRa \leftrightarrow h(b) \in h(a)$, by the isomorphism-property of h, and hence $h(a) = \{h(b) \mid bRa\}$, since $\mathrm{Ran}(h)$ is transitive. Next we shall construct h.

The recursion theorem (put $t(a,b,f) := \mathrm{Ran}(f)$) states that exactly one operation σ exists for which $\sigma(a) = \{\sigma(b) \mid bRa\}$. So take $B := \{\sigma(a) \mid a \in A\}$ and $h := \{\langle a,\sigma(a)\rangle \mid a \in A\}$ then the argument above shows that the only transitive epsilon-structure, isomorphic to $\mathbf{A}$, must be $\mathbf{B} := \langle B,\in_B\rangle$ and the only isomorphism between $\mathbf{A}$ and $\mathbf{B}$ must be h.

We finally show that $\mathbf{B}$ and h satisfy the requirements indeed. The transitivity of B immediately follows from the definition and the recursion-equation for σ (which holds for h also). We show that h is injective using R-induction. Suppose that $\forall b \in A\ [h(c) = h(b) \to c = a]$ when cRa (induction-hypothesis) and $h(a) = h(b)$. We have to show that $a=b$; but for this it is sufficient to show that a and b have the same R-predecessors since R is extensional.

(*i*) if cRa then $h(c) \in h(a)$; therefore $h(c) \in h(b)$. So for some c' with $c'Rb$, $h(c) = h(c')$. Using the induction-hypothesis, $c'=c$, therefore cRb.

(*ii*) if cRb then $h(c) \in h(b)$; therefore $h(c) \in h(a)$. So for some c' with

$c'Ra$, $h(c) = h(c')$. Using the induction-hypothesis once more, $c'=c$, hence cRa. Finally, h maps R onto ϵ_B: if aRb then $h(a) \epsilon h(b)$, and, conversely, if $h(a) \epsilon h(b)$ then $h(a) = h(a')$ for some a' with $a'Rb$. But since h is injective it follows that $a=a'$ and aRb. □

Corollary 2.10: Isomorphic transitive epsilon structures are equal.
Proof: Transitive epsilon structures are well-founded and extensional. Apply the second half of 2.9 . □

Corollary 2.11: Well-founded extensional structures only have trivial automorphisms.
Proof: Suppose that $\mathbf{A}$ is a well-founded extensional structure with a non-trivial automorphism g. Let h map $\mathbf{A}$ isomorphically onto a transitive epsilon-structure. Then $h \circ g$ is an isomorphism between $\mathbf{A}$ and a transitive epsilon-structure different from h which is impossible. □

Corollary 2.12: There is at most one isomorphism between well-founded extensional structures.
Proof: If h and g are different isomorphisms between $\mathbf{A}$ and $\mathbf{B}$ then $g^{-1} \circ h$ is a non-trivial automorphism of $\mathbf{A}$. Apply 2.11. □

Suppose that $\mathbf{A}$ is well-founded and extensional. 2.9 implies that the collection of structures isomorphic to $\mathbf{A}$ contains exactly one transitive epsilon-structure $\mathbf{B} = \langle B, \epsilon_B \rangle$. We call B the *type* of $\mathbf{A}$.

Definition 2.13: The *type* of a well-founded extensional structure $\mathbf{A}$ is the (unique) transitive set B for which $\mathbf{A} \cong \langle B, \epsilon_B \rangle$.

If $\mathbf{A}$ is an extensional epsilon structure which happens to be non-transitive then the type of $\mathbf{A}$ is also called the *collapse* of $\mathbf{A}$. For instance, the collapse of the epsilon-structure on: $\{a,\{a\}, \{a,\{a\}\},\{\{a\},\{a,\{a\}\}\}\}$ (where a is an arbitrary set) is the set $\{0,1,2,\{1,2\}\}$. Exercise 15 gives an other example.

Exercises

1. S, R and t are such that $aSb \rightarrow t(a)\ R\ t(b)$.
   a) If R is well-founded, so is S.
   b) Is it true that S is local whenever R is?

2. Prove the assertion made following definition 2.3.

3. $R \cup (S \circ R) \subseteq S \rightarrow \bar{R} \subseteq S$ (cf. 2.4).

4. Prove lemma 2.4.

5. $\emptyset$ is an element of every non-empty transitive set.

6. If a is transitive so are $\bigcup a$ and $\mathcal{P}a$.

7. If the elements of A are transitive then both $\bigcup A$ and $\bigcap A$ are transitive.

8. $\{\emptyset\}$ is an element of every transitive set containing at least two elements.

9. Prove the principle of R-induction for local well-founded transitive R without using the replacement-axiom.

10. Suppose that R is a relation on the set A. $B \subseteq A$ is called a *tail* if $b \in B \wedge bRc \rightarrow c \in B$. Prove that R is well-founded if every non-empty tail has an R-minimal element.

11. Give direct proofs for 2.10 and 2.11, that is, without using 2.9 .

12. The type of a well-ordering is a transitive set of transitive sets.

13. Transitive, extensional, well-founded relations are well-orderings.

14. For any relation R the following two assertions are equivalent:
    (*i*) $\bar{R}$ is antisymmetric;
    (*ii*) There is no finite sequence $a_1, \ldots, a_n$ $(n > 1)$ of pairwise different elements for which $a_i R a_{i+1}$ if $1 \leqslant i < n$ and $a_n R a_1$.

15. Is the relation $R := \{\langle 1,2\rangle, \langle 3,2\rangle, \langle 4,2\rangle, \langle 4,1\rangle, \langle 4,3\rangle, \langle 5,4\rangle, \langle 5,3\rangle\}$ extensional and well-founded on $\{1,2,3,4,5\}$? If so, determine the collapse of the corresponding structure.

16. Define R by $aRb := \bigcup a \in b$. Is R local? well-founded? extensional?

17. (*i*) Give examples of collections of pairs R and S for which $S = R \cup (S \circ R)$ and $S \neq \bar{R}$.
    (*ii*) If R is well-founded and local and $S = R \cup (S \circ R)$ then $S = \bar{R}$.

3 ORDINAL NUMBERS

Well-orderings play a rôle of central importance in set theory. They give rise to generalisations of the natural numbers such as the counting of objects and the principle of mathematical induction. Since in most cases we do not need to distinguish between isomorphic well-orderings we may as well restrict ourselves to well-ordering *types* which constitute a special subcategory. The fact that every well-ordering has a type is due mainly to the replacement-axiom. The type of a well-ordering is called an *ordinal number* (*ordinal*, for short). Hence these are transitive sets being well-ordered by their epsilon-relation. Examples are 0, 1, 2, ..., ω, ω^+, ... etc. However, we prefer to take a different definition of the concept of an ordinal which is possible thanks to the regularity-axiom. We shall show that this definition is, in fact, adequate.

<u>Definition 3.1</u>: An *ordinal number* is a transitive set of transitive sets.

In the sequel small greek letters usually denote ordinals. OR is the collection of all ordinals. The next lemma says in effect that OR is transitive:

<u>Lemma 3.2</u>: Elements of ordinals are ordinals.

Proof: Suppose that a is an element of an ordinal α. Then a is transitive since elements of ordinals are transitive. And if $b \in a$ then $b \in \alpha$ (for α is transitive) and hence b is transitive too. □

OR cannot be a set: if so, it would be an ordinal by 3.1 and 3.2 and therefore OR $\in$ OR. But this is excluded by the regularity-axiom.

A proof of the (provably incorrect) statement OR $\in$ OR yields, what is usually referred to as the *Paradox of Burali-Forti*; this could be carried out in Cantor's naive set theory.

One of the most important properties of ordinals is their comparability:

<u>Theorem 3.3</u>: For any two ordinals α and β, either $\alpha \in \beta$ or $\beta \in \alpha$ or $\alpha = \beta$.

Proof: We use epsilon-induction to show that for all α :
$\forall\beta[\alpha \in \beta \vee \beta \in \alpha \vee \alpha = \beta]$. So we can assume as an induction-hypothesis that

$$\forall\gamma[\gamma \in \alpha \to \forall\beta\ (\gamma \in \beta \vee \beta \in \gamma \vee \gamma = \beta)] \tag{1}$$

on the basis of which we have to show that $\forall\beta(\alpha \in \beta \vee \beta \in \alpha \vee \alpha = \beta)$. To this end we apply a second epsilon-induction, so we may assume moreover

$\forall\gamma[\gamma \in \beta \to (\alpha \in \gamma \vee \gamma \in \alpha \vee \alpha = \gamma)]$ (2)

and we have to show now that $\alpha \in \beta \vee \beta \in \alpha \vee \alpha = \beta$.

So suppose that finally

$\alpha \notin \beta$ and $\beta \notin \alpha$; (3)

We then show that $\alpha = \beta$ using the extensionality-axiom:

(*i*) If $\gamma \in \alpha$ then $\beta \notin \gamma$ and $\gamma \neq \beta$ by (3) so $\gamma \in \beta$ using (1).

(*ii*) If $\gamma \in \beta$ then $\alpha \notin \gamma$ and $\alpha \neq \gamma$ again by (3) hence $\gamma \in \alpha$, this time using (2). □

Theorem 3.3 implies that our definition of an ordinal is adequate. Indeed, according to exercise 12 of §2, the type of a well-ordering is an ordinal, and conversely, every ordinal α is the type of a well-ordering, viz. $\langle\alpha, \epsilon_\alpha\rangle$: 3.3 says that this is a well-ordering.

Because $\in$ behaves as a well-ordering on OR (apart from the fact that it is no relation) one often writes '$\alpha < \beta$' or '$\beta > \alpha$' in stead of '$\alpha \in \beta$' and '$\alpha \leqslant \beta$' ('$\beta \geqslant \alpha$') in stead of '$\alpha \subseteq \beta$'.

Note that $a^+ := a \cup \{a\}$.

<u>Theorem 3.4</u>: (*i*) $0 \in \mathrm{OR}$;

(*ii*) $\alpha \in \mathrm{OR} \to \alpha^+ \in \mathrm{OR}$;

(*iii*) $a \subseteq \mathrm{OR} \to \bigcup a \in \mathrm{OR}$.

Proof: exercise. □

<u>Theorem 3.5</u>: $\omega \in \mathrm{OR}$.

Proof: From 3.4(*i*), (*ii*) and mathematical induction (I.20.9) it follows that $\omega \subseteq \mathrm{OR}$. Hence the elements of ω are transitive. So it suffices to show that ω is transitive: $\forall n \in \omega(n \subseteq \omega)$. This again trivially follows by mathematical induction. □

For an ordinal α we distinguish two possibilities: either it has a greatest element β (in the sense of $<$) -and clearly then $\alpha = \beta^+$- or it has no greatest element, i.e., $\alpha \subseteq \bigcup\alpha$.

<u>Definition 3.6</u>: (*i*) $\alpha \in \mathrm{OR}$ is called a *successor*,

notation: $\mathrm{Suc}(\alpha)$, if $\exists\beta \in \alpha(\alpha = \beta^{+})$;

(*ii*) $\alpha \in \mathrm{OR}$ is called a *limit*, notation: $\mathrm{Lim}(\alpha)$, if $\alpha \subseteq \cup\alpha$ and $\alpha \neq 0$.

The natural numbers 1,2,3,... are examples of successors, ω is the first limit. Then a series of successors follows: ω^{+}, ω^{++}, etc., after which a next limit occurs.

Restricting $\in$ to OR one immediately obtains induction- and recursion principles for ordinals ('*transfinite*' induction resp. recursion). Often these are applied in a somewhat modified form. The ordinary form looks as follows:

$\forall\alpha[\forall\beta(\beta < \alpha \rightarrow P(\beta)) \rightarrow P(\alpha)] \rightarrow \forall\alpha\, P(\alpha)$.

The modified version now is:

$P(0) \wedge \forall\alpha[P(\alpha) \rightarrow P(\alpha^{+})] \wedge \forall\alpha[\mathrm{Lim}\,\alpha \wedge \forall\beta < \alpha[P(\beta)] \rightarrow P(\alpha)] \rightarrow \forall\alpha\, P(\alpha)$.

This immediately follows from the ordinary form noting that every ordinal either is 0 or a successor or a limit.

The modified recursion principle asserts that for given t_1, t_2 and t_3 the next triple of equations has exactly one solution for σ:

$$\sigma(a,0) = t_1(a)$$
$$\sigma(a,\alpha^{+}) = t_2(a,\alpha,\sigma(a,\alpha))$$
$$\sigma(a,\gamma) = t_3(a,\gamma,\{\langle\xi,\sigma(a,\xi)\rangle \mid \xi<\gamma\}) \quad \text{(where Lim } \gamma).$$

To see this we define t by

$$t(a,\alpha,f) = \begin{cases} t_1(a) \text{ if } \alpha = 0 \\ t_2(a,\beta,f(\beta)) \text{ if } \alpha = \beta^{+} \\ t_3(a,\alpha,f) \text{ if Lim } \alpha. \end{cases}$$

The recursion theorem says that the equation $\sigma(a,\alpha) = t(a,\alpha,\{\langle\xi,\sigma(a,\xi)\rangle \mid \xi<\alpha\})$ has exactly one solution; but clearly this one has the same solutions as the preceding triple. In practice it is often the case that an operation defined by recursion in this modified way has a meaning which is easier to grasp than in the case of just one equation.

Every set of ordinals is well-ordered by the membership-relation; it therefore has a smallest element, which happens to be the *intersection* of the ordinals in the set, cf. exercise 9. According to 3.4 (*iii*), the sum-set of a set of ordinals is again an ordinal; this happens to be the smallest ordinal greater than or equal to the ordinals in the set.

These facts are frequently being used without explicit reference.

Exercises

1. A set a is an ordinal iff it is element of every set A for which $0 \in A \wedge \forall b \in a \cap A(b^+ \in A) \wedge \forall b \subseteq a \cap A(\bigcup b \in A)$.

2. The type of a well-ordering $\langle A,R\rangle$ in which A has no R-greatest element is a limit.

3. Define $R \subseteq (2 \times \omega)^2$ by
$\langle i,n\rangle R \langle j,m\rangle := i < j \vee (i = j \wedge n < m)$.
Verify that $\langle 2 \times \omega,R\rangle$ is a well-ordering the type of which is the first limit greater than ω.

4. If the collection A contains every one of its transitive subsets as an element then OR $\subseteq A$.

5. A set a is an ordinal iff it is element of every set A for which $\forall b \subseteq a \cap A$ (b transitive $\rightarrow b \in A$).

6. Not all transitive sets are ordinals; not all sets of ordinals are ordinals.

7. Transitive sets with a connected epsilon relation are ordinals.

8. Construct a well-founded collection of pairs which is not local.

9. If $a \subseteq$ OR and $a \neq \emptyset$, then $\bigcap a$ is the smallest element of a.

10. $\alpha^+ = \beta^+ \rightarrow \alpha = \beta$.

11. Define *finiteness* for ordinals as follows:
Fin(a) := $\forall\beta \leqslant \alpha$ ($\neg$ Lim β).
Prove: (a) Fin$(\alpha) \leftrightarrow$ Fin(α^+);
(b) Fin$(\alpha) \leftrightarrow \alpha \in \omega$;
(c) $\neg$ Fin(ω).

4 THE CUMULATIVE HIERARCHY

In set theory, ordinals are often used to measure things. Such measurements are intuitively attractive since the concept of an ordinal seems to be simpler than the general idea of a set: the ordinals are well-ordered so OR has a (relatively) simple structure. The *rank 'function'* ρ is an operation which measures, in a sense, the number of steps in which a set is built up from the empty set.

Definition 4.1: ρ is the unique operation satisfying the recursive equation $\rho(a) = \{\rho(b) \mid b \bar{\in} a\}$.

Since $b \bar{\in} a$ iff $b \in TC(a)$ we can write this condition also as $\rho(a) = \{\rho(b) \mid b \in TC(a)\}$. As an illustration we compute some values of ρ.

$TC(\emptyset) = \emptyset$; hence $\rho(\emptyset) = \emptyset = 0$.

$TC(\{\emptyset\}) = \{\emptyset\}$; hence $\rho(\{\emptyset\}) = \{\rho(\emptyset)\} = \{\emptyset\} = 1$.

$TC(\{\{\emptyset\}\}) = \{\emptyset,\{\emptyset\}\}$; hence $\rho(\{\{\emptyset\}\}) = \{\rho(\emptyset), \rho(\{\emptyset\})\} = \{\emptyset,\{\emptyset\}\} = 2$.

There also exists a recursion equation for ρ in terms of $\in$, as the following computation shows:

$$\begin{aligned} \rho(a) &= \{\rho(c) \mid c \in TC(a)\} && (4.1)\\ &= \{\rho(c) \mid c \in a \cup \bigcup_{b\in a} TC(b)\} && (2.4)\\ &= \{\rho(c) \mid c \in a\} \cup \bigcup_{b\in a} \{\rho(c) \mid c \in TC(b)\} \\ &= \{\rho(b) \mid b \in a\} \cup \bigcup \{\rho(b) \mid b \in a\} && (4.1)\end{aligned}$$

We use this one in the proof of the next

Lemma 4.2: (*i*) $\rho(a) \in OR$

(*ii*) $\rho(\alpha) = \alpha$

(*iii*) $\rho(a) = \bigcup\{\rho(b)^+ \mid b \in a\}$.

(*i*) states that the values of ρ are ordinals; (*ii*) that ρ is the identity on OR and (*iii*) that $\rho(a)$ is the least ordinal greater than ranks of elements of a.

Proof: (*i*) Apply $\bar{\in}$ - induction. Suppose that all $\rho(b)$ for which $b \in TC(a)$, are ordinals. By 4.1, $\rho(a)$ is a set of ordinals; so we have to prove it is a transitive set. Hence let $x \in \beta$ and $\beta \in \rho(a)$. By 4.1, $\beta = \rho(b)$ for some $b \in TC(a)$. Again by 4.1, $x = \rho(c)$ for some $c \in TC(b)$. Thus $c \in TC(a)$ and $x \in \rho(a)$.

(*ii*) Again we apply $\bar{\in}$ - induction: suppose that $\rho(\beta) = \beta$ for all $\beta \in \alpha$. Notice that $TC(\alpha) = \alpha$, hence $\rho(\alpha) = \{\rho(\beta) \mid \beta \in TC(\alpha)\} = \{\beta \mid \beta \in \alpha\} = \alpha$.

(*iii*) Verify that $A \cup \bigcup A = \bigcup\{\beta^+ \mid \beta \in A\}$ if $A \subseteq OR$ and use the recursive equation for ρ in terms of $\in$ derived just before 4.2. □

Some more simple properties of rank can be found in exercises 1-5; another $\in$ -recursive equation in exercise 6.

One can view rank also as a measure of the height of a set in the so-called *cumulative hierarchy*. This is an operation V defined on OR (not to be confused with the *collection* V of all sets) by means of the following recursive equation:

Definition 4.3: $V_\alpha = \bigcup\{\mathcal{P}V_\beta \mid \beta < \alpha\}$.

The hierarchy was introduced first by Von Neumann. We compute some of the V_α for small α: $V_0 = \bigcup\{\mathcal{P}V_\beta \mid \beta < 0\} = \bigcup\emptyset = \emptyset$;

$V_1 = \bigcup\{\mathcal{P}V_\beta \mid \beta < 1\} = \bigcup\{\mathcal{P}\emptyset\} = \mathcal{P}\emptyset$;

$V_3 = \bigcup\{\mathcal{P}V_0, \mathcal{P}V_1\} = \bigcup\{\{\emptyset\},\{\emptyset,\{\emptyset\}\}\} = \{\emptyset,\{\emptyset\}\} = \mathcal{P}\mathcal{P}\emptyset$.

Furthermore $V_3 = \mathcal{P}\mathcal{P}\mathcal{P}\emptyset$ etc., hence $V_\omega = \emptyset \cup \mathcal{P}\emptyset \cup \mathcal{P}\mathcal{P}\emptyset \cup \ldots$.

The process will become somewhat clearer after 4.5. A set A is called *supertransitive* if $a \in A \wedge b \subseteq a \rightarrow b \in A$.

Lemma 4.4: (*i*) every V_α is supertransitive;

(*ii*) every V_α is transitive;

(*iii*) $\alpha < \beta \rightarrow V_\alpha \in V_\beta$ (and $V_\alpha \subseteq V_\beta$).

Proof: (*i*) If $a \in V_\alpha$ then by definition $a \in \mathcal{P}V_\beta$ for some $\beta < \alpha$, i.e., $a \subseteq V_\beta$. If furthermore $b \subseteq a$ then $b \subseteq V_\beta$ too and, again by definition, $b \in V_\alpha$.

(*ii*) Transfinite induction ("on α"). Induction hypothesis: all V_β are transitive for $\beta < \alpha$. Let $a \in V_\alpha$ and $b \in a$. By 4.3, $a \subseteq V_\beta$, for some $\beta < \alpha$. Hence $b \in V_\beta$ and by the induction hypothesis, $b \subseteq V_\beta$. Therefore again by 4.3, $b \in V_\alpha$. (*iii*) If $\alpha < \beta$ then $V_\alpha \in \mathcal{P}V_\alpha \subseteq V_\beta$. □

Next we give some alternative recursive equations for the V_α.

Lemma 4.5: (*i*) $V_0 = \emptyset$;

(*ii*) $V_{\alpha+1} = \mathcal{P}V_\alpha$;

(*iii*) $V_\gamma = \bigcup_{\xi<\gamma} V_\xi$ if Limγ.

Proof: We have already seen that $V_0 = \emptyset$. (*ii*) follows from 4.3, by noting that if $\beta < \alpha$, (since by 4.4 (*iii*) $V_\beta \subseteq V_\alpha$) then $\mathcal{P}V_\beta \subseteq \mathcal{P}V_\alpha$. As to (*iii*), we always have $\bigcup_{\xi<\gamma} V_\xi \subseteq V_\gamma$. And if Lim$\gamma$ then also $V_\gamma = \bigcup_{\xi<\gamma} \mathcal{P}V_\xi = \bigcup_{\xi<\gamma} V_{\xi+1} \subseteq \bigcup_{\xi<\gamma} V_\xi$. □

We now make the (surprisingly simple) connection between the rank-operation and the cumulative hierarchy by means of

Theorem 4.6: $V_\alpha = \{a \mid \rho(a) < \alpha\}$.

Proof: Induction on α. The next five assertions are equivalent: $a \in V_\alpha$ $\exists\beta < \alpha\ (a \subset V_\beta)$ (4.3); $\exists\beta < \alpha\ \forall b \in a\ \ \rho(b) < \beta$ (induction hypothesis); $\exists\beta < \alpha\ (\rho(a) \leqslant \beta)$ (4.2(*iii*)); $\rho(a) < \alpha$. □

Corollary 4.7: (*i*) $V = \bigcup_{a \in OR} V_a$;

(*ii*) $\{\langle a,b\rangle \mid \rho(a) < \rho(b)\}$ is well-founded and local.

4.7 (*i*) states that every set is an element of some V_a. Induction using the collection in 4.7(*ii*) is called *induction with respect to rank*. It is a bit stronger than the usual induction w.r.t $\in$ or even $\bar{\in}$ since its inductive hypothesis is stronger: $a \in b$ implies $a \in TC(b)$ and $a \in TC(b)$ implies $\rho(a) < \rho(b)$, but neither implication can be reversed.

Some trivial consequences of 4.6 are stated in 4.8:

Corollary 4.8: (*i*) $\rho(V_\alpha) = \alpha$;

(*ii*) $a \notin V_{\rho(a)}$; in particular $\alpha \notin V_\alpha$;

(*iii*) $a \subseteq V_{\rho(a)}$; in particular $\alpha \subseteq V_\alpha$;

(*iv*) $a \in V_{\rho(a)+}$; in particular $\alpha \in V_{\alpha+}$.

In the theory *ZF* without the regularity-axiom one can show as follows that the collection **R** of regular sets introduced in §1 equals the collection $\bigcup_{a \in OR} V_a$ of sets from the cumulative hierarchy. To show that $\mathbf{R} \subseteq \bigcup_{a \in OR} V_a$ we use the fact that every class containing all of its subsets as elements has **R** as a subclass (this was proved at the end of §1). This certainly is true of $\bigcup_{\alpha \in OR} V_\alpha$: when $a \subseteq \bigcup_{\alpha \in OR} V_\alpha$, we can associate with every $b \in a$ the smallest $\xi \in$ OR, say, ξ_b, such that $b \in V_{\xi_b}$. Then $a \subseteq V_\alpha$ where $\alpha = \bigcup_{b \in a} \xi_b$. Hence $a \in V_{\alpha^+}$. Conversely, since $\emptyset \in \mathbf{R}$ and **R** is closed under powersets and sums it follows using transfinite induction that every V_a is a member of **R**. Therefore $\bigcup_{a \in OR} V_a \subseteq \mathbf{R}$ also.

It should be admitted that in the theory **ZF** minus regularity the notion of an ordinal has to be defined in a slightly different way in order to insure the validity of, for instance, the principle of transfinite induction. One could define an ordinal in this theory as an ordinal in the old sense which is, moreover, an element of **R**.

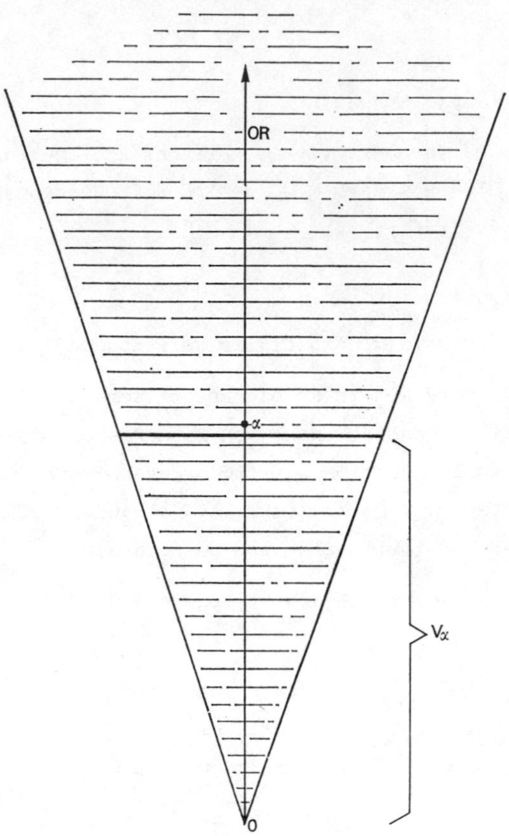

Fig. 1

The cumulative hierarchy gives a rather clear picture of the universe of all sets and how it is built if the regularity-axiom holds. Look at figure 1: one can think of the ordinals as being ordered on a vertical line; the universe being an extending mass around this axis of which the "partial universes", the V_a , are initial segments. The figure can represent a bottom-piece of the universe only since we cannot imagine an end to the series of partial universes. In §§7 and 11 we shall study more extensively the topics of the cumulative hierarchy and partial universes. We finally present the *bottom-operation*, which in a uniform way associates with every non-empty collection a non-empty subset of that collection. Suppose that K is such a collection. Since K is non-empty, for instance $a \in K$, there exists a $\alpha \in$ OR (for instance $\rho(a)^+$) such that $K \cap V_a \neq \emptyset$. Let β be the smallest such ordinal. The *bottom* of K is the set $K \cap V_\beta$. Somewhat different but equivalent to this we make the

Definition 4.9: Bottom $(K) := \{a \in K \mid \forall b \in K\ \rho(a) \leqslant \rho(b)\}$.

Notice that the bottom-operation is not an operation in the proper sense since its arguments are collections. 4.9 should be read as a *definition scheme* which associates a class-abstract with every class-abstract, the value of which is a set. The relevant properties are stated in:

Lemma 4.10 (*i*) Bottom $(K) \in V$;
(*ii*) Bottom $(K) \subseteq K$;
(*iii*) $K \neq \emptyset \rightarrow$ Bottom $(K) \neq \emptyset$.

Applications of 4.10 can be found in exercise 8 and 9 and in §10.

Exercises

1. $\rho(a) = 0 \leftrightarrow a = 0$; $\rho(a) = 1 \leftrightarrow a = 1$.
2. $a \in b \rightarrow \rho(a) < \rho(b)$ (but not conversely); $\alpha < \rho(a) \rightarrow \exists b \in a\ (\alpha \leqslant \rho(b))$.
3. $a \subseteq b \rightarrow \rho(a) \leqslant \rho(b)$.
4. $\mathrm{Suc}\ (\rho(a)) \rightarrow \rho(a) = \rho(\bigcup a)^{+}$; $\mathrm{Lim}(\rho(a)) \rightarrow \rho(a) = \rho(\bigcup a)$.
5. $\rho(\mathcal{P}a) = \rho(a)^{+}$.
6. $\rho(a) = \mathrm{TC}(\{\rho(b) \mid b \in a\})$.
7. Give an example of a collection for which $\{a \in K \mid a$ minimal in $K\}$ is not a set; same question w.r.t. $\{a \in K \mid a$ is $\bar{\bar{\in}}$ -minimal in $K\}$.
8. If R is well-founded, every non-empty collection has an R-minimal element. (Compare the discussion immediately before 2.6. We do not require here that R is local). (Hint: suppose that $a \in K$ and define recursively $A_0 = \{a\}$ and $A_{n+1} = \bigcup\{\mathrm{Bottom}(\{c \in K \mid c\ R\ b\}) \mid b \in A_n\}$. Every R-minimal element of $\bigcup_{n \in \omega} A_n$ is R-minimal in K).
9. Prove that the transitive closure of a well-founded collection of ordered pairs is again well-founded. (Compare the remark just after 2.5. Again, we do not presuppose locality here.) (Hint: rewrite the proof of 2.5(*ii*) using exercise 8.)
10. $\mathcal{P}(\mathrm{TC}(A))$ and $\bigcup_{a \in TC(A)} \mathcal{P}(a)$ are examples of transitive, supertransitive sets containing A (the second example even is the least such set) in general not having the form V_a.

11. R is reflexive, symmetrical and transitive; $[a] := \text{Bottom}(\{b|b\ R\ a\})$. Prove that $[a] = [b] \leftrightarrow aRb$.

12. Prove 4.10.

5 ORDINAL ARITHMETIC

In this section we extend the operations of addition, multiplication and exponentiation to the class of all ordinals. The following recursive definition of addition uses the idea that addition is iteration of the successor-operation.

Definition 5.1: (*i*) $\alpha + 0 = \alpha$;

(*ii*) $\alpha + \beta^+ = (\alpha + \beta)^+$;

(*iii*) $\alpha + \gamma = \bigcup_{\xi<\gamma} (\alpha + \xi)$ if Limγ.

An easy way to unify these equations into one (and a simplification) is given by the

Lemma 5.2: (*i*) $\alpha + \beta = \alpha \cup \bigcup\{(\alpha + \xi)^+|\xi < \beta\}$;

(*ii*) $\alpha + \beta = \alpha \cup \{(\alpha + \xi)|\xi < \beta\}$.

Proof: (*i*) Induction w.r.t. β. The case β=0 is trivial.

$\alpha \cup \bigcup\{(\alpha + \xi)^+|\xi < \beta^+\} =$

$\alpha \cup \bigcup\{(\alpha + \xi)^+|\xi < \beta\} \cup (\alpha + \beta)^+ = (\alpha + \beta) \cup (\alpha + \beta)^+$ (inductionhypothesis)

$= (\alpha + \beta)^+ = \alpha + \beta^+$ (5.1(*ii*)). And if Limβ then it is not difficult to see that

$\bigcup_{\xi<\beta} (\alpha + \xi) = \alpha \cup \bigcup\{(\alpha + \xi)^+|\xi < \beta\}$, using 5.1(*i*) and (*ii*). Hence the result follows by 5.1(*iii*).

(*ii*) It follows from (*i*) that $\alpha \cup \{\alpha + \xi|\xi < \beta\} \subseteq \alpha + \beta$. To see that, conversely, $\alpha + \beta \subseteq \alpha \cup \{\alpha + \xi|\xi < \beta\}$, we use induction w.r.t. β. Suppose that $\delta \in \alpha + \beta$. Then according to (*i*) either $\delta \in \alpha$ or $\delta \in (\alpha + \xi)^+$ for some $\xi < \beta$. In the first case we are done. In the second either $\delta = \alpha + \xi$ and we are through again, or $\delta \in \alpha + \xi$ but then the induction-hypothesis helps us out. □

We give some examples.

$\alpha + 1 = \alpha + 0^+ = (\alpha + 0)^+ = \alpha^+$ (5.1(*i*),(*ii*)).

$0 + \alpha = 0 \cup \{0 + \xi|\xi < \alpha\} = \{\xi|\xi < \alpha\} = \alpha$ (5.2(*ii*) using induction w.r.t.α).

α is *absorbed* by β if $\alpha + \beta = \beta$; this phenomenon occurs frequently. For instance ω absorbs every natural number: $m + \omega = m \cup \{m + n | n < \omega\} = \omega$.

A binary operation τ is *commutative* if for all x,y: $\tau(x,y) = \tau(y,x)$. Ordinal addition does *not* commute since $1 + \omega = \omega \neq \omega^+ = \omega + 1$. It is *not left-monotonic* in the sense that if $\alpha < \beta$ then for all γ $\alpha + \gamma < \beta + \gamma$ because, for instance, $0 < 1$ but also $0 + \omega \nless 1 + \omega$. However compare 5.3(*ii*). Some convenient properties are listed in the next lemma.

Lemma 5.3: (*i*) $\alpha + (\beta + \gamma) = (\alpha + \beta) + \gamma$;

(*ii*) $\alpha + \beta < \alpha + \gamma \leftrightarrow \beta < \gamma$;

(*iii*) $\alpha + \beta = \alpha + \gamma \leftrightarrow \beta = \gamma$;

(*iv*) $\alpha < \beta \rightarrow \alpha + \gamma \leqslant \beta + \gamma$;

(*v*) $\alpha + \gamma < \beta + \gamma \rightarrow \alpha < \beta$.

Proof: (*i*) From 5.2(*ii*), using induction w.r.t. γ:

$$\begin{aligned}\alpha + (\beta + \gamma) &= \alpha \cup \{\alpha + \xi | \xi < \beta + \gamma\} \\ &= \alpha \cup \{\alpha + \xi | \xi < \beta\} \cup \{\alpha + (\beta + \delta) | \delta < \gamma\} \\ &= (\alpha + \beta) \cup \{(\alpha + \beta) + \delta | \delta < \gamma\} \\ &= (\alpha + \beta) + \gamma.\end{aligned}$$

(*ii*) From right to left this is immediate from 5.2(*ii*); the converse follows trivially. (*iii*) is immediate from (*ii*). (*iv*) Induction w.r.t. γ. Suppose that $\alpha < \beta$ and $\delta < \alpha + \gamma$. Then (5.2(*ii*)) $\delta < \alpha$ (hence $\delta < \beta \leqslant \beta + \gamma$) or for some $\xi < \gamma$, $\delta = \alpha + \xi$. Inductionhypothesis and (*ii*): $\alpha + \xi \leqslant \beta + \xi < \beta + \gamma$. (*v*) is immediate from (*iv*). □

Conceiving of multiplication as iterated addition, one obtains the following recursion equations:

Definition 5.4: (*i*) $\alpha.0 = 0$;

(*ii*) $\alpha.\beta^+ = (\alpha.\beta) + \alpha$;

(*iii*) $\alpha.\gamma = \bigcup_{\xi < \gamma} (\alpha.\xi)$ (if Lim γ).

Corresponding to 5.2 we have the

Lemma 5.5: (*i*) $\alpha.\beta = \bigcup\{\alpha.\xi + \alpha | \xi < \beta\}$;

(*ii*) $\alpha.\beta = \{\alpha.\xi + \delta | \xi < \beta \wedge \delta < \alpha\}$.

Proof: (*i*) Induction w.r.t. β. The case $\beta = 0$ presents no difficulties. Next, $\cup\{\alpha.\xi + \alpha | \xi < \beta^+\} =$
$\cup\{\alpha.\xi + \alpha | \xi < \beta\} \cup (\alpha.\beta + \alpha) = (\alpha.\beta) \cup (\alpha.\beta + \alpha) = \alpha.\beta + \alpha = \alpha.\beta^+$. We leave the limit-case to the reader.

(*ii*) It follows from (*i*) that $\alpha.\xi + \delta \in \alpha.\beta$ whenever $\xi < \beta$ and $\delta < \alpha$. Conversely, suppose that $\eta < \alpha.\beta$. According to (*i*) $\eta < \alpha.\xi + \alpha$ for some $\xi < \beta$. By 5.2(*ii*) either $\eta < \alpha.\xi$ or for some $\delta < \alpha$, $\eta = \alpha.\xi + \delta$. Hence, apply induction w.r.t. β. □

Some examples. From 5.5(*ii*), $\alpha.0 = 0.\alpha = 0$; also, $\alpha.1 = 1.\alpha$ (using an induction on α). From 5.4 we see that $\alpha.2 = \alpha + \alpha$; hence the *second* factor functions as the *multiplier*. This is essential since multiplication does not commute due to absorption-phenomena. For instance, $2.\omega = \bigcup_{n<\omega}(2.n) = \omega$, while $\omega.2 = \omega + \omega > \omega$.

Some properties are listed below.

<u>Lemma 5.6</u>: (*i*) $\alpha.(\beta.\gamma) = (\alpha.\beta).\gamma$;

(*ii*) $\alpha.(\beta + \gamma) = \alpha.\beta + \alpha.\gamma$;

(*iii*) $\alpha.\beta < \alpha.\gamma \leftrightarrow (\beta < \gamma \wedge \alpha \neq 0)$;

(*iv*) $\alpha.\beta = \alpha.\gamma \rightarrow (\beta = \gamma \vee \alpha = 0)$;

(*v*) $\alpha < \beta \rightarrow \alpha.\gamma \leqslant \beta.\gamma$;

(*vi*) $\alpha.\gamma < \beta.\gamma \rightarrow \alpha < \beta$.

Proof: One may prove (*ii*) by means of an induction w.r.t. γ using 5.5(*ii*) and 5.2(*ii*):

$$\begin{aligned}
\alpha.(\beta + \gamma) &= \{\alpha.\xi + \eta | \xi < \beta + \gamma \wedge \eta < \alpha\} \\
&= \{\alpha.\xi + \eta | \xi \in \beta \cup \{\beta + \delta | \delta < \gamma\} \wedge \eta < \alpha\} \\
&= \{\alpha.\xi + \eta | \xi \in \beta \wedge \eta < \alpha\} \cup \\
&\quad \cup\{\alpha\ \xi + \eta | \xi \in \{\beta + \delta | \delta < \gamma\} \wedge \eta < \alpha\} \\
&= \alpha.\beta \cup \{\alpha.(\beta + \delta) + \eta | \delta < \gamma \wedge \eta < \alpha\} \\
&= \alpha.\beta \cup \{\alpha.\beta + (\alpha.\delta + \eta) | \delta < \gamma \wedge \eta < \alpha\} \\
&= \alpha.\beta \cup \{\alpha.\beta + \xi | \xi \in \{\alpha.\delta + \eta | \delta < \gamma \wedge \eta < \alpha\}\} \\
&= \alpha.\beta \cup \{\alpha.\beta + \xi | \xi < \alpha.\gamma\} = \alpha.\beta + \alpha.\gamma.
\end{aligned}$$

(*i*) is proved in more or less the same fashion using 5.5(*ii*) and 5.6(*ii*).

(*iii*): it follows from 5.5(*ii*) that $\beta < \gamma \wedge \alpha \neq 0 \rightarrow \alpha.\beta < \alpha.\gamma$; the converse follows trivially.

(iv) follows immediately from *(iii)*.

(v) follows from 5.2*(ii)* using induction w.r.t. γ(and 5.3*(iv)*).

(vi) finally is an immediate consequence from *(v)*. □

Though multiplication distributes over addition from the left (5.6*(ii)*) it does not distribute from the right: for instance, we have $(1+1).\omega = 2.\omega = \omega$ but $1.\omega + 1.\omega = \omega + \omega \neq \omega$.

Finally, exponentiation iterates multiplication, that is to say we make the

<u>Definition 5.7</u>: *(i)* $\alpha^0 = 1$;

(ii) $\alpha^{\beta+1} = \alpha^\beta . \alpha$;

(iii) $\alpha^\gamma = \bigcup_{0<\xi<\gamma} \alpha^\xi$ if Lim γ.

Corresponding to 5.2 and 5.5 we obtain

<u>Lemma 5.8</u>: if $\alpha > 0$ then

(i) $\alpha^\beta = 1 \cup \bigcup\{\alpha^\xi.\alpha \mid \xi < \beta\}$

(ii) $\alpha^\beta = 1 \cup \{\alpha^\xi.\delta + \eta \mid \xi < \beta \wedge \delta < \alpha \wedge \eta < \alpha^\xi\}$

Proof: *(i)* induction w.r.t. β. For the successor-case:

$$1 \cup \bigcup\{\alpha^\xi.\alpha \mid \xi < \beta^+\} = 1 \cup \bigcup\{\alpha^\xi.\alpha \mid \xi < \beta\} \cup \alpha^\beta.\alpha$$
$$= \alpha^\beta \cup \alpha^\beta.\alpha = \alpha^\beta.\alpha.$$

The cases $\beta = 0$ and Lim β are trivial.

(ii) The right-hand-side is a subset of the left-hand-side: $0 < \alpha^\beta$; and if $\xi < \beta$, $\delta < \alpha$ and $\eta < \alpha^\xi$ then $\alpha^\xi.\delta + \eta < \alpha^\xi.\delta + \alpha^\xi = \alpha^\xi.(\delta + 1) \leqslant \alpha^\xi.\alpha = \alpha^{\xi+1} \leqslant \alpha^\beta$ according to resp. 5.3*(ii)*, 5.4*(ii)*, 5.6*(iii)* and 5.8*(i)*. Conversely, the left-hand-side is a subset of the right-hand-side too: suppose that $\gamma < \alpha^\beta$. Then a smallest $\xi' < \beta$ exists such that $\gamma < \alpha^{\xi'}$. By 5.7*(iii)*, ξ' is not a limit. If $\xi' = 0$ then $\alpha^{\xi'} = 1$, $\gamma = 0$ and we are done. Otherwise $\xi' = \xi + 1$, $\xi < \beta$, and $\alpha^\xi \leqslant \gamma < \alpha^\xi.\alpha$. Hence a smallest $\delta' \leqslant \alpha$ exists such that $\gamma < \alpha^\xi.\delta'$.

By 5.4*(i)* and *(iii)* δ' has to be a successor, say, $\delta' = \delta + 1$. Thus $\delta < \alpha$ and $\alpha^\xi.\delta \leqslant \gamma < \alpha^\xi.\delta + \alpha^\xi$. Finally take the smallest η' for which $\gamma < \alpha^\xi.\delta+\eta'$. Again $\eta' = \eta + 1$ and $\eta < \alpha^\xi$ and now $\alpha^\xi.\delta + \eta \leqslant \gamma < \alpha^\xi.\delta + \eta'$. But then $\gamma = \alpha^\xi.\delta + \eta$. □

It is easy to see that $\alpha^{\beta} = 0$ iff $\alpha = 0$; $\alpha^{1} = \alpha$ and $1^{\alpha} = 1$.
Again there are absorption-phenomena such as $2^{\omega} = \omega$. Exponentiation does not distribute over multiplication: $(\omega.2)^{2} = (\omega.2).(\omega.2) = (\omega.(2.\omega)).2 = \omega^{2}.2$; on the other hand $\omega^{2}.2^{2} = \omega^{2}.4 > \omega^{2}.2$. Some properties are listed in the next lemma:

<u>Lemma 5.9</u>: (*i*) $\alpha^{\beta+\gamma} = \alpha^{\beta}.\alpha^{\gamma}$;

(*ii*) $(\alpha^{\beta})^{\gamma} = \alpha^{\beta.\gamma}$;

(*iii*) $\alpha^{\beta} < \alpha^{\gamma} \leftrightarrow (\beta < \alpha \wedge \alpha > 1)$;

(*iv*) $\alpha^{\beta} = \alpha^{\gamma} \rightarrow (\beta = \gamma \vee \alpha \leqslant 1)$;

(*v*) $\alpha < \beta \rightarrow \alpha^{\gamma} \leqslant \beta^{\gamma}$;

(*vi*) $\alpha^{\gamma} < \beta^{\gamma} \rightarrow \alpha < \beta$.

Proof: (*i*) and (*ii*) are proved by induction w.r.t. γ using 5.8(*ii*). (*iii*): using induction w.r.t. γ one may show that if $\beta < \gamma$ and $\alpha > 1$ then $\alpha^{\beta} < \alpha^{\gamma}$ for instance from 5.7. The converse follows immediately. (*iv*) follows from (*iii*). (*v*) is proved by induction w.r.t. γ using 5.7; this has (*vi*) as a consequence. □

We close this paragraph with *Cantor's normal-form theorem* which generalizes decimal notation for natural numbers.

<u>Theorem 5.10</u>: Suppose that α is some ordinal greater than 1. Then every ordinal $\gamma > 0$ has a unique representation

$$\gamma = \alpha^{\beta_0}.\alpha_0 + \ldots + \alpha^{\beta_n}.\alpha_n \tag{1}$$

where n is some natural number, $0 < \alpha_i < \alpha$ ($i \leqslant n$) and $\beta_0 > \beta_1 > \ldots > \beta_n$.

Proof: We first show that $\gamma < \alpha^{\beta_0+1}$ if the representation (1) satisfies the conditions required. For this purpose, it suffices to show that if $k \leqslant n$

$$\gamma_k := \alpha^{\beta_0}.\alpha_0 + \ldots + \alpha^{\beta_{k-1}}.\alpha_{k-1} + \alpha^{\beta_k+1} \leqslant \alpha^{\beta_0+1} \tag{2}$$

since clearly $\gamma < \gamma_n$. (2) is proved by induction on k.
If k=0, $\gamma_0 = \alpha^{\beta_0+1}$ and (2) holds trivially.
For the successor-case we get

$$\begin{aligned}\gamma_{k+1} &= \alpha^{\beta_0}.\alpha_0 + \ldots + \alpha^{\beta_k}.\alpha_k + \alpha^{\beta_{k+1}+1}\\ &\leqslant \alpha^{\beta_0}.\alpha_0 + \ldots + \alpha^{\beta_k}.\alpha_k + \alpha^{\beta_k}\end{aligned}$$

$= \alpha^{\beta_0}.\alpha_0 + \ldots + \alpha^{\beta_k}.(\alpha_k + 1)$

$\leqslant \alpha^{\beta_0}.\alpha_0 + \ldots + \alpha^{\beta_k}.\alpha$

$= \alpha^{\beta_0}.\alpha_0 + \ldots + \alpha^{\beta_k+1} = \gamma_k$, and the result follows using the induction-hypothesis $\gamma_k \leqslant \alpha^{\beta_0+1}$. Uniqueness of the representation (1) is proved using induction on γ. Suppose that (1) holds but also that $\gamma \equiv \alpha^{\beta'_0}.\alpha'_0 + \ldots + \alpha^{\beta'_m}.\alpha'_m$ $(0 < \alpha'_i < \alpha;\ \beta'_0 > \beta'_1 > \ldots > \beta'_m)$. By the previous remark, if $\eta := \alpha^{\beta_1}.\alpha_1 + \ldots + \alpha^{\beta_n}.\alpha_n$ then $\eta < \alpha^{\beta_1+1} \leqslant \alpha^{\beta_0}$; in the same way, if $\eta' := \alpha^{\beta'_1}.\alpha'_1 + \ldots + \alpha^{\beta'_m}.\alpha'_m$ then $\eta' < \alpha^{\beta'_0}$. If we had, for instance, $\beta_0 < \beta'_0$ then

$\gamma = \alpha^{\beta_0}.\alpha_0 + \eta < \alpha^{\beta_0}.\alpha_0 + \alpha^{\beta_0} = \alpha^{\beta_0}.(\alpha_0 + 1)$

$\leqslant \alpha^{\beta_0}.\alpha = \alpha^{\beta_0+1} \leqslant \alpha^{\beta'_0} \leqslant \alpha^{\beta'_0}.\alpha'_0 \leqslant \alpha^{\beta'_0}.\alpha'_0 + \eta' = \gamma$ which is impossible, and the same holds for $\beta'_0 < \beta_0$. Therefore, $\beta_0 = \beta'_0$. And if we had, say, $\alpha_0 < \alpha'_0$ then

$\gamma = \alpha^{\beta_0}.\alpha_0 + \eta < \alpha^{\beta_0}.\alpha_0 + \alpha^{\beta_0} = \alpha^{\beta_0} = \alpha^{\beta_0}.(\alpha_0 + 1)$

$\leqslant \alpha^{\beta_0}.\alpha'_0 \leqslant \alpha^{\beta_0}.\alpha'_0 + \eta' = \gamma$; hence $\alpha_0 = \alpha'_0$. Therefore $\eta = \eta'$. Since $\eta < \alpha^{\beta_0} \leqslant \gamma$ we can use now the induction-hypothesis which says that the representations for η and η' are identical therefore the representations for γ are identical as well. Finally we show that a representation of the form (1) exists using induction on γ.

In exercise 2 it is stated that $\gamma \leqslant \alpha^\gamma$; therefore $\gamma < \alpha^{\gamma+1}$. Let β be the smallest ordinal for which $\gamma < \alpha^\beta$. Then β cannot be 0 since $\gamma > 0$ and Lim β is excluded since then by 5.7(*iii*) there would be a $\xi < \beta$ such that $\gamma < \alpha^\xi$. Hence $\beta = \beta_0 + 1$ and $\alpha^{\beta_0} \leqslant \gamma$. If $\alpha^{\beta_0} = \gamma$ we are done so suppose that $\alpha^{\beta_0} < \gamma$. The smallest η such that $\gamma < \alpha^{\beta_0}.\eta$ can also be shown to be of the form $\eta = \alpha_0 + 1$ where $\alpha_0 < \alpha$. If $\alpha^{\beta_0}.\alpha_0 = \gamma$ we are through and otherwise $\alpha^{\beta_0}.\alpha_0 < \gamma$, hence $\alpha^{\beta_0}.\alpha_0 + \gamma' = \gamma$ for some γ'. Clearly, $\gamma' < \alpha^{\beta_0} \leqslant \gamma$ hence according to the induction-hypothesis γ' possesses a representation $\gamma' = \alpha^{\beta_1}.\alpha_1 + \ldots \alpha^{\beta_n}.\alpha_n$. Then $\beta_1 < \beta_0$ and $\gamma = \alpha^{\beta_0}.\alpha_0 + \ldots + \alpha^{\beta_n}.\alpha_n$ is the required representation. □

Notice that if $1 < \alpha < \omega$ then $\alpha^\omega = \omega$. But also if $\omega \leqslant \alpha$ there are many γ for which $\alpha^\gamma = \gamma$ as we shall see in the next paragraph. For such γ the normalform-representation is trivial.

Exercises

1. If $\alpha < \beta$ there exists exactly one γ such that $\alpha + \gamma = \beta$.

2. If $1 < \alpha$ then $\beta \leqslant \alpha^{\beta}$.

3. If $\langle A,R\rangle$ and $\langle B,S\rangle$ are well-orderings with type α and β respectively and $A \cap B = \emptyset$ then $\langle A \cup B,\ R \cup (A \times B) \cup S\rangle$ is a well-ordering with type $\alpha + \beta$.

4. If $\langle A,R\rangle$ and $\langle B,S\rangle$ are well-orderings of type α and β respectively and $\prec$ is the relation defined on $A \times B$ by $\langle a_1,b_1\rangle \prec \langle a_2,b_2\rangle := b_1\ S\ b_2 \vee (b_1 = b_2 \wedge a_1\ R\ a_2)$ then $\langle A \times B, \prec\rangle$ is a well-ordering of type $\alpha . \beta$.

5. Prove 5.9(*i*) and (*ii*).

6. Give counterexamples for the missing implications in 5.3, 5.6 and 5.9.

7. Prove the assertion made at the beginning of §5, i.e., that the operations introduced in this paragraph extend the corresponding arithmetical operations.

6 NORMAL OPERATIONS

Let $\langle A,R\rangle$ be a (partial) ordering. A set $B \subseteq A$ is called *cofinal* in A (w.r.t. R) if for every $a \in A$, $a \in B$ or $\exists b \in B(a\ R\ b)$. In case A does not have an R-maximal element this condition is equivalent with $\forall a \in A\ \exists b \in B(a\ R\ b)$ even if R is irreflexive.

As an example, the natural numbers are cofinal in the set of reals (w.r.t. the natural ordering). In the sequel we only consider sets of ordinals under the usual ordering.

Examples: every ordinal is cofinal in itself; $\{\alpha\}$ is cofinal in $\alpha + 1$. ω has several cofinal subsets, in fact, every infinite subset of ω is cofinal in ω. $\{\omega^{3.n} \mid 10^{10} < n < \omega\}$ is an example of a set cofinal in ω^{ω}.

Every set a of ordinals is well-ordered by its epsilon-relation ϵ_a, hence the type of $\langle a,\epsilon_a\rangle$ is an ordinal. We now make the

Definition 6.1: $cf(\alpha)$, the *cofinality* of α, is the smallest ordinal which is type of the epsilon-structure of a cofinal subset of α.

The examples imply that $cf(\alpha) \leqslant \alpha$, in particular $cf(0) = 0$, and $cf(\alpha + 1) = 1$. Hence the operation cf is interesting only for limit-arguments. If Limα, α does not have finite cofinal subsets; therefore $\omega \leqslant cf(\alpha)$. In particular $cf(\omega) = \omega$.

Definition 6.2: α is called *regular* if $\omega \leqslant \alpha$ and $cf(\alpha) = \alpha$.

This notion of regularity is not to be confused with that of §1.
We have seen that ω is regular. There exists a conjecture that it might be consistent to assume that there are no more regular ordinals. (Jech 1973 p. 165 exercise 26.) The *axiom of choice* (§9) implies the existence of lots of regular ordinals (10.17; 10.18).
If ω is the only regular ordinal then cf would be a rather trivial operation witness the

Lemma 6.3: $cf(cf(\alpha)) = cf(\alpha)$.

Proof: Evidently, $cf(cf(\alpha)) \leqslant cf(\alpha)$. So let $\beta := cf(\alpha)$ and suppose $cf(\beta) < \beta$. Take $b \subseteq \beta$ cofinal in β of type $cf(\beta)$ and suppose that $h: cf(\beta) \to b$ is the corresponding epsilon-isomorphism. Since $\beta = cf(\alpha)$ there also exists an epsilon-isomorphism $g: \beta \to \alpha$ onto a cofinal subset of α. It is not difficult to show that $g \circ h$ is an epsilon-isomorphism of $cf(\beta)$ onto a cofinal subset of α, therefore $\beta = cf(\alpha) \leqslant cf(\beta)$ contradicting $cf(\beta) < \beta$. □

Hence if Lim α then $cf(\alpha)$ is regular.
Another characterisation of $cf(\alpha)$ is given by

Lemma 6.4: $\beta < cf(\alpha)$ iff there is no $f: \beta \to \alpha$ for which Ran(f) is cofinal in α (unless $\alpha = 0$).

Proof: If $cf(\alpha) \leqslant \beta$ and $\alpha \neq 0$ we can define an $f: \beta \to \alpha$ with Ran(f) cofinal in α as follows. Let h map $cf(\alpha)$ onto a cofinal subset of α and define f by

$$f(\xi) := \begin{cases} h(\xi) & \text{if } \xi < cf(\alpha) \\ 0 & \text{if } cf(\alpha) \leqslant \xi < \beta. \end{cases}$$

Conversely, suppose that f maps β onto a cofinal subset of α. Notice that f need not be an epsilon isomorphism! Nevertheless part of f is an epsiloniso-morphism onto a cofinal subset of α. For let $B := \{\xi \in \beta \mid \forall\delta < \xi(f(\delta) < f(\xi))\}$. Clearly $f \upharpoonright B$ preserves the ordering. Also Ran($f \upharpoonright B$) is cofinal in α: let $\eta < \alpha$ and $\xi_0 := \bigcap \{\xi \in \beta \mid \eta \leqslant f(\xi)\}$. The reader, who has no practical experience with ordinals, may prefer to think of ξ_0 as " the smallest ordinal ξ such that $\eta \leqslant f(\xi)$ ", cf. exercise 3.9. Then $\xi_0 \in B$ and $f(\xi_0) \geqslant \eta$. Now let h map B order-isomorphically onto the type β' of $\langle B, \epsilon_B \rangle$. Using induction on ξ it follows that $h(\xi) \leqslant \xi$ if $\xi \in B$. Therefore $\beta' \leqslant \beta$. By definition, $cf(\alpha) \leqslant \beta'$.□

$B \subseteq \alpha$ is called *bounded* in α if B is not cofinal in α, in other words, if for some $\alpha' \in \alpha$, $B \subseteq \alpha'$. Hence 6.4 states among other things that if $\beta < cf(\alpha)$ and f: $\beta \to \alpha$ then Ran(f) is bounded in α. In particular, if α is regular and $\beta < \alpha$ then $\beta \underset{1}{<} \alpha$ (cf. Theorem 8.8). The countable ordinals (such as $\omega + \omega$, $\omega.\omega$, ω^{ω}, $\omega^{\omega^{\omega}}$; there are in fact uncountable many) thus all have cofinality ω.

Notice that the replacement-axiom implies a kind of regularity for the collection OR: if σ: OR $\to$ OR is an operation and $\alpha \in$ OR then $\{\sigma(\xi) | \xi < \alpha\}$ is bounded in OR.

Let A be some collection of ordinals and σ its *collapsing-isomorphism*, i.e. σ is defined on A and satisfies the recursive equation $\sigma(\alpha) = \{\sigma(\beta) | \beta \in \alpha \cap A\}$. Then $\sigma(A) := \{\sigma(\alpha) | \alpha \in A\}$ is a set if and only if A is a set. The inverse σ^{-1} of σ satisfies the recursive equation $\sigma^{-1}(\alpha) = \bigcap\{\beta \in A | \forall\xi < \alpha\ \sigma^{-1}(\xi) < \beta\}$ on $\sigma(A)$. σ^{-1} is called the *enumeration* of A. In general, an *enumeration* is an ordinal-valued operation defined on an ordinal or on OR which preserves membership.

A collection $A \subseteq$ OR is called *closed* if the sum-sets of non-empty bounded subsets of A are members of A, in other words if $\bigcup a \in A$ whenever $\emptyset \neq a \subseteq A$ and $\exists\alpha \in A(a \subseteq \alpha)$.

We are interested in enumerations of closed collections. The following pages show that they provide an interesting theory.

Lemma 6.5: A is closed iff for every $\alpha \in A$, $\alpha \cap A \neq \emptyset \to \bigcup(\alpha \cap A) \in A$.

Proof: If $\alpha \in A$ and $\alpha \cap A \neq \emptyset$ then $\alpha \cap A$ is a non-empty bounded subset of A. Conversely, if $a \subseteq A$ is non-empty and bounded in A there exists a smallest $\alpha \in A$ for which $a \subseteq \alpha$. But then $\bigcup a = \bigcup(\alpha \cap A)$. □

The collection of limit-ordinals is closed, but the collection of all successors is not.

There is a reason for borrowing the term "closed" from topology: a set A is closed precisely if A is closed in the sense of the "order"-topology on $\alpha := \bigcap\{\beta | A \subseteq \beta\}$ which has as a basis the set of initialsegments $\{\xi | \xi < \delta\}$ and the intervals $\{\xi | \gamma < \xi < \delta\}$ where $\gamma, \delta \leqslant \alpha$.

Lemma 6.6: If σ enumerates A the following three assertions are equivalent:

(i) A is closed;

(ii) if Lim α and $\alpha \in$ Dom σ then $\sigma(\alpha) = \bigcup_{\xi < \alpha} \sigma(\xi)$;

(iii) if Lim α, $\alpha \in$ Dom σ and $a \subseteq \alpha$ is cofinal in α then $\sigma(\alpha) = \bigcup_{\xi \in a} \sigma(\xi)$.

Proof: (i) $\rightarrow$ (ii): $a := \{\sigma(\xi) | \xi < \alpha\} \subseteq \sigma(\alpha) \in A$ is a non-empty bounded subset of A hence $\bigcup a \in A$. Now $\bigcup a \leqslant \sigma(\alpha)$ and hence $\bigcup a = \sigma(\alpha)$ since σ enumerates A.

(ii) $\rightarrow$ (iii): $\bigcup_{\xi \in a} \sigma(\xi) \leqslant \bigcup_{\xi < \alpha} \sigma(\xi)$. But $\bigcup_{\xi < \alpha} \sigma(\xi) \leqslant \bigcup_{\xi \in a} \sigma(\xi)$ also, for if $\eta < \sigma(\xi)$, $\xi < \alpha$, then $\delta \in a$ exists such that $\xi \leqslant \delta$; therefore $\eta < \sigma(\xi) \leqslant \sigma(\delta)$.

(iii) $\rightarrow$ (i): Suppose that $\emptyset \neq b \subseteq A$, b bounded in A and $a := \{\xi | \sigma(\xi) \in b\}$. If a has a greatest element then so has b and $\bigcup b \in A$. Otherwise $\alpha := \bigcup a$ is a limit, a is cofinal in α and hence $\sigma(\alpha) = \bigcup_{\xi \in a} \sigma(\xi) = \bigcup b \in A$. □

The *closure* of a collection $A \subseteq$ OR is $A \cup \{\bigcup a | a$ non-empty bounded subset of $A\}$.

Lemma 6.7: The closure of A is the smallest closed collection containing A.

Proof: Let B be the closure of A. It is clear that B is a subclass of every closed class containing A. So we need to show that B is closed. Therefore, let $b \subseteq B$ be non-empty and bounded in B. We may as well assume that $b \subseteq \bigcup b$ since otherwise b has a greatest element. Define for $\alpha \in b$
$f(\alpha) := \bigcap\{\gamma \in A | \alpha \leqslant \gamma \leqslant \bigcup b\}$. Then $f(\alpha) \in A$, $\alpha \leqslant f(\alpha) \leqslant \bigcup b$ and
$\bigcup b = \bigcup\{f(\alpha) | \alpha \in b\} \in B$. □

Lemma 6.8: Suppose that σ and τ enumerate A and its closure, respectively, and $\sigma(\xi) = \tau(\delta)$. Then either $\xi = \delta$ or $\delta = \xi + 1$.

Proof: Induction w.r.t. ξ. cf. exercice 4. □

So the *types* of a set of ordinals and its closure differ very little. Nevertheless, for some results closedness of the sets involved is essential as we shall have occasion to note several times.

Definition 6.9 : Enumerations of closed collections are called *normal*.

Hence a normal operation is an ordinal-valued strictly increasing operation defined on an ordinal or on OR, which is *continuous* in the sense of the topology described just before lemma 6.6 (i.e. which satisfies 6.6(*ii*)).

If σ is an enumeration and $\alpha \in$ Dom σ then $\alpha \leqslant \sigma(\alpha)$: otherwise there would be a *smallest* $\alpha \in$ Dom σ for which $\sigma(\alpha) < \alpha$; but then $\sigma(\sigma(\alpha)) < \sigma(\alpha)$ contradicting the minimality of α. An $\alpha \in$ Dom σ for which $\sigma(\alpha) = \alpha$ is called a *fixed point* of σ. Enumerations not necessarily have fixed points, as the enumeration of all successor-ordinals (defined by $\sigma(\alpha) := \alpha + 1$) shows. This operation is not normal either since the ω-th successor is $\omega+1$ while the limit of the first ω successors is ω. To simplify the discussion we shall deal from now on only with normal operations defined on all ordinals; they enumerate closed collections cofinal in OR. For these operations we have the following *fixed-point theorem*:

Theorem 6.10 : The fixed-points of a normal operation form a closed collection which is cofinal in OR.

Proof: Closed: suppose that a is a set of fixed-points of the normal operation σ and $\alpha = \bigcup a$. We have to prove that $\sigma(\alpha) = \alpha$. If $\alpha \in a$ this is trivial. Otherwise, Lim α and by 6.6(*iii*) and the cofinality of a in α:
$\sigma(\alpha) = \bigcup_{\xi \in a} \sigma(\xi) = \bigcup_{\xi \in a} \xi = \bigcup a = \alpha$.
Also, there are arbitrarily large fixed-points: take any $\alpha \in$ OR; we construct a fixed-point $\beta \geqslant \alpha$ as follows. If $\sigma(\alpha) = \alpha$ we may take $\beta = \alpha$. Otherwise, $\alpha < \sigma(\alpha)$. Define f by recursion to satisfy $f(0) = \alpha$, $f(n + 1) = \sigma(f(n))$ and take $\beta := \bigcup_{n \in \omega} f(n)$. Using an induction one sees that $f(n) < f(n + 1)$ for all $n \in \omega$ hence Lim β and $\{f(n) \mid n < \omega\}$ is cofinal in β. From 6.6(*iii*) it follows that $\sigma(\beta) = \bigcup_{n \in \omega} \sigma(f(n)) = \bigcup_{n \in \omega} f(n + 1) = \beta$. □

By the *derivative* of a normal operation σ we mean the enumeration of the fixed-points of σ; hence 6.10 states that the derivative of a normal operation again is a normal operation.

Examples of normal operations are σ, τ and η defined by $\sigma(\xi) := \alpha + \xi$; $\tau(\xi) := \alpha . \xi$; $\eta(\xi) := \alpha^{\xi}$ (where α is some fixed ordinal $\geqslant 2$).

Sometimes it may come as a surprise to discover that a quickly-growing operation is normal and hence has fixed-points; examples are exponentiation on the basis ω and the enumeration of initial-numbers defined in 8.6.

<u>Corollary 6.11</u>: If α is a fixed-point of the normal operation σ then

(*i*) if $\omega < \mathrm{cf}(\alpha)$ then $\{\beta < \alpha \mid \sigma(\beta) = \beta\}$ is cofinal in α and

(*ii*) if α is regular then α is a fixed-point of the derivative of σ.

Proof: For (*i*) we can almost copy the second half of the proof of 6.10: if $\gamma < \alpha$ let $f(0) = \gamma$, $f(n+1) = \sigma(f(n))$ and $\beta := \bigcup_{n\in\omega} f(n)$. Then $\sigma(\beta) = \beta$ and if $f(n) < \alpha$ then $f(n+1) = \sigma(f(n)) < \sigma(\alpha) = \alpha$; by induction $\beta \leqslant \alpha$ and hence $\beta < \alpha$ since $\mathrm{cf}(\beta) = \omega < \mathrm{cf}(\alpha)$.
In case (*ii*) α even is regular, so $\{\beta < \alpha \mid \sigma(\beta) = \beta\}$ must be order-isomorphic with α; this isomorphism is given by the derivative of σ and since this is normal it has α as a fixed-point. □

<u>Lemma 6.12</u>: (*i*) intersections of closed collections are closed;

(*ii*) if ω, $\alpha < \mathrm{cf}(\beta)$ and $A_\xi \subseteq \beta$ is closed and cofinal in β $(\xi < \alpha)$ then $\bigcap_{\xi<\alpha} A_\xi$ is (closed and) cofinal in β.

(*iii*) if $A_\xi \subseteq \mathrm{OR}$ is closed and cofinal in OR $(\xi < \alpha)$ then $\bigcap_{\xi<\alpha} A_\xi$ is (closed and) cofinal in OR.

Proof: (*i*) is trivial. For (*ii*), suppose that $\gamma_0 < \beta$ and define recursively $\gamma_{n+1} = \bigcup_{\xi<\alpha} \bigcap (A_\xi - \gamma_n)$. If $\gamma_n < \beta$ then $\gamma_{n+1} < \beta$ also, since $\alpha < \mathrm{cf}(\beta)$; hence $\gamma := \bigcup_{n\in\omega} \gamma_n < \beta$ since $\omega < \mathrm{cf}(\beta)$. Now $\gamma_n \leqslant \bigcap (A_\xi - \gamma_n) \leqslant \gamma_{n+1}$; therefore $\gamma = \bigcup_{n\in\omega} \bigcap (A_\xi - \gamma_n)$ $(\xi < \alpha)$. Hence $\gamma \in A_\xi$, since $\bigcap(A_\xi - \gamma_n) \in A_\xi$ and A_ξ is closed. Thus, $\gamma_0 \leqslant \gamma \in \bigcap_{\xi<\alpha} A_\xi$. The proof for (*iii*) is similar. □

We really need $\alpha < \mathrm{cf}(\beta)$ in 6.12(*ii*) as, for instance, $\bigcap_{n\in\omega} (\omega - n) = \emptyset$, but each ω-n is closed and cofinal in ω.

For every normal operation σ it is possible to prove the existence of a binary operation τ such that if one defines σ_α by $\sigma_\alpha(\beta) := \tau(\alpha,\beta)$ the following conditions are satisfied: (*i*) $\sigma_0 = \sigma$; (*ii*) $\sigma_{\sigma+1}$ is the derivative of σ_α and (*iii*) if Lim α then σ_α enumerates $\{\delta \mid \forall\xi < \alpha\ \sigma_\xi(\delta) = \delta\}$, i.e., the fixed points common to all σ_ξ, $\xi < \alpha$. Notice that the existence of τ does *not* follow from the recursion theorem 2.7 because the σ_α are not sets. From 6.10 and 6.12 it follows that all σ_α are normal and any regular fixed-point $\beta > \omega$ of σ is fixed-point of all σ_α when $\alpha < \beta$. σ_α is called the α-*th derivative* of σ.

<u>Lemma 6.13</u>: (*i*) If $A_\alpha \subseteq \mathrm{OR}$ is closed and cofinal in OR for every $\alpha \in \mathrm{OR}$ then so is $\{\beta | \beta \in \bigcap_{\alpha<\beta} A_\alpha\}$.

(*ii*) If $\gamma > \omega$ is regular and $A_\alpha \subseteq \gamma$ is closed and cofinal in γ for every $\alpha < \gamma$, then so is $\{\beta | \beta \in \bigcap_{\alpha<\beta} A_\alpha\}$.

Proof: (*i*) Cofinal: define σ by: $\sigma(\xi)$ is the smallest element of $\bigcap_{\delta<\xi} A_\delta$ greater than all $\sigma(\delta)$ where $\delta < \xi$. (This is possible by 6.12 (*iii*)).
Then σ is normal, for if Lim γ then $\xi < \delta < \gamma \to \sigma(\delta) \in A_\xi$, hence $\bigcup_{\delta<\gamma} \sigma(\delta) \in A_\xi$ for every $\xi < \gamma$ since A_ξ is closed and thus $\bigcup_{\delta<\gamma} \sigma(\delta) \in \bigcap_{\xi<\gamma} A_\xi$. Therefore, $\sigma(\gamma) = \bigcup_{\delta<\gamma} \sigma(\delta)$. From the fixed-point theorem we get arbitrarily large fixed-points of σ. But if β is such a fixed-point then $\beta = \sigma(\beta) \in \bigcap_{\alpha<\beta} A_\alpha$.
Closed: let σ be the enumeration of $A := \{\beta | \beta \in \bigcap_{\alpha<\beta} A_\alpha\}$; we show that $\sigma(\gamma) = \bigcup_{\xi<\gamma} \sigma(\xi)$ if Lim γ. For this purpose it clearly suffices to show that $\beta \in A$, where $\beta := \bigcup_{\xi<\gamma} \sigma(\xi)$. Now if $\delta < \beta$, then for some $\xi_0 < \gamma$ we have $\delta < \sigma(\xi)$ whenever $\xi_0 < \xi$. Thus $\xi_0 < \xi < \gamma \to \sigma(\xi) \in A_\delta$. Since A_δ is closed we get $\beta \in A_\delta$ for any $\delta < \beta$; therefore $\beta \in A$.
(*ii*) Same proof, but now we remain inside γ. □

Using the results above we can construct normal operations which increase faster and faster, starting with some normal operation σ not being the identity on OR: if σ_α is the α-th derivative of σ, it increases faster than σ_β if $\beta < \alpha$. If σ_α enumerates A_α, eventually the enumeration of the "diagonal" $A := \{\beta | \beta \in \bigcap_{\alpha<\beta} A_\alpha\}$ increases faster than A_α since $A - \alpha \subseteq A_\alpha$. Next we can iterate the process, leading from σ to the enumeration of A over all ordinals, diagonalize again, etc.

<u>Exercises</u>

1. The construction in the second part of the proof of 6.10 yields the *least* fixed point greater than or equal to a given ordinal α.
2. If R is a total ordering on $\alpha \in \mathrm{OR}$ (not necessarily $R = \in_\alpha$!) then there exists $a \subseteq \alpha$ cofinal in α w.r.t. R and well-ordered by $R \cap a^2$.

3. Every ordinal α has a cofinal, *closed* subset order-isomorphic to $cf(\alpha)$.
4. Prove 6.8 .
5. If σ is normal, $\sigma(\alpha)$ regular and Lim α then $\sigma(\alpha) = \alpha$.
6. Let σ enumerate OR-$\{0\}$. Determine some derivatives of σ.
7. If α is type of (the epsilon-structure of) a cofinal part of β then $cf(\alpha) = cf(\beta)$.
8. $cf(\alpha) = cf(\beta)$ iff α and β have order-isomorphic cofinal subsets.

7 THE REFLECTION-PRINCIPLE

Let Θ be any formula from the ZF-formalism. Θ^A, the *relativization* of Θ w.r.t. the collection A , is the formula obtained from Θ by restricting bound variables in Θ to membership in A, that is to say: quantifier-expressions of the form '$\exists x$' and '$\forall x$' in Θ are replaced by '$\exists x \in A$' and '$\forall x \in A$' respectively. Intuitively one may think of Θ^A as expressing with respect to A what Θ expresses with respect to the universe of all sets.

For the sequel of this section we fix a sequence of sets W_a $(\alpha \in \text{OR})$ such that

(*i*) $\alpha \leqslant \beta \rightarrow W_a \subseteq W_\beta$ and

(*ii*) $\text{Lim } \gamma \rightarrow W_\gamma = \bigcup_{\xi < \gamma} W_\xi$,

and we define $W := \bigcup_{a \in OR} W_a$. W does *not* need to be a set - in non-trivial applications it actually will not be one (cf. exercise 1). The main example of a sequence of sets satisfying (*i*) and (*ii*) is the cumulative hierarchy. But one may also divide any given collection W into a hierarchy satisfying (*i*) and (*ii*) by putting $W_a := W \cap V_a$ for example. There are more ways of devising hierarchies like this, however we shall not have occasion to apply these.

We now associate with every formula Θ (having the free variables, say, $x_0, \ldots, x_{n-1}$) the collection $|\Theta|$ of Θ-*stable* ordinals defined by
$|\Theta| := \{\alpha \in \text{OR} \mid \forall x_0, \ldots, x_{n-1} \in W_a (\Theta^{W_a} \leftrightarrow \Theta^W)\}$.

With respect to the cumulative hierarchy we have the following examples. The ordinals, stable w.r.t. the formula $\exists y\ (x \in y)$, are exactly the limit ordinals.

A more difficult example: α is stable w.r.t. the formula saying that every well-ordering has a type (which as a matter of fact happens to be theorem 2.9 using definition 2.13) exactly if every well-ordering $\langle A,R\rangle \in V_\alpha$ has a type $\beta < \alpha$ (the verification of this statement is tiresome but not particularly difficult). It looks as if such ordinals are rare. ω is an easy example, but one easily verifies that the next one must be rather big, for instance, it cannot be countable, neither can it be the first uncountable ordinal nor have the same power etc.; one even may come to think that ω is the only example. Nevertheless, the main result of this paragraph implies that there are a great many of such ordinals:

Theorem 7.1: Let Ψ be any formula of the ZF-formalism.
Then $\|\Psi\| := \bigcap \{|\Phi| \mid \Phi \text{ subformula of } \Psi\}$ is a closed collection cofinal in OR.

Proof: We start by making five remarks.

1. Every formula has only finitely many subformulas, hence $\|\Psi\|$ is a finite intersection of collections.
2. If Ψ happens to be quantifier-free then $\Psi^A = \Psi$ hence $|\Psi| = \|\Psi\| =$ OR.
3. Trivially, $|\neg\Psi| = |\Psi|$; therefore $\|\neg\Psi\| = \|\Psi\|$.
4. Also, $|\Phi_1| \cap |\Phi_2| \subseteq |\Phi_1 \wedge \Phi_2|$; hence $\|\Phi_1 \wedge \Phi_2\| = \|\Phi_1\| \cap \|\Phi_2\|$.
5. We may as well assume that $\neg$ $\wedge$ and $\exists$ are the only logical signs in Ψ since the others can be defined in terms of those.

We now prove that $\|\Psi\|$ is closed. Suppose that $a \subseteq \|\Psi\|$ and $a \subseteq \alpha = \bigcup a$; we show that $\alpha \in |\Phi|$ for every subformula Φ of Ψ using induction on the number of occurrences of logical connectives in Φ. If Φ does not have any logical signs or if Φ happens to be a negation or a conjunction the result follows from 2, 3 or 4 respectively, using the induction-hypotheses for the last two cases. So assume that $\Phi = \exists y\ \Phi'(y,x_1,\ldots,x_n)$. Suppose that $a_1,\ldots,a_n \in W_\alpha$ and that $[\exists y\ \Phi'(y,a_1,\ldots,a_n)]^{W_\alpha}$ holds. Then for some $b \in W_\alpha$ $[\Phi'(b,a_1,\ldots,a_n)]^{W_\alpha}$. By inductionhypothesis, $\alpha \in |\Phi'|$, hence $[\Phi'(b,a_1,\ldots,a_n)]^W$.
Therefore $[\exists y\ \Phi'(y,a_1,\ldots,a_n)]^W$. Conversely, suppose that $[\exists y\ \Phi'(y,a_1,\ldots,a_n)]^W$. Since α is a limit, $W_\alpha = \bigcup_{\xi<\alpha} W_\xi = \bigcup_{\xi\in a} W_\xi$. As $a_1,\ldots,a_n \in W_\alpha$ there exists a $\xi \in a$ such that $a_1,\ldots,a_n \in W_\xi$. Now $\xi \in a \subseteq \|\Psi\| \subseteq |\Phi|$, therefore $[\exists y\ \Phi'(y,a_1,\ldots,a_n)]^{W_\xi}$; in other words, $b \in W_\xi$ exists such that $[\Phi'(b,a_1,\ldots,a_n)]^{W_\xi}$. Also, $\xi \in a \subseteq \|\Psi\| \subseteq |\Phi'|$, hence $[\Phi'(b,a_1,\ldots,a_n)]^W$ and

since $\alpha \in |\Phi'|$, by induction-hypothesis it follows that $[\Phi'(b,a_1,\ldots,a_n)]^{W_\alpha}$. Therefore $[\exists y\ \Phi'(y,a_1,\ldots,a_n)]^{W_\alpha}$.

Next we show that $\|\Psi\|$ is cofinal in OR. According to 6.12(*iii*) it suffices to show that $|\Psi|$ has a closed subcollection which is cofinal in OR. For this we use induction w.r.t. the number of occurrences of logical connectives in Ψ. Using remarks 2,3,4 and 5 we see that the only non-trivial case is $\Psi = \exists y\ \Phi(y,x_1,\ldots,x_n)$. Define the operation τ by: $\tau(\alpha)$ is the least β greater than all $\tau(\xi)\,(\xi < \alpha)$ such that, if $x_1,\ldots,x_n \in W_\alpha$ and $(\exists y\ \Phi)^W$, then $\exists y \in W_\beta\ \Phi^W$. Such a β exists by the replacement-axiom. Using properties (*i*) and (*ii*) of the W_a we see that τ is normal.

By the induction-hypothesis there exists a collection $A \subseteq |\Phi|$, closed and cofinal in OR; from 6.10 and 6.12 it follows that $B := \{\alpha \in A \mid \tau(\alpha) = \alpha\}$ is again closed and cofinal in OR. We show that $B \subseteq |\Psi|$: suppose that $x_1,\ldots,x_n \in W_\beta$, $\beta \in B$. If $(\exists y\ \Phi)^{W_\beta}$ then for some $y \in W_\beta$, Φ^{W_β}. Since $\beta \in A$, Φ^W holds. Thus $(\exists y\ \Phi)^W$. Conversely, if $(\exists y\ \Phi)^W$ then by definition of τ, $\exists y \in W_{\tau(\beta)}\ \Phi^W$. But $\beta \in B$ hence $\tau(\beta) = \beta$. Also, $\beta \in A$, therefore $(\exists y\ \Phi)^{W_\beta}$. □

By the *reflection-principle* (in set theory) one usually means the

Corollary 7.2: If $\Phi(x_1,\ldots,x_n)$ is a formula of the ZF-formalism we can prove as a ZF-theorem the sentence $\forall\alpha\ \exists\beta > \alpha\ \forall x_1,\ldots,x_n \in V_\beta\ [\Phi \leftrightarrow \Phi^{V_\beta}]$.

This expresses that properties of the universe (expressible by ZF-formulas) reflect to arbitrarily great partial universes V_β.

Exercises

1. If $\bigcup_{a \in OR} W_a \in V$ then for some $\gamma \in$ OR, $\bigcup_{a \in OR} W_a = \bigcup_{a < \gamma} W_a$. (Hence if $\bigcup_{a \in OR} W_a \in V$ then 7.1 is trivial).
2. Does there exist a formula Ψ for which $|\Psi|$ is not closed?
3. Suppose that $\alpha > \omega$ and $\alpha \in |$'every well-ordering has a type'$|$. Prove that α is uncountable. Prove that $\beta < \alpha \rightarrow \beta \underset{1}{<} \alpha$.

8 INITIAL NUMBERS

In the previous section we showed (cf. exercise 3) the existence of uncountable ordinals in a rather roundabout way; we do this again a bit handier using the *operation* Γ *of Hartogs:*

Definition 8.1: $\Gamma(a) := \{\alpha \in \mathrm{OR} \mid \alpha \leqslant_1 a\}$.

The remarkable thing about Γ is that its values are sets:

Lemma 8.2: $\Gamma(a) \in \mathrm{OR}$.

Proof: If $\alpha \leqslant_1 a$ and $\beta < \alpha$ then $\beta \leqslant_1 a$; therefore $\Gamma(a)$ is a transitive collection of ordinals. So it suffices to show that $\Gamma(a)$ is a *set*. Consider the set of all pairs $\langle A\ R\rangle \in \mathcal{P}(a) \times \mathcal{P}(a \times a)$ for which R well-orders A and notice that $\Gamma(a)$ exactly is the collection of types of such pairs: if α is type of $\langle A,R\rangle$, $A \subseteq a$, then $\alpha =_1 A \leqslant_1 a$, hence $\alpha \leqslant_1 a$ and, conversely, if $\alpha \leqslant_1 a$, for instance by $f:\alpha \to a$, then $\langle f[\alpha], \{\langle f(\xi), f(\delta)\rangle \mid \xi < \delta < \alpha\}\rangle$ is a well-ordering of the required form of type α. As $\Gamma(a)$ is the image of a set, the replacement-axiom implies that it is a set. □

For natural-number-arguments the values of Γ are easily described: $\Gamma(n) = n+1$. The main properties of Γ are listed by the

Theorem 8.3: (*i*) $\neg\, \Gamma(a) \leqslant_1 a$, hence

(*ii*) $\alpha <_1 \Gamma(\alpha)$ if $\alpha \in \mathrm{OR}$;

(*iii*) $\alpha < \Gamma(a) \to \alpha <_1 \Gamma(a)$.

Proof: (*i*) If $\Gamma(c) \leqslant_1 a$ then $\Gamma(a) \in \Gamma(a)$ by 8.1 and 8.2 contradiction.

(*ii*) $\alpha < \Gamma(\alpha)$ hence $\alpha \leqslant_1 \Gamma(\alpha)$. And $\alpha =_1 \Gamma(\alpha)$ is excluded by (*i*).

(*iii*) If $\alpha < \Gamma(a)$ then $\alpha \leqslant_1 \Gamma(a)$. If $\alpha =_1 \Gamma(a)$ then (since $\alpha \leqslant_1 a$) $\Gamma(a) \leqslant_1 a$ contradicting (*i*). □

8.3(*i*) implies *Hartogs' lemma* : to every set a there exists an ordinal α which cannot be embedded in a. 8.3(*iii*) motivates the next

Definition 8.4: An ordinal $\alpha \geqslant \omega$ is called an *initial number* if for all $\beta < \alpha$, $\beta <_1 \alpha$.

Hence an initial number is an infinite ordinal minimal among the ordinals of a certain power. By 8.3(*iii*), this is true of $\Gamma(a)$ when $\omega \leqslant \Gamma(a)$; 8.3(*ii*) and

8.1 state that $\Gamma(\alpha)$ is the first initial-number greater than α (if $\alpha \geqslant \omega$).

Theorem 8.5: The collection of initial numbers is closed and cofinal in OR.

Proof: Cofinal: if α is any infinite ordinal, $\Gamma(\alpha)$ is an initial number greater than α.
Closed: suppose that A is a set of initial numbers for which $A \subseteq \alpha = \cup A$. Then $\alpha \geqslant \omega$ and if $\beta < \alpha$ then for some $\gamma \in A$, $\beta < \gamma$, hence, $\beta \underset{1}{<} \gamma$ and since $\gamma < \alpha$ we get $\beta \underset{1}{<} \alpha$, that is to say, α is an initial-number also. □

The recursion equations of 8.6 define the enumeration of the collection of initial numbers:

Definition 8.6: 1. $\omega_0 = \omega$,
2. $\omega_{\alpha+1} = \Gamma(\omega_\alpha)$,
3. $\omega_\gamma = \bigcup_{\xi<\gamma} \omega_\xi$ if Lim γ.

The ω-operation increases very fast; nevertheless 6.10 implies the existence of many ordinals α for which $\omega_\alpha = \alpha$.

Lemma 8.7: Initial numbers are limits.
Proof: cf. exercise 1. □

Theorem 8.8: Regular ordinals are initial numbers.
Proof: Suppose α is regular. By definition, $\omega \leqslant \alpha$. If $\beta < \alpha$ then $\beta \underset{1}{\leqslant} \alpha$ anyway. If $\beta \underset{1}{=} \alpha$ would hold then some f: $\beta \to \alpha$ with Ran(f) $= \alpha$ would exist, hence $\mathrm{cf}(\alpha) \leqslant \beta$ by Lemma 6.4. But $\mathrm{cf}(\alpha) = \alpha > \beta$, contradiction. □

If Lim α then $\mathrm{cf}(\omega_\alpha) = \mathrm{cf}(\alpha)$, so there are lots of non-regular initial numbers, for instance ω_ω, $\omega_{\omega+\omega}$ ω_{ω_1} etc.

Definition 8.9: An initial number ω_α is called *weakly inaccessible* if it is regular and Lim α.

We shall see later on that such initial numbers are rather rare; in fact it even is impossible to prove their existence on the basis of the ZF-axioms.

Theorem 8.10: ω_a is weakly inaccessible iff α is a regular fixed-point of the ω-operation.

Proof: If ω_a is weakly inaccessible, then $cf(\alpha) \leqslant \alpha \leqslant \omega_a = cf(\omega_a) = cf(\alpha)$ and hence $\alpha = \omega_a$.

Conversely, if $\omega_a = \alpha$ then Lim α by 8.7. □

Notice that the proof of 6.10 only produces fixed-points of cofinality ω. To facilitate the proof of 8.13, we introduce the following

Definition 8.11: If $\alpha, \beta, \gamma, \delta \in$ OR, then $\langle \alpha, \beta \rangle \prec \langle \gamma, \delta \rangle$ holds just in case one of the following conditions is satisfied:

(*i*) $\alpha \cup \beta < \gamma \cup \delta$;

(*ii*) $\alpha \cup \beta = \gamma \cup \delta$ and $\alpha < \gamma$;

(*iii*) $\alpha \cup \beta = \gamma \cup \delta$, $\alpha = \gamma$ and $\beta < \delta$.

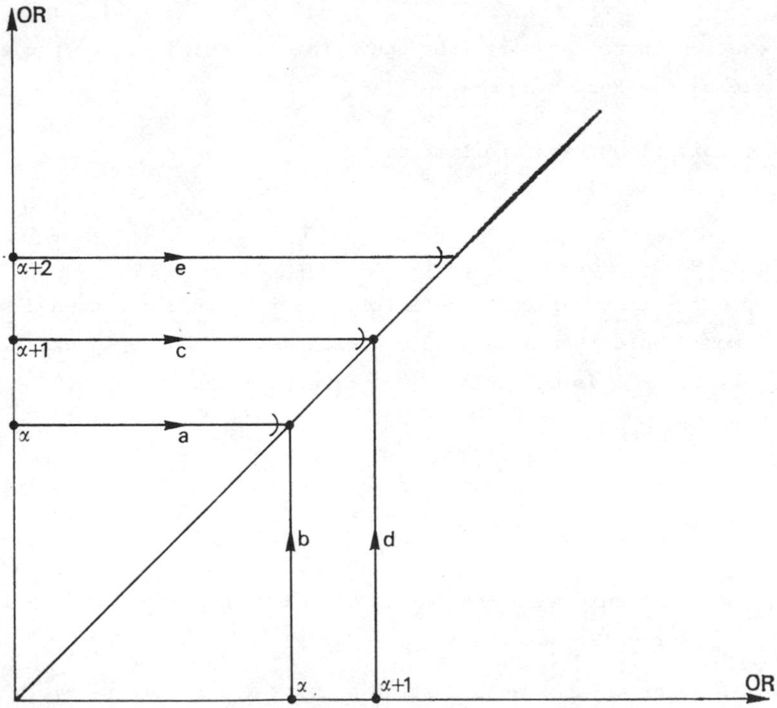

Fig. 2

Figure 2 illustrates the ordering hereby defined on the collection OR × OR: follow the lines in the figure in the order a-b-c-d-e.

Theorem 8.12: (*i*) $\prec$ is a well-ordering of OR × OR;

(*ii*) sets of the form $\eta \times \eta$ ($\eta \in$ OR) are *initialsegments* of OR × OR w.r.t. $\prec$, i.e., if $<\gamma,\delta> \in \eta \times \eta$ and $<\alpha,\beta> \prec <\gamma,\delta>$ then $<\alpha,\beta> \in \eta \times \eta$.

Proof: The second part follows trivially from 8.11. For the first part: let $\emptyset \neq A \subseteq$ OR × OR; first take the pairs $<\alpha,\beta> \in A$ for which $\alpha \cup \beta$ is minimal; among those take the pairs $<\alpha,\beta>$ in which α is minimal and finally from these the pair $<\alpha,\beta>$ with smallest β - this pair clearly is $\prec$-minimal in A. □

Theorem 8.13: $\omega \leqslant \alpha \to \alpha \times \alpha \underset{1}{=} \alpha$.

Proof: If not, we take the smallest $\alpha \geqslant \omega$ such that $\alpha \times \alpha \underset{1}{\neq} \alpha$. Since certainly $\alpha \underset{1}{\leqslant} \alpha \times \alpha$ we get $\alpha \underset{1}{<} \alpha \times \alpha$. Also $\omega < \alpha$, for $\omega \times \omega \underset{1}{=} \omega$, as we have seen in chapter I. Finally, α is an initial number: otherwise there would be some $\beta < \alpha$ for which $\omega \leqslant \beta$ and $\beta \underset{1}{=} \alpha$, but then also $\beta \times \beta \underset{1}{=} \beta$ by the choice of α, and we would have $\alpha \times \alpha \underset{1}{=} \beta \times \beta \underset{1}{=} \beta \underset{1}{=} \alpha$, contradiction.
Let γ be the type of $\alpha \times \alpha$ under the well-ordering $\prec$. Clearly, $\alpha < \gamma$ since $\alpha \underset{1}{<} \alpha \times \alpha \underset{1}{=} \gamma$. Hence there exist an order-isomorphism h between α and some proper initial segment (w.r.t. $\prec$) of $\alpha \times \alpha$. Take $<\xi,\eta> \prec$-minimal in $\alpha\times\alpha$-Ran(h) and $\delta := (\xi \cup \eta)^+$ then h maps α injectively into $\delta \times \delta$ while $\delta < \alpha$.
By the choice of α we have $\delta \times \delta \underset{1}{=} \delta$ and $\alpha \underset{1}{\leqslant} \delta \times \delta \underset{1}{=} \delta \underset{1}{\leqslant} \alpha$; therefore $\delta \underset{1}{=} \alpha$. Since α is an initial number, we arrive at a contradiction. □

Corollary 8.14: If A has a well-ordering and $\omega \underset{1}{\leqslant} A$ then $A \times A \underset{1}{=} A$.

Corollary 8.15: $\omega_a \times \omega_a \underset{1}{=} \omega_a$.

This last corollary plays an important rôle in §10 devoted to cardinal numbers and their arithmetic.

Exercises

1. Prove lemma 8.7.
2. Determine the first 20 elements of OR × OR under the well-ordering of 8.11.

3. Let τ enumerate the elements of $OR \times OR$ in the ordering given by $\prec$, i.e., $\tau(\langle\alpha,\beta\rangle) = \{\tau(\langle\gamma,\delta\rangle) \mid \langle\gamma,\delta\rangle \prec \langle\alpha,\beta\rangle\}$.
   Prove that $\tau(\langle 0,\omega_a\rangle) = \omega_a$.

9 THE AXIOM OF CHOICE

Imagine an infinite set of pairs of shoes A. There are several sets A' the elements of which are shoes with the property that A' contains of each pair in A exactly one element. One could take for example the set A' of right-hand shoes from the pairs in A. But now, if A would consist of an infinite amount of pairs of socks no way of defining a similar set A' seems to be possible since unfortunately the sock-manufacturers usually produce indistinguishable socks for both feet. To get such an A' one cannot do better than just *choosing* a sock from each pair in A in an unsystematic way. The above example is due to Russell.

In I.15 we have seen an attempt to characterize the "number of elements" of sets via bijections and "greater" or "smaller" via injections. The least one should want (either for pragmatic reasons, or extrapolating our experience with finite sets) is that the relation "greater than" is a total order. Put differently: for any two sets A and B we should like to have $A \leqslant_1 B$ or $B \leqslant_1 A$. However, in general there is no way to get the required injection. Intuitively, the best thing we can do is to make successive choices of pairs $\langle a,b\rangle$ ($a \in A$, $b \in B$) until one of the sets is exhausted by this process, for we then have the injection asked for (or its inverse).

To be able to solve these and similar problems we may introduce the *axiom of choice*, denoted by AC.

Though in general people feel this principle to be intuitively evident we always shall indicate its use in proofs. The reason is that AC seems to be different in some way, from the other ZF-axioms. It deals with the existence of sets in a novel way. The sets whose existence is asserted by the ZF-axioms, such as the sum-set, power-set or image of a set under an operation, always can be described (defined) in an obvious way; on the other hand (as in the examples given above) sets obtained using the axiom of choice usually cannot

be defined - in fact if they can (as in the case of the shoes) AC is not needed to show their existence.
In this way, applications of *AC* stand out among applications of the other ZF-axioms (though applications of AC went unnoticed some time before Zermelo's explicit statement of the axiom in 1904; cf. [van Dalen - Monna 1972], Ch. I §3). In mathematical contexts, even of a definite set-theoretic flavour, such as topology, one will never find references to set theoretic axioms other than *AC* (in one form or the other; often the "Zorn-lemma"-formulation is used).

In order to have a convenient formulation of the axiom of choice, we introduce the following

Definition 9.1: A function f is called *choice-function* for the set A if $\mathrm{Dom}(f) = A - \{\emptyset\}$, and for all $a \in \mathrm{Dom}(f)$, $f(a) \in a$.

Hence a choice-function for A picks an element from every non-empty set in A. The *axiom of choice* now has the following simple formulation:

AC: Every set has a choice function.

It already follows from the ZF-axioms that several sets do have choice-functions, for instance *finite* sets have (cf. exercise 1). Also, *AC* appears to be rather plausible: of course we can choose an element C_a in every non-empty set $a \in A$; why should there be no function $f: A \to \cup A$ making all such choices all at once?
Whatever one's fundamental views concerning *AC* may be, the fact is that *AC* considerably adds to the strength of *ZF*. There are several variants of *AC* (cf. Rubin and Rubin, 1963); we present four of them. (N.B. there are lots of *equivalents*, but the four, listed below, are almost linguistic reformulations-their equivalence with *AC* is obvious).

AC 1: Every power-set has a choice function.

AC 2: Every binary relation contains a function with the same domain.

AC 3: Every function contains an injection with the same range.

AC 4: Every partition has a set of representatives.

The equivalence with *AC* is easily established:

AC $\to$ *AC 1*: Trivial. □

AC 1 → *AC 2* : Assume $R \subseteq A \times A$ and (*AC* 1) let f be a choice function for $\mathcal{P}(A)$. The required function $h \subseteq R$ with $\mathrm{Dom}(h) = \mathrm{Dom}(R)$ can be defined by $h(a) := f(\{b \in A \mid a\ R\ b\})$. □

AC 2 → *AC 3*: Let f be some function and (*AC 2*) $h \subseteq f^{-1}$ a function with $\mathrm{Dom}(h) = \mathrm{Ran}(h)$. Then h^{-1} is injective and $\mathrm{Ran}(h^{-1}) = \mathrm{Ran}(f)$. □

AC 3 → *AC 4*: Suppose that P is a partition (i.e., if $A, B \in P$ and $A \neq B$ then $A \cap B = \emptyset$). Define $f\colon \bigcup P \to P$ by $f(a) := \bigcup\{A \in P \mid a \in A\}$ (so $a \in f(a) \in P$). *AC 3* asserts the existence of an injection $h \subseteq f$ with $\mathrm{Ran}(h) = \mathrm{Ran}(f) = P$. Now $\mathrm{Dom}(h)$ is the required set of representatives. □

AC 4 → *AC*: If A is any set, $\{\{a\} \times a \mid a \in A\}$ is a partition (i.e., a set of disjoint sets) hence (*AC 4*) it has a set of representatives; and this is at the same time a choice function for A. □

The first explicit formulation of the axiom of choice was provided by Zermelo (in 1904, when he listed his axioms for set theory). His main purpose, in making the assumption about the set theoretic universe explicit, was to provide a framework for proving his *well-ordering theorem*:

Theorem 9.2: (*AC*): Every set has a well-ordering.

Proof: Suppose that A is a set for which we want to construct a well-ordering; take (*AC*) some choice function f for $\mathcal{P}(A)$. Using the recursion theorem there is an operation τ defined on OR satisfying

$$\tau(\alpha) = \begin{cases} f(A - \{\tau(\beta) \mid \beta < \alpha\}) & \text{if } A - \{\tau(\beta) \mid \beta < \alpha\} \neq \emptyset \\ A & \text{if } A - \{\tau(\beta) \mid \beta < \alpha\} = \emptyset. \end{cases}$$

If $\beta_1 < \beta_2 < \alpha$ and $A - \{\tau(\beta) \mid \beta < \alpha\} \neq \emptyset$ then $\tau(\beta_1), \tau(\beta_2) \in A$ and $\tau(\beta_1) \neq \tau(\beta_2)$; hence $\{<\beta, \tau(\beta)> \mid \beta < \alpha\}$ embeds α into A. By theorem 8.3(*i*), not every ordinal can be embedded into A. So there is a smallest α_0 for which $A - \{\tau(\beta) \mid \beta < \alpha_0\} = \emptyset$. But then $h := \{<\beta, \tau(\beta)> \mid \beta < \alpha_0\}$ is a bijection between α_0 and A and $\{<h(\beta_1), h(\beta_2)> \mid \beta_1 < \beta_2 < \alpha_0\}$ is a well-ordering of A. □

The simple proof of 9.2 uses the recursion theorem hence the replacement-axiom. We now present Zermelo's proof which is purely combinatorial and very elegant, not using recursion, replacement-axiom or ordinals.
Technically speaking Zermelo's proof shows that $Z \vdash AC \to$ well-ordering theorem. This is an improvement compared to theorem 9.2 which showed

ZF $\vdash$ AC $\to$ well-ordering theorem. (Z is *Zermelo-set theory*, that is, ZF minus replacement.) The reader who is not primarily interested in these metamathematical refinements can safely skip the following proof.

Suppose again that f is a choice function for the set A, which is to be well-ordered. We write $B^- := B - \{f(B)\}$ if $\emptyset \neq B \subseteq A$; clearly $B^- \subseteq C \subseteq B$ implies that $B^- = C$ or $C = B$. Define G to be the collection of $\Gamma \subseteq \mathcal{P}(A)$ for which

1° $A \in \Gamma$;

2° $\emptyset \neq \Gamma' \subseteq \Gamma \to \bigcap\Gamma' \in \Gamma$;

3° $\emptyset \neq B \in \Gamma \to B^- \in \Gamma$.

Notice that $\mathcal{P}(A) \in G$ hence $\Gamma_0 := \bigcap G \subseteq \mathcal{P}(A)$. It is easily verified that $\Gamma_0 \in G$. We next show that Γ_0 is well-ordered by $\supseteq$ and that $A \leqslant_1 \Gamma_0$; this clearly establishes a well-ordering of A.

a) The inclusion-relation on Γ_0 is connected, i.e., $\forall B \in \Gamma_0\ \forall C \in \Gamma_0 (B \subseteq C$ or $C \subseteq B)$. To prove this, consider the set $\Delta := \{B \in \Gamma_0 \mid \forall C \in \Gamma_0 (B \subseteq C$ or $C \subseteq B)\}$ and note that it suffices to show $\Delta \in G$, since this implies $\Gamma_0 = \bigcap G \subseteq \Delta$ as required. That Δ satisfies 1° and 2° is easily seen. As to 3°, suppose that $\emptyset \neq B \in \Delta$ (1); we have to show that $B^- \in \Delta$, or ($B \in \Delta \subseteq \Gamma_0$ hence $B^- \in \Gamma_0$) in other words $\forall C \in \Gamma_0 (B^- \subseteq C$ or $C \subseteq B^-)$. To this end consider $\Sigma := \{C \in \Gamma_0 \mid B^- \subseteq C$ or $C \subseteq B^-\}$. As before, it suffices to show that $\Sigma \in G$ and again verification of 1° and 2° is easy. Finally 3°: assume that $\emptyset \neq C \in \Sigma$ (2), we now need to show that $C^- \in \Sigma$, i.e., ($C^- \in \Gamma_0$) $B^- \subseteq C^-$ or $C^- \subseteq B^-$.

Since (1) $B \in \Delta$ we have $B \subseteq C^-$ or $C^- \subseteq B$; in the first case $B^- \subseteq C^-$ and we are done. So we assume now that $C^- \subseteq B$ (3).

Since (2) $C \in \Sigma$, we have $B^- \subseteq C$ or $C \subseteq B^-$; in the second case $C^- \subseteq B^-$ and we are finished, so suppose that $B^- \subseteq C$ (4).

Finally by (1) $B \in \Delta$ we get $B \subseteq C$ or $C \subseteq B$. In case $B \subseteq C$ it follows from (3) that $C^- = B$ or $B = C$ and in both cases $B^- \subseteq C^-$. If $C \subseteq B$ using (4) it follows that $B^- = C$ or $C = B$ thus in both cases $C^- \subseteq B^-$.

b) We now show that $\supseteq$ well-orders Γ_0. Suppose that $\emptyset \neq \Delta \subseteq \Gamma_0$ and consider $\Sigma := \{B \in \Gamma_0 \mid \bigcup\Delta \subseteq B\}$. Trivially, $A \in \Sigma$, hence, since $\emptyset \neq \Sigma \subseteq \Gamma_0 \in G$, we have $D := \bigcap\Sigma \in \Gamma_0$. Clearly, $\bigcup\Delta \subseteq D$. If $D = \emptyset$ then $\bigcup\Delta = \emptyset$, hence $\Delta = \{\emptyset\}$ and Δ has a $\supseteq$-minimal element. If $D \neq \emptyset$ then $D^- \in \Gamma_0 - \Sigma$ (by definition of D) therefore

$C \not\subseteq D^-$ for some $C \in \Delta$ and hence $D^- \subseteq C \neq D^-$ since $\supseteq$ is connected on Γ_0. Now $C \subseteq \bigcup\Delta \subseteq D$, therefore $C = D = \bigcup\Delta$ and $\bigcup\Delta \in \Delta$. Thus $\bigcup\Delta$ is $\supseteq$-minimal in Δ.
c) Finally we inject A in Γ_0 by means of the function h defined by $h(a) := \bigcap\{B \in \Gamma_0 | a \in B\}$. This is an injection indeed: for $a \in h(a)$ and $a \notin (h(a))^-$ - that is to say, $a = f(h(a))$ which makes h injective. The well-ordering of Γ_0 induces a well-ordering of A via h: namely, $\{\langle a,b\rangle \in A \times A | h(b) \subseteq h(a) \neq h(b)\}$. The definition of Γ_0 in this proof forms an interesting example of a so-called *inductive definition* (cf. chapter III).

We have proved the well-ordering theorem, using the axiom of choice; conversely this theorem implies the axiom of choice - i.e., these principles are equivalent in the system ZF. For suppose that we want a choice function for $\mathcal{P}(A)$. The well-ordering theorem provides us with a well-ordering R of A. Since R is connected, every non-empty $B \subseteq A$ has a unique R-minimal element. The function $f(B) := \bigcup\{b \in B | \forall a \in B \neg\, a R b\}$ which associates to every non-empty $B \subseteq A$ its R-minimal element (note that $\{b \in B | \forall a \in B \neg\, a R b\}$ is a singleton) is a choice function for $\mathcal{P}(A)$.

Comparability of sets with respect to their cardinality is stated by the so-called *trichotomy-theorem*:

Theorem 9.3:(AC): For all sets A and B, $A \leqslant_1 B$ or $B \leqslant_1 A$.

Proof: Using AC in the form of the well-ordering theorem, let R and S be well-orderings of A resp. B, and suppose that α and β are the types of $\langle A,R\rangle$ and $\langle B,S\rangle$ respectively. Then $A =_1 \alpha$ and $B =_1 \beta$. If $\alpha \leqslant \beta$ then clearly $A \leqslant_1 B$; otherwise $\beta \leqslant \alpha$ and $B \leqslant_1 A$. □

The trichotomy-theorem, conversely, implies the axiom of choice (Hartogs (1915) for instance in the form of the well-ordering theorem. Let A be some set we want to well-order. By 8.2, $\Gamma(A)$ is an ordinal for which (8.3(*i*)) not $\Gamma(A) \leqslant_1 A$. Hence $A \leqslant_1 \Gamma(A)$, using the trichotomy theorem. Now the well-ordering of $\Gamma(A)$ induces one on A.

Next we formulate a form of the axiom of choice known under the name *Zorn's lemma*, which is extremely useful in pure mathematics.
Let R partially order A. $B \subseteq A$ is called a *chain* (w.r.t. R) if $R \cap (B \times B)$ is a total ordering of B. $a \in A$ is called *upper bound* of $B \subseteq A$ if $b R a$ for all

$b \in B$ and $a \in A$ is called *maximal* if for all $a' \in A$, aRa' implies $a' = a$. Zorn's lemma is the

Theorem 9.4:(AC): If every chain in the partial ordering $\langle A,R \rangle$ has an upper bound, then A has a maximal element.

Proof: It suffices to find a chain $M \subseteq A$ maximal w.r.t. inclusion, i.e., not proper part of some other chain. For if m is an upper bound of M it must be maximal too: otherwise there would be some $m' \in A$ for which mRm' and $m \neq m'$ and $M \cup \{m'\}$ would be a chain properly containing M. Constructing such an M is easy using recursion. Use AC in the form of the well-ordering theorem; suppose that α is type of A w.r.t. to some well-ordering and $h{:}\alpha \to A$ the corresponding bijection. Define M_ξ by

$$M_\xi = \begin{cases} \bigcup_{\delta<\xi} M_\delta \cup \{h(\xi)\} & \text{if this is a chain;} \\ \bigcup_{\delta<\xi} M_\delta & \text{otherwise.} \end{cases}$$

It is easily verified that $\bigcup_{\xi<\alpha} M_\xi$ is a maximal chain. □

One can construct proofs for 9.3 and 9.4 using AC in the choice-function-formulation which can be compared to the Zermelo-proof of 9.2 we have given.

The usefulness of Zorn's lemma as an equivalent of AC can be illustrated by observing how smoothly it proves most equivalents of AC. Take for instance AC *4*. Assume P to be a partition for which we want to have a set of representatives. Take A to be the set of sets of representatives of *subsets* of P. (Hence $A \subset \wp(\bigcup P)$; for instance, all sets of representatives of *finite* subsets of P are in A.) Consider A, partially ordered by $\subseteq$. If $K \subseteq A$ is a chain then $\bigcup K \in A$ is an upper bound for K hence Zorn's lemma gives us a maximal element of A; this clearly is a set of representatives for the whole of P.

Finally we discuss one more equivalent of AC: the powerset of an ordinal can be well-ordered.

If the axiom of choice holds, this follows as a subcase of the well-ordering theorem. Conversely we have the

Theorem 9.5: If power sets of ordinals are well-orderable, AC holds.

Proof: (H. Rubin, 1960). If suffices to well-order partial universes since according to 4.8(*iii*) every set is part of one. We show how to well-order V_γ. Let $\lambda := \Gamma(V_\gamma)$ (cf. 8.1, 8.2) and let S well-order $\mathcal{P}(\lambda)$. We recursively define well-orderings R_ξ of V_ξ $(\xi \leqslant \gamma)$ as follows. Take $R_0 = \emptyset$, and $R_\alpha = \bigcup_{\xi<\alpha} R_\xi$ if Lim α. Supposing finally that R_β well-orders V_β $(\beta < \gamma)$ we show how to extend it to a well-ordering of $V_{\beta+1}$. Let h: $V_\beta \to \lambda'$ be the order-isomorphism between $\langle V_\beta, R_\beta \rangle$ and its type λ'. Then $\lambda' < \lambda$ for $\lambda' =_1 V_\beta \lesssim_1 V_\gamma$. Thus, S induces a well-ordering S^β of $V_{\beta+1} - V_\beta$ via h. The required extension $R_{\beta+1}$ of R_β is defined by $R_{\beta+1} = R_\beta \cup S^\beta \cup (V_\beta \times (V_{\beta+1} - V_\beta))$. □

In the presence of the axiom of choice there is an operation τ satisfying $\mathcal{P}(\omega_\alpha) =_1 \omega_{\tau(\alpha)}$. One may wonder what τ looks like. The results of Cohen have shown that on the basis of ZF (or even ZF + AC) the operation τ cannot be determined. There are many possible candidates for τ. Of course, $\alpha < \tau(\alpha)$ since by Cantor's theorem $\omega_\alpha \lesssim_1 \mathcal{P}(\omega_\alpha)$. One version of the *generalized continuum-hypothesis* says that $\tau(\alpha) = \alpha + 1$, in other words, that $\mathcal{P}(\omega_\alpha) =_1 \omega_{\alpha+1}$ for all α - this is the smallest possible value in all cases. (Cantor's continuum-hypothesis says that $\tau(0) = 1$, that is, $\mathcal{P}(\omega) =_1 \omega_1$.)

Since every ordinal either is finite or is equivalent to an initial number we get from 9.5 the

Corollary 9.6: The generalized continuum-hypothesis implies the axiom of choice.

The proof of 9.6 by means of 9.5 may be questioned for two reasons: first, 9.5 uses the axiom of regularity and second, there exist different formulations of the generalized continuumhypothesis not immediately implying the assumption of 9.5 (for instance: there is no set of cardinality strictly between ω_α and $\mathcal{P}(\omega_\alpha)$). Both objections can be dealt with at the cost of complicating the proof of 9.6.

Finally we present some set-theoretic assertions weaker than the axiom of choice but still not provable from the ZF-axioms alone.

In analysis one can often use *DC*, the *axiom of dependent choices*, in stead of *AC*. This is formulated in the

Theorem 9.7: (*AC*): If $A \neq \emptyset$, $R \subseteq A \times A$ and $\forall a \in A\ \exists b \in A\ (aRb)$ then a function $f{:}\,\omega \to A$ exists such that $\forall n \in \omega\ [f(n)\ R\ f(n+1)]$.

Proof: Take a choice-function h for $\mathcal{P}(A)$ and define f recursively by $f(0) = h(A)$; $f(n+1) = h(\{a \in A | f(n)R\ a\})$ then f satisfies the requirements. □

DC opens the possibility of characterizing well-foundedness in terms of sequences instead of subsets:

Theorem 9.8: (*DC*): R is well-founded iff there is no function f defined on ω such that $\forall n \in \omega\ [f(n+1)Rf(n)]$.

Proof: If $\forall n \in \omega\ [f(n+1)R\ f(n)]$ than $\{f(n) \mid n \in \omega\}$ has no R-minimal element; conversely if R is not well-founded there exists a non-empty A failing an R-minimal element, that is, $\forall a \in A\ \exists\ b \in A\ (bRa)$.
DC now accounts for the sequence required. □

The version of well-foundedness, introduced by 9.8 says that *there are no infinite descending sequences*.

Several weakenings of the axiom of choice are obtained by restricting the cardinality of the sets involved. Since finite sets have choice functions anyway (no need for *AC*) one of the weakest non-trivial variants is the *axiom of countable choice CC* stating that countable sets do have choice functions.

Theorem 9.9: $DC \to CC$.
Proof: Assume $h: \omega \to A$ is bijective. Without loss of generality we may suppose that $\emptyset \notin A$ and $a \cap b = \emptyset$ whenever $a,b \in A$ and $a \neq b$. (Otherwise we consider $A' := \{\{a\} \times a | a \in A\}$.)
Define $R := \bigcup_{n \in \omega} (h(n) \times h(n+1))$; *DC* supplies us with a function f for which $a \cap \text{Ran}\ (f)$ consists of one element when $a \in A$ with the exception of at most finitely many $a \in A$. But we do not need *DC* to choose finitely many times (cf. exercise 1). □

Theorem 9.10: (*CC*): $A \underset{1}{\leqslant} \omega$ or $\omega \underset{1}{\leqslant} A$.

Proof: If A cannot be embedded into ω it cannot be finite, therefore it has n-element subsets for every $n \in \omega$ (use induction on n). Define $U := \{\{B \subseteq A | B \underset{1}{=} n\} | n \in \omega\}$ then $U \underset{1}{=} \omega$ hence (*CC*) it has a choice function.
Transform this into a function $f: \omega \to \mathcal{P}(A)$ with $f(n) \underset{1}{=} n$.

The sets $f(n) - \bigcup_{m<n} f(m)$ $(n \in \omega)$ form a countable set of pairwise disjoint sets; applying *CC* once again we find $\omega \leqslant_1 A$, since there are infinitely many non-empty sets among them. □

Dedekind called a set A *infinite* if it had a $B \subseteq A$, $B \neq A$, such that $B =_1 A$. In such a case we call A *Dedekind-infinite*. If $\omega \leqslant_1 A$ A is Dedekind-infinite, as A inherits this property from ω. Conversely, if A is Dedekind-infinite, say, $f\colon A \to A$ is an injection and $a \in A - \mathrm{Ran}(f)$, then a, $f(a)$, $f(f(a))$, ... are all different elements of A and hence $\omega \leqslant_1 A$.

Dedekind-infinite sets certainly are infinite in the sense of not being of equal power as some natural number. Using 9.10 we can show that *CC* implies the converse: if A is infinite then we cannot have $A <_1 \omega$. From 9.10 it follows that in that case, $\omega \leqslant_1 A$, which according to the previous remarks means that A is Dedekind-infinite.

Large parts of analysis become impossible without something like *DC* or *CC*. Hence it is not too surprising that the axiom of choice was in use long before 1904 without being noticed. It was Zermelo's ingenious construction of an actual well-ordering out of an innocent-looking choice function that roused feelings of distrust against the axiom of choice.

Gödel, in 1938, established a certain harmlessness for *AC*, at least in a formal sense, by proving that if the ZF-axioms are consistent (free from contradiction) they remain so after the addition of *AC* (and, by the way, the generalized continuum hypothesis) (cf [Gödel, 1940], [Cohen, 1966]). Though consistency and truth are different things, nowadays most people do not doubt *AC* anymore; there is only one more or less serious set theoretic proposition known which contradicts it, the so-called *axiom of determinateness AD* (cf. chapter III 8).

Fraenkel in 1922 proved that *AC* cannot be derived from the ZF-axioms if one weakens the axiom of extensionality in order to admit objects called *individuals* (or urelements) - elementless objects different from the empty set. The method has been polished by Mostowski (in 1939). (Instead of weakening extensionality one can also weaken regularity and let sets a for which $a = \{a\}$ take over the rôle of the individuals.) It was only in 1963, that Cohen proved the underivability of *AC*, *DC* and *CC* from the full ZF-axiomatics.

To be specific, we have ZF $\nvdash$ CC $\not\to$ DC $\not\to$ AC $\not\to$ GCH.

<u>Exercises</u>

1 Prove that finite sets have choice functions, without using *AC*. (Hint: use induction on the number of elements.)

2 Define $A \leqslant_s B$:= $A = \emptyset$ or there exists a surjection of B onto A. Prove:

(*i*) $A \leqslant_s B \to \mathcal{P}(A) \leqslant_1 \mathcal{P}(B)$;

(*ii*) $A \leqslant_1 B \to A \leqslant_s B$;

(*iii*) (*AC*) $A \leqslant_s B \to A \leqslant_1 B$.

3 $A \nless_1 \omega \to \omega <_1 \mathcal{P}(\mathcal{P}(A))$.

4 Let f and g be choice functions for $\mathcal{P}(A)$ and $\mathcal{P}(B)$ respectively, and G the collection of all $\Gamma \subseteq \mathcal{P}(A \times B)$ for which $\emptyset \in \Gamma$; $\Gamma' \subseteq \Gamma \to \bigcup\Gamma' \in \Gamma$ and if $h \in \Gamma$, Dom $h \neq A$, Ran $h \neq B$ then $h^+ := h \cup \{< f(A-\text{Dom } h), g(B - \text{Ran } h)>\} \in \Gamma$. Define $\Gamma_0 := \bigcap G$ and prove that Γ_0 is an injection either embedding A into B or mapping part of A onto the whole of B (thereby giving another proof for 9.3.)

5 Give proofs of *AC*, *AC*1-3, the well-ordering theorem and trichotomy, using Zorn's lemma.

6 The following assertions are equivalent:

(*i*) *AC*.

(*ii*) Every partial order $\langle A,R \rangle$ has a chain $K \subseteq A$ containing all of its upperbounds.

(*iii*) (Hansdorff's maximum-principle). Every partial order $\langle A,R \rangle$ has a chain $K \subseteq A$ not being a proper part of other chains.

(*iv*) Every partial order $\langle A,R \rangle$ has a chain $K \subseteq A$ well-ordered by R and containing all of its upperbounds.

7 Define the relation $\prec$ on $\mathcal{P}(\alpha)$ ($\alpha \in$ OR) by $A \prec B$:= $A \neq B$ and $\bigcap((A - B) \cup (B - A)) \in B$.

Prove that $\prec$ is a total ordering of $\mathcal{P}(\alpha)$. If $\omega \leqslant \alpha$ then every subset of $\mathcal{P}(\alpha)$ which is well-ordered by $\prec$ has power at most α.

8 S is called a *selector* for A iff for all $a \in A$, $a \cap S$ is a one-element set. *AC 4* states that *partitions* have selectors. Prove (*AC*) that a set A of finite sets has a selector iff every finite subset of A has one.

9 (AC) If $\langle A,R \rangle$ is a partial order there exists a connected ordering S of A with $R \subseteq S$ (the *order extension principle*.)

10 AC is equivalent with the *multiplicative principle* (Russell): if for all $x \in \mathrm{Dom}(f)$, $f(x) \neq \emptyset$ then $\underset{x \in Dom(f)}{\mathsf{X}} f(x) \neq \emptyset$.

11 $Q \subset \mathcal{P}(A)$ is said to have *finite character* over A iff a subset X of A is member of Q iff every finite subset of X is. Prove the equivalence of AC with the *Lemma of Teichmüller and Tukey*: if Q has finite character over A and is non-empty, it has a maximal element (i.e., not properly included in some other element of Q).

10 CARDINAL NUMBERS

Natural numbers are usually cast for two different rôles, namely that of order type of a well-ordering (derived from the process of counting) and that of measure of quantity. The first notion generalizes to that of an *ordinal*, the second to that of a *cardinal*. In the realm of the finite these two notions are extensionally the same. For infinite sets, however, they behave differently. Intuitively the cardinal of a set A is something A has in common exactly with the sets equivalent with A, i.e., the sets, the elements of which can be brought in a one to one-correspondence with the elements of A. If we denote this cardinal by $|A|$ then clearly the following condition should be satisfied:

Criterion of adequacy: $|A| = |B| \leftrightarrow A \underset{I}{=} B$. [*]

Since $\underset{I}{=}$ is a "relation" of equivalence, the mathematically simple-minded solution would be to define $|A|$ as the collection of sets equivalent with A -and this has been done in fact by Frege and Russell. The problem, however, arises that by this definition, $|A|$ never will be a set- unless $A = \emptyset$ (cf. I.19.3).

Nevertheless one can keep close to the idea of equivalence-classes by taking as a

<u>Definition 10.1</u>: $|A| := \mathrm{Bottom}(\{C | C \underset{I}{=} A\})$.

The bottom-operation was introduced in 4.9 (cf. also exercise 4.11). 10.1 has been proposed by Scott. According to 4.10(i) we then have $|A| \in V$. Also, [*] is satisfied: if $A \underset{I}{=} B$ then obviously $\{C | C \underset{I}{=} A\} = \{C | C \underset{I}{=} B\}$ hence

$|A| = |B|$. Conversely, if $|A| = |B|$ then since $A \in \{C | C =_I A\}$ by 4.10(*ii*) $|A| \neq \emptyset$, for instance, $C \in |A|$. From 4.10 (*ii*) $C =_I A$. Since $|A| = |B|$, we get $C \in |B|$, hence $C =_I B$. But then $A =_I B$.
The bottom-operation exists thanks to the regularity-axiom, hence the possibility to save the Frege-Russell idea via 10.1 also depends on this axiom.
If one admits the axiom of choice there is an alternative: then every set is equivalent with an ordinal and we could choose one element of $\{C | C =_I A\}$ in a canonical way by defining $|A| := \bigcap\{\alpha | \alpha =_I A\}$.
As we do not want to commit ourselves to *AC* we nevertheless prefer 10.1.
In the absence of both axioms of regularity and choice it is (provably) impossible to find a definition of cardinality adequate in the sense of [*] and there is no other way left than postulating the existence of a cardinality-operation satisfying [*] as an axiom.

For the rest it is quite irrelevant which direction is taken (and to ask what cardinals "really are") because the use of cardinals can be eliminated by simply translating statements concerning cardinals into statements about sets and bijections.
We shall never use any properties of $|A|$ than those given by [*].

These preliminaries out of the way, we shall base the future development on 10.1. $|A|$ is called the *cardinal* or the *power* of A and a *cardinal* is a set of the form $|A|$.

<u>Definition 10.2</u>: $\aleph_\alpha := |\omega_\alpha|$ $(\alpha \in \mathrm{OR})$.

An *aleph* is the cardinal of an initial number or, equivalently, the cardinal of an infinite well-ordered set. If the axiom of choice holds every cardinal is either finite (i.e., cardinal of a finite set) or an aleph.
We shall use the letters $\mathfrak{m}$, $\mathfrak{n}$, $\mathfrak{p}$, $\mathfrak{q}$, ... for cardinals.

<u>Definition 10.3</u>: (*i*) $\mathfrak{m} \leqslant \mathfrak{n} := \exists A \exists B\, [A \subseteq B \wedge |A| = \mathfrak{m} \wedge |B| = \mathfrak{n}]$;
(*ii*) $\mathfrak{m} < \mathfrak{n} := \mathfrak{m} \leqslant \mathfrak{n} \wedge \mathfrak{m} \neq \mathfrak{n}$.

<u>Theorem 10.4</u>: (*i*) $|A| \leqslant |B| \leftrightarrow A \leqslant_I B$; (*iii*) $\mathfrak{m} \leqslant \mathfrak{n} \wedge \mathfrak{n} \leqslant \mathfrak{p} \rightarrow \mathfrak{m} \leqslant \mathfrak{p}$;
(*ii*) $\mathfrak{m} \leqslant \mathfrak{m}$; (*iv*) $\mathfrak{m} \leqslant \mathfrak{n} \wedge \mathfrak{n} \leqslant \mathfrak{m} \rightarrow \mathfrak{m} = \mathfrak{n}$.

Proof: Trivial. (*iv*) follows from the Cantor-Bernstein-theorem in chapter I. □

So $\leqslant$ partially orders the cardinals. The assertion that $\leqslant$ is connected clearly amounts to the trichotomy-theorem 9.3 and hence to the axiom of choice; this in turn implies that $<$ is well-founded. We now introduce the operations of sum, product and exponentiation for cardinals; it is easy to show that these extend the corresponding arithmetical operations identifying natural numbers and their cardinals.

Definition 10.5: $\mathfrak{m}+\mathfrak{n}=\mathfrak{p}$ if and only if there exist disjoint sets A and B of cardinality $\mathfrak{m}$ and $\mathfrak{n}$ respectively such that $\mathfrak{p} = |A \cup B|$.

Notice that $\mathfrak{m} + \mathfrak{n}$ does not depend on the choice of the representatives A and B: if $|A'| = |A|$, $|B'| = |B|$ and $A' \cap B' = \emptyset$ then $|A' \cup B'| = |A \cup B|$. Also, disjoint representatives always can be found:

$|A \times \{0\}| = |A|$, $|B \times \{1\}| = |B|$ and $A \times \{0\} \cap B \times \{1\} = \emptyset$. Since $\cup$ satisfies the commutative and associative laws, the same holds for cardinal addition, cf. 10.8.

Definition 10.6: (AC): $\sum_{i\in I} \mathfrak{m}_i = \mathfrak{p}$ if there exists a function f defined on I for which $|f(i)| = \mathfrak{m}_i$ $(i \in I)$ and $\mathfrak{p} = |\bigcup_{i\in I} (f(i) \times \{i\})|$.

Transfinite sums, as defined in 10.6, in general may be used only in contexts assuming the axiom of choice. Both the existence of the function f in 10.6 and the independence of $\mathfrak{p}$ from the actual choice of f require AC:

Existence (AC): Assume $\mathfrak{m}$ to be a function associating a cardinal $\mathfrak{m}_i = \mathfrak{m}(i)$ to every $i \in I$ then there exists a function f defined on I such that for all $i \in I$, $|f(i)| = \mathfrak{m}_i$.

Proof: $f(i) := \bigcap \{\alpha \in OR \mid |\alpha| = \mathfrak{m}_i\}$. □

Often the $\mathfrak{m}_i$ are given in such a way that a function f, as required, may be found without the help of AC (for instance, the $\mathfrak{m}_i$ may be alephs); but even then we shall be in need of the

Uniqueness (AC): If f and g are functions defined on I such that for all $i \in I$,

$|f(i)| = |g(i)|$ then $|\bigcup_{i \in I} (f(i) \times \{i\})| = |\bigcup_{i \in I} (g(i) \times \{i\})|$.

Proof: Choose a bijection h_i between $f(i)$ and $g(i)$ for every $i \in I$. (Note that $h_i \in \mathcal{P}(\bigcup_{i \in I} (f(i) \times g(i)))$ hence a choice function for $\mathcal{PP}(\bigcup_{i \in I} (f(i) \times g(i)))$ is required.) Then the function h, defined by $h(\langle a,i \rangle) := \langle h_i(a), i \rangle$, is the required bijection. □

The last proof is simple enough. That *AC* is really needed here follows from the next example. In the second proof of 10.18 only transfinite sums of alephs are considered and nothing more of *AC* is demanded there than the fact that transfinite sums be uniquely defined. Nevertheless the result of 10.18 is known to be improvable without the use of the axiom of choice (cf. [Jech, 1973], p. 142)

Definition 10.7: $\mathfrak{m} \cdot \mathfrak{n} = \mathfrak{p}$ if sets A and B exist of cardinality $\mathfrak{m}$ and $\mathfrak{n}$ respectively for which $\mathfrak{p} = |A \times B|$.

As with cardinal sums, the cardinal product always exists and is uniquely determined. 10.8 lists some properties of these operations:

Theorem 10.8: (*i*) $\mathfrak{m} + \mathfrak{n} = \mathfrak{n} + \mathfrak{m}$; $\mathfrak{m} \cdot \mathfrak{n} = \mathfrak{n} \cdot \mathfrak{m}$;

(*ii*) $\mathfrak{m} + (\mathfrak{n} + \mathfrak{p}) = (\mathfrak{m} + \mathfrak{n}) + \mathfrak{p}$; $\mathfrak{m} \cdot (\mathfrak{n} \cdot \mathfrak{p}) = (\mathfrak{m} \cdot \mathfrak{n}) \cdot \mathfrak{p}$;

(*iii*) $\mathfrak{m} \leqslant \mathfrak{n} \rightarrow \mathfrak{m} + \mathfrak{p} \leqslant \mathfrak{n} + \mathfrak{p} \wedge \mathfrak{m} \cdot \mathfrak{p} \leqslant \mathfrak{n} \cdot \mathfrak{p}$;

(*iv*) $\mathfrak{m} \cdot (\mathfrak{n} + \mathfrak{p}) = \mathfrak{m} \cdot \mathfrak{n} + \mathfrak{m} \cdot \mathfrak{p}$ and

(*AC*) $\mathfrak{m} \cdot \sum_{i \in I} \mathfrak{n}_i = \sum_{i \in I} (\mathfrak{m} \cdot \mathfrak{n}_i)$;

(*v*) (*AC*) $\forall i \in I (\mathfrak{m}_i = \mathfrak{m}) \rightarrow \sum_{i \in I} \mathfrak{m}_i = |I| \cdot \mathfrak{m}$.

Besides 10.8(*iii*) there also exist monotonicity-properties for transfinite sums; we leave the formulation to the reader. Addition and multiplication of *alephs* is a trivial business due to the absorption-property:

Theorem 10.9: $\aleph_\alpha + \aleph_\beta = \aleph_\alpha \cdot \aleph_\beta = \aleph_{\alpha \cup \beta}$.

Proof: If for instance $\beta \leqslant \alpha$ then according to 8.13

$\omega_\alpha \leqslant_1 \omega_\alpha \times \{0\} \cup \omega_\beta \times \{1\} \subseteq \omega_\alpha \times \omega_\beta \subseteq \omega_\alpha \times \omega_\alpha =_1 \omega_\alpha$. □

In particular, if the axiom of choice is valid then $\mathfrak{m}^2 = \mathfrak{m}$ for every infinite

cardinal. It is well worth noticing that, conversely, this assertion implies the axiom of choice as do a number of other consequences of *AC* in cardinal arithmetic (cf. in general [Rubin and Rubin 1963]).
We now present a result which although seemingly trivial cannot be proved without the axiom of choice:

Lemma 10.10 (*AC*): $\bigcup_{i \in I} A_i \leqslant_I \bigcup_{i \in I} (A_i \times \{i\})$.

Proof: The map $h(<a,i>) := a$ is a surjection from $\bigcup_{i \in I} (A_i \times \{i\})$ onto $\bigcup_{i \in I} A_i$. Now apply *AC* 3. □

Corollary 10.11 (*AC*): $|\bigcup_{i \in I} A_i| \leqslant \sum_{i \in I} |A_i|$.

Proof: Immediate from 10.10. □

Note that this is an inflated version of I.17.9.

Lemma 10.12: If $\emptyset \neq a \subseteq$ OR then $\bigcup_{\xi \in a} (\omega_\xi \times \{\xi\}) =_I \omega_{\bigcup a}$.

Proof: Let $\alpha := \bigcup a$. Obviously, $\bigcup_{\xi \in a} (\omega_\xi \times \{\xi\}) \subseteq \omega_a \times (\omega_a + 1) =_I \omega_a$. Conversely, the map h defined by $h(\delta) := <\delta, \bigcap\{\xi \in a | \delta \in \omega_\xi\}>$ injects ω_a into $\bigcup_{\xi \in a} (\omega_\xi \times \{\xi\})$. □

The *least upper bound* of a set of cardinals is the least cardinal greater than or equal to every cardinal in the set; it exists if the axiom of choice holds.

Corollary 10.13 (*AC*): The cardinal sum of a set of cardinals equals the least upper bound of the set.

Proof: Immediate from 10.12. □

We now introduce the concepts of regularity and cofinality for cardinals. In the absence of the axiom of choice, we can define cofinality only for alephs:

Definition 10.14: $\mathrm{cf}(\aleph_a)$ is the smallest cardinal $\mathfrak{m}$ such that a partition of ω_a into $\mathfrak{m}$ parts of power $< \aleph_a$ exists.

Notice that $\mathrm{cf}(\aleph_a)$ is a well-defined cardinal $\leqslant \aleph_a$ and hence an aleph. For if

P partitions ω_a, the map associating with $p \in P$ its smallest element injects P into ω_a; hence P is well-ordered and $|P| \leqslant \aleph_a$. The link with ordinal-cofinality (def. 6.1) is given by the

Theorem 10.15: $\mathrm{cf}(\aleph_a) = |\mathrm{cf}(\omega_a)|$.

Proof: Assume that h maps $\mathrm{cf}(\omega_a)$ onto a cofinal subset of ω_a. Then $\{h(\xi) \times \{\xi\} \mid \xi < \mathrm{cf}(\omega_a)\}$ partitions a set of power $\aleph_a$ into smaller pieces and therefore $\mathrm{cf}(\aleph_a) \leqslant |\mathrm{cf}(\omega_a)|$.
For the converse, if $\mathrm{cf}(\aleph_a) = \aleph_a$, then $|\mathrm{cf}(\omega_a)| \leqslant |\omega_a| = \aleph_a = \mathrm{cf}(\aleph_a)$ and we are finished. So assume $\mathrm{cf}(\aleph_a) < \aleph_a$ and let P partition ω_a into $\mathrm{cf}(\aleph_a)$ pieces of power $< \aleph_a$. Let $\bar{p}$ be the order-type of $p \in P$. As $|p| < \aleph_a$, $\bar{p}$ is an ordinal in ω_a.
$\bar{P} := \{\bar{p} \mid p \in P\}$ cannot be bounded in ω_a: if $\bar{p} \subseteq \gamma < \omega_a$, we can associate with each $\delta < \omega_a$ the pair $\langle p_\delta, \xi_\delta \rangle$ consisting of the element $p_\delta \in P$ for which $\delta \in p_\delta$ and the ordinal $\xi_\delta < \gamma$ which is the image of δ under the order-isomorphism between p_δ and $\bar{p}_\delta$. This correspondence injects ω_a into $P \times \gamma$; but then $\aleph_a \leqslant_1 |P \times \gamma| = |P|.|\gamma| < \aleph_a$ - contradiction. Thus, $\bar{P}$ is cofinal in ω_a. As we have noticed before, P can be well-ordered and hence $\bar{P} \leqslant_1 P$. Also, $|\mathrm{cf}(\omega_a)| \leqslant |\bar{P}|$. Thus, $|\mathrm{cf}(\omega_a)| \leqslant |\bar{P}| \leqslant |P| = \mathrm{cf}(\aleph_a)$. □

Definition 10.16: $\mathfrak{m}$ is called *regular* if there is no way of partitioning a set of power $\mathfrak{m}$ into less than $\mathfrak{m}$ pieces of smaller power.

Correlating this to ordinal-regularity (6.2) we get

Corollary 10.17: $\aleph_a$ is regular just in case ω_a is.

Proof: Note that $\aleph_a$ is regular iff $\mathrm{cf}(\aleph_a) = \aleph_a$. By 10.15, this amounts to $|\mathrm{cf}(\omega_a)| = \aleph_a$; and this in turn clearly is equivalent to $\mathrm{cf}(\omega_a) = \omega_a$ which by definition means that ω_a is regular. □

Assuming the axiom of choice, $\mathrm{cf}(\mathfrak{m})$ is defined for every infinite cardinal; by definition it is the smallest cardinal $\mathfrak{n}$ for which a representation $\mathfrak{m} = \sum_{i \in I} \mathfrak{m}_i$ exists satisfying $|I| = \mathfrak{n}$ and $\mathfrak{m}_i < \mathfrak{m}$ for all $i \in I$.

As we have noticed shortly after 6.2 it may be the case that, in the absence of the axiom of choice, $\aleph_0$ is the only regular cardinal; but assuming *AC* lots of cardinals are regular :

Theorem 10.18 (*AC*): $\aleph_{\alpha+1}$ is regular.

Proof: We shall use 10.17. Suppose that there exists an $a \subseteq \omega_{\alpha+1}$ such that a is cofinal in $\omega_{\alpha+1}$ and has order-type $< \omega_{\alpha+1}$ and hence $|a| \leqslant \aleph_{\alpha}$. As every ordinal $\xi \in a$ has power $\leqslant \aleph_{\alpha}$ we can assume (*AC*) the existence of injections $h_{\xi} : \xi \to \omega_{\alpha}$ for each $\xi \in a$. Now the assignment of the pair $<\xi, h_{\xi}(\delta)>$ to the ordinal $\delta \in \omega_{\alpha+1}$ where $\xi := \bigcap\{\eta \in a | \delta \in \eta\}$ clearly injects $\omega_{\alpha+1}$ into $a \times \omega_{\alpha}$. As this set has power $\leqslant \aleph_{\alpha}$ a contradiction is obtained. □

A simpler proof of 10.18 in which the role of *AC* is less clearly visible is the following

2nd proof: Assume $|I| < \aleph_{\alpha+1}$ and $\mathfrak{m}_i < \aleph_{\alpha+1}$ $(i \in I)$. Then $|I| \leqslant \aleph_{\alpha}$, $\mathfrak{m}_i \leqslant \aleph_{\alpha}$ $(i \in I)$ and $\sum_{i \in I} \mathfrak{m}_i \leqslant \sum_{i \in I} \aleph_{\alpha} = \aleph_{\alpha} \cdot |I| \leqslant \aleph_{\alpha} \cdot \aleph_{\alpha} = \aleph_{\alpha} < \aleph_{\alpha+1}$; therefore $\mathrm{cf}(\aleph_{\alpha+1}) = \aleph_{\alpha+1}$. □

The concepts of cofinality and regularity are important in connection with transfinite products and exponentiation of cardinals which we now introduce:

Definition 10.19 (*AC*): $\prod_{i \in I} \mathfrak{m}_i = \mathfrak{m}$ if there exists a function f defined on I such that for all $i \in I$ $|f(i)| = \mathfrak{m}_i$ and $\mathfrak{m} = |\mathop{\mathsf{X}}_{i \in I} f(i)|$.

Just as transfinite sums, transfinite products can be used only in contexts presupposing the axiom of choice: only under this condition existence and uniqueness of the product are guaranteed. cf. also exercise 10 of §9.

Cardinal exponentiation is a special case of the transfinite product:

Definition 10.20: $\mathfrak{m}^{\mathfrak{n}} = \mathfrak{p}$ if there exist sets A and B with cardinality $\mathfrak{m}$ and $\mathfrak{n}$ respectively for which $\mathfrak{p} = |A^B|$.

That exponentiation is uniquely defined is proved as in the case of sum and product. The next theorem lists some fundamental properties:

Theorem 10.21: (*i*) (*AC*) $\prod_{i \in I} \mathfrak{m} = \mathfrak{m}^{|I|}$;

(*ii*) $2^{|A|} = |\wp(A)|$;

(*iii*) $\mathfrak{m}^{\mathfrak{n}+\mathfrak{p}} = \mathfrak{m}^{\mathfrak{n}} \mathfrak{m}^{\mathfrak{p}}$; in general (*AC*) $\mathfrak{m}^{\sum_{i \in I} \mathfrak{n}_i} = \prod_{i \in I} \mathfrak{m}^{\mathfrak{n}_i}$;

$(iv)\quad (m^n)^p = m^{n.p}$;

$(v)\quad (m.n)^p = m^p.n^p$; in general

$(AC)\quad (\prod_{i\in I} m_i)^p = \prod_{i\in I} m_i^p$.

Proof: Exercise 4. - As to (ii) notice that we identify the natural numbers with their cardinals: $2 = |2|$ etc. □

Formulation of the obvious monotonicity-laws is left to the reader.

The following law, *König's inequality*, is rather fundamental:

Theorem 10.22 (AC): $\forall i \in I\ (m_i < n_i) \rightarrow \sum_{i\in I} m_i < \prod_{i\in I} n_i$.

Proof: Assume that $|A_i| = m_i$, $|B_i| = n_i$ and $h: \bigcup_{i\in I}(A_i \times \{i\}) \rightarrow \mathsf{X}_{i\in I} B_i$. It suffices to show that h cannot be surjective (leaving it to the reader to verify that it can be injective). Since $A_i \underset{1}{<} B_i$ $f \in \mathsf{X}_{i\in I} B_i$ exists $(AC!)$ such that $f(i) \in B_i - \{h(\langle a,i\rangle)(i)\,|\,a \in A_i\}$; clearly $f \notin \mathrm{Ran}(h)$. □

König's inequality in the cardinal-free version-if $A_i \underset{1}{<} B_i$ $(i \in I)$ then $\bigcup_{i\in I} A_i \underset{1}{<} \mathsf{X}_{i\in I} B_i$ - implies Cantor's theorem (take $|A_i| = 1$, $|B_i| = 2$) as well as the axiom of choice (take $|A_i| = 0$ and use exercise 9.10).

One of the most important corollaries of 10.22 is the

Theorem 10.23 (AC) : $\aleph_a < \mathrm{cf}(2^{\aleph_a})$.

Proof: Assume we had, nevertheless, a partition P of $\mathcal{P}(\omega_a)$ (cf. 10.21 (ii)) of cardinality $\aleph_a$ in pieces of power less then $2^{\aleph_a}$. Then

$|\mathcal{P}(\omega_a)| = |\bigcup P| = \sum_{p\in P} |p| < \prod_{p\in P} 2^{\aleph_a} = (2^{\aleph_\alpha})^{\aleph_\alpha} = 2^{\aleph_\alpha}$. □

The importance of 10.23 is due to the fact that the set-theoretic axioms introduced up to now really do not give any information about the position of $2^{\aleph_a}$ among the alephs except that $\aleph_a < \mathrm{cf}(2^{\aleph_a})$. (Of course we also know that $\aleph_a < 2^{\aleph_a}$ - but this is being implied by 10.23.) For instance, it is possible to assume without getting into inconsistencies that $2^{\aleph_0} = \aleph_1$ (the so-called *continuum-hypothesis*); but also that $2^{\aleph_0} = \aleph_2$, or $2^{\aleph_0} = \aleph_{100}$, or $2^{\aleph_0} = \aleph_{\omega_1}$. On the other hand, $2^{\aleph_0} \neq \aleph_\omega$ as $\mathrm{cf}(\aleph_\omega) = \aleph_0$. Much the same holds for powers of other *regular* cardinals.

Possible forms of the operation τ for which $2^{\aleph_a} = \aleph_{\tau(a)}$ were studied by Easton.

In more or less unprecise terms, his result states that if τ is such that (i) $\alpha \leqslant \beta \to \tau(\alpha) \leqslant \tau(\beta)$ (which is an obvious requirement) and (ii) $cf(\aleph_{\tau(\alpha)}) > \aleph_\alpha$ (this is 10.23) then we can consistently assume $2^{\aleph_\alpha} = \aleph_{\tau(\alpha)}$ for all *regular* $\aleph_\alpha$.

The *singular cardinal problem* asks whether this can be extended to hold not only for regular, but also for *singular* (= non-regular) cardinals also; this question is yet partly unsolved. Obviously, powers of singular cardinals sometimes depend on others. If, for instance, $2^{\aleph_n} = \aleph_{\omega+1}$ for all $n < \omega$ (a possibility not to be excluded by Easton's result) we obtain

$$2^{\aleph_\omega} = 2^{\sum_{n\in\omega}\aleph_n} = \prod_{n\in\omega} 2^{\aleph_n} = \prod_{n\in\omega} 2^{\aleph_0} = (2^{\aleph_0})^{\aleph_0} = 2^{\aleph_0} = \aleph_{\omega+1} .$$

A few years ago, Jensen (1974) proved that *all* values $\tau(\alpha)$ for $\aleph_\alpha$ singular depend on the behaviour of τ on the indices of regular alephs - that is, as long as a certain strong consequence of the existence of a very large cardinal (such as a measurable one, cf. §11) happens to be false. And a bit earlier Silver (1974) proved some results outright without such an assumption; see 10.27 for an example. (The proof is taken from [Baumgartner-Prikry 1977] which also has a nice general discussion and indicates generalisations.) In case the *generalized continuum-hypothesis* holds (i.e., $\tau(\alpha) = \alpha+1$ for all α), we can evaluate all powers $\aleph_\alpha^{\aleph_\beta}$, cf. 10.26.

But first, we present two lemmas, which enable us to reduce powers to powers of smaller basis.

Lemma 10.24 (AC): $\aleph_\beta \leqslant \aleph_\alpha \to \aleph_{\alpha+1}^{\aleph_\beta} = \aleph_\alpha^{\aleph_\beta}.\aleph_{\alpha+1}$.

Proof: Since $\omega_{\alpha+1}$ is regular and $\omega_\beta \leqslant \omega_\alpha$, every function f: $\omega_\beta \to \omega_{\alpha+1}$ has a bounded range in $\omega_{\alpha+1}$; therefore $\omega_{\alpha+1}^{\omega_\beta} = \bigcup_{\gamma<\omega_{\alpha+1}} \gamma^{\omega_\beta}$. So

$$\aleph_{\alpha+1}^{\aleph_\beta} = |\omega_{\alpha+1}^{\omega_\beta}| = |\bigcup_{\gamma<\omega_{\alpha+1}} \gamma^{\omega_\beta}| \leqslant \sum_{\gamma<\omega_{\alpha+1}} |\gamma^{\omega_\beta}| =$$

$$= \sum_{\gamma<\omega_{\alpha+1}} |\gamma|^{\aleph_\beta} \leqslant \sum_{\gamma<\omega_{\alpha+1}} \aleph_\alpha^{\aleph_\beta} = \aleph_\alpha^{\aleph_\beta}.\aleph_{\alpha+1} .$$

On the other hand, both $\aleph_\alpha^{\aleph_\beta}$ and $\aleph_{\alpha+1}$ are at most $\aleph_{\alpha+1}^{\aleph_\beta}$, hence $\aleph_\alpha^{\aleph_\beta}.\aleph_{\alpha+1} \leqslant \aleph_{\alpha+1}^{\aleph_\beta}$. □

Lemma 10.25 (AC): If Lim α and $\aleph_\beta < cf(\aleph_\alpha)$ then $\aleph_\alpha^{\aleph_\beta} = \sum_{\gamma<\alpha} \aleph_\gamma^{\aleph_\beta}$.

Proof: From the hypotheses it follows that $\omega_\alpha^{\omega_\beta} = \bigcup_{\gamma<\alpha} \omega_\gamma^{\omega_\beta}$ since $\omega_\alpha = \bigcup_{\gamma<\alpha} \omega_\gamma$.
Therefore

$$\aleph_\alpha^{\aleph_\beta} = |\omega_\alpha^{\omega_\beta}| = |\bigcup_{\gamma<\alpha} \omega_\gamma^{\omega_\beta}| \leqslant \sum_{\gamma<\alpha} |\omega_\gamma^{\omega_\beta}| = \sum_{\gamma<\alpha} \aleph_\gamma^{\aleph_\beta} \leqslant \sum_{\gamma<\alpha} \aleph_\alpha^{\aleph_\beta} = \aleph_\alpha^{\aleph_\beta} \cdot |\alpha| = \aleph_\alpha^{\aleph_\beta}. \qquad \square$$

Finally the next theorem completely answers the question how to evaluate $\aleph_\alpha^{\aleph_\beta}$ if the generalised continuum-hypothesis (GCH) holds:

<u>Theorem 10.26</u> (GCH):

(*i*) if $\aleph_\beta < \mathrm{cf}\,(\aleph_\alpha)$ then $\aleph_\alpha^{\aleph_\beta} = \aleph_\alpha$;

(*ii*) if $\mathrm{cf}\,(\aleph_\alpha) \leqslant \aleph_\beta < \aleph_\alpha$ then $\aleph_\alpha^{\aleph_\beta} = \aleph_{\alpha+1}$;

(*iii*) if $\aleph_\alpha \leqslant \aleph_\beta$ then $\aleph_\alpha^{\aleph_\beta} = \aleph_{\beta+1}$.

Proof: (*i*) If $\alpha = \gamma+1$ then $\mathrm{cf}\,(\aleph_\alpha) = \aleph_\alpha$ and
$\aleph_\alpha^{\aleph_\beta} = \aleph_{\gamma+1}^{\aleph_\beta} = (2^{\aleph_\gamma})^{\aleph_\beta} = 2^{\aleph_\gamma \cdot \aleph_\beta} = 2^{\aleph_\gamma} = \aleph_{\gamma+1} = \aleph_\alpha$; if Lim α then (by 10.25)
$\aleph_\alpha \leqslant \aleph_\alpha^{\aleph_\beta} = \sum_{\gamma<\alpha} \aleph_\gamma^{\aleph_\beta} \leqslant \sum_{\gamma<\alpha} (2^{\aleph_\gamma})^{\aleph_\beta} = \sum_{\gamma<\alpha} 2^{\aleph_\gamma \cdot \aleph_\beta} =$

$= \sum_{\gamma<\alpha} 2^{\aleph_{\gamma\cup\beta}} = \sum_{\gamma<\alpha} \aleph_{(\gamma\cup\beta)+1} \leqslant \sum_{\gamma<\alpha} \aleph_\alpha = \aleph_\alpha \cdot |\alpha| = \aleph_\alpha$.

(*ii*) Assume that $A \subseteq \omega_\alpha$ is cofinal in ω_α and has type $\mathrm{cf}\,(\omega_\alpha)$. Then
$|A| = |\mathrm{cf}(\omega_\alpha)| = \mathrm{cf}(\aleph_\alpha) \leqslant \aleph_\beta$ and $\aleph_\alpha = |\bigcup_{\gamma\in A} \gamma| \leqslant \sum_{\gamma\in A} |\gamma| < \prod_{\gamma\in A} \aleph_\alpha =$
$= \aleph_\alpha^{|A|} \leqslant \aleph_\alpha^{\aleph_\beta} \leqslant (2^{\aleph_\alpha})^{\aleph_\beta} = 2^{\aleph_\alpha \cdot \aleph_\beta} = 2^{\aleph_\alpha} = \aleph_{\alpha+1}$.
(*iii*) $\aleph_{\beta+1} = 2^{\aleph_\beta} \leqslant \aleph_\alpha^{\aleph_\beta} \leqslant (2^{\aleph_\alpha})^{\aleph_\beta} = 2^{\aleph_\alpha \cdot \aleph_\beta} = 2^{\aleph_\beta} = \aleph_{\beta+1}$. □

The reader may have gathered the impression that the generalised continuum hypothesis presupposes the theory of alephs since we have used the standard-formulation

GCH : $2^{\aleph_\alpha} = \aleph_{\alpha+1}$ (also called *aleph-hypothesis*).

But one may also give an "aleph-free" formulation:
GCH' : $\mathfrak{n} \leqslant \mathfrak{m} < 2^{\mathfrak{n}} \rightarrow \mathfrak{n} = \mathfrak{m}$,or even avoiding cardinals altogether
$A \underset{1}{\leqslant} B \underset{1}{<} \mathcal{P}(A) \rightarrow A \underset{1}{=} B$.

(Assuming the sets and cardinals occurring to be infinite.)
Clearly if we assume the axiom of choice, both formulations are equivalent. But the proof of corollary 9.6 cannot be used to show that GCH' implies *AC*. For a proof that it does, cf. [Cohen 1966, p. 148]; further information can be

obtained from [Drake (1974)] and [Fraenkel et al. (1973)].

We end this paragraph by presenting one remarkable case of Silver's theorem saying that the GCH holds at $\aleph_{\omega_1}$ as soon as it holds below $\aleph_{\omega_1}$. Though the proof given is the longest one up to now in this book, it still belongs to axiomatic set theory in the narrow sense - a subject in which new results really are quite rare.

Theorem 10.27 (*AC*): If, for all $\alpha < \omega_1$, $2^{\aleph_\alpha} = \aleph_{\alpha+1}$, then $2^{\aleph_{\omega_1}} = \aleph_{\omega_1+1}$.

Proof: The main step depends on notions and results from §6 which we review. As we assume *AC*, ω_1 is regular (10.18; 10.17). $A \subseteq \omega_1$ is *bounded* (in ω_1) if for some $\alpha < \omega_1$, $A \subseteq \alpha$; otherwise A is *cofinal*. $A \subseteq \omega_1$ is *closed* if limits of bounded sequences from A belong to A. Intersections of countable many closed cofinal sets are again closed cofinal (6.12) and the "diagonal intersection" $\{\alpha | \forall\beta < \alpha\ (\alpha \in A_\beta)\}$ of an ω_1-sequence A_β $(\beta < \omega_1)$ of closed cofinal sets also is closed cofinal (6.13 (*ii*)). Finally, $A \subseteq \omega_1$ is called *stationary* if it intersects every closed cofinal set.
We now start the proof. Hence assume $2^{\aleph_\alpha} = \aleph_{\alpha+1}$ for all $\alpha < \omega_1$. We then can enumerate $\mathcal{P}(\omega_\alpha)$ as $\{A^\alpha_\xi | \xi < \omega_{\alpha+1}\}$ $(\alpha < \omega_1)$. Now associate to every $A \subset \omega_{\omega_1}$ the function $f_A : \omega_1 \to \omega_{\omega_1}$ by putting $f_A(\alpha) = \xi$ if $A \cap \omega_\alpha = A^\alpha_\xi$ and define the relation R putting $R(A,B)$ if $\{\alpha | f_A(\alpha) < f_B(\alpha)\}$ is stationary.
Then R is *connected*, i.e., for all $A,B \subseteq \omega_{\omega_1}$: $R(A,B)$ or $R(B,A)$ or $A = B$. For consider the three sets $I = \{\alpha | f_A(\alpha) < f_B(\alpha)\}$, $\text{II} = \{\alpha | f_B(\alpha) < f_A(\alpha)\}$ and $\text{III} = \{\alpha | f_A(\alpha) = f_B(\alpha)\}$. At least one of them is stationary. (Otherwise pick closed cofinal sets X,Y,Z from with they are disjoint respectively. As $X \cap Y \cap Z$ is closed cofinal also it has an element α. But α has to be in one of I, II and III.) If this happens to be I or II we are done. And if III is stationary it is also cofinal (why?) but then $A = B$ (why?). As we have to prove $2^{\aleph_{\omega_1}} = \aleph_{\omega_1+1}$ we may as well assume $2^{\aleph_{\omega_1}} > \aleph_{\omega_1+1}$, thus heading for a contradiction. From this, by a cardinality-argument, it follows that some $B \subseteq \omega_{\omega_1}$ has at least $\aleph_{\omega_1+1}$ - many R-predecessors: if not, take any $X \subseteq \mathcal{P}(\omega_{\omega_1})$ of power $\aleph_{\omega_1+1}$, then $X \cup \bigcup\{\{B | R(B,A)\} | A \in X\}$ has power $\aleph_{\omega_1+1}$ and as $2^{\aleph_{\omega_1}} > \aleph_{\omega_1+1}$ we may pick $B \subseteq \omega_{\omega_1}$ outside this set. Since R is connected it follows that $X \subseteq \{A | R(A,B)\}$. Fix such a B.
If $\alpha < \omega_1$ then $f_B(\alpha) < \omega_{\alpha+1}$; hence a one-one map $g_\alpha : f_B(\alpha) \to \omega_\alpha$ can be choosen.

If $R(A,B)$ then by definition of R, $S_A := \{\alpha | f_A(\alpha) < f_B(\alpha)\}$ is stationary and $\alpha \in S_A \to g_a(f_A(\alpha)) < \omega_a$. We argue that for some $\gamma = \gamma_A < \omega_1$, $g_a(f_A(\alpha)) < \omega_\gamma$ for all α in a stationary subset $T = T_A$ of S_A. For suppose not. Let $\gamma < \omega_1$ be arbitrary. By assumption $\{\alpha \in S_A | g_a(f_A(\alpha)) < \omega_\gamma\}$ is not stationary, hence disjoint from some closed cofinal C_γ.

Form $C := \{\alpha | \forall \gamma < \alpha \ (\alpha \in C_\gamma)\}$, the diagonal intersection of the C_γ - this again is closed cofinal. Notice that the limit-ordinals below ω_1 form a closed cofinal set also; therefore a limit $\alpha \in C \cap S_A$ exists.

As $g_a(f_A(\alpha)) < \omega_a = \bigcup_{\gamma<a} \omega_\gamma$ we may pick $\gamma < \alpha$ such that $g_a(f_A(\alpha)) < \omega_\gamma$. By choice of C_γ, $\alpha \notin C_\gamma$. But as $\alpha \in C$, $\alpha \in C_\gamma$ for all $\gamma < \alpha$; contradiction. By assumption of 10.27 we see that $\{<T_A, \gamma_A> | R(A,B)\}$ has power at most $2^{\aleph_1} \cdot \aleph_1 = \aleph_2$; as $|\{A | R(A,B)\}| = \aleph_{\omega_1+1}$ is regular we find T and γ such that for $\aleph_{\omega_1+1}$ - many A with $R(A,B)$, $T_A = T$ and $\gamma_A = \gamma$.

The number of functions mapping T into ω_γ is $\aleph_\gamma^{\aleph_1} \leqslant 2^{\aleph_\gamma \cdot \aleph_1} \leqslant \aleph_{\gamma+2} < \aleph_{\omega_1+1}$; by the regularity of $\aleph_{\omega_1+1}$ we find h: $T \to \omega_\gamma$ such that for $\aleph_{\omega_1+1}$ - many A with $R(A,B)$, $T_A = T$ and $\gamma_A = \gamma$: $\forall \alpha \in T \ (g_a(f_A(\alpha)) = h(\alpha))$.

Pick two different ones, A_1 and A_2.

Then $\forall \alpha \in T \ (g_a(f_{A_1}(\alpha)) = g_a(f_{A_2}(\alpha)))$ and since the g_a are one-one, $\forall \alpha \in T(f_{A_1}(\alpha) = f_{A_2}(\alpha))$. As every stationary set is a fortiori cofinal this implies $A_1 = A_2$ as before - contradiction. □

Though the question what happens to $2^{\aleph_\omega}$ if $2^{\aleph_n} = \aleph_{n+1}$ for all n seems to be simpler, this still is unanswered. The method of 10.27 applies to cardinals of uncountable cofinality only as the notion of a closed cofinal subset of ω is not satisfactory.

Exercises

1 It has been maintained before 10.2 that the notion of a cardinal can be eliminated from theorems. Give some examples (as we did for 10.22.)

2 Prove 10.4.

3 Prove 10.8.

4 Prove 10.21.

5 (AC) If Lim α and $\xi < \delta < \alpha$ implies $\mathfrak{m}_\xi < \mathfrak{m}_\delta$ then $\sum_{\xi<a} \mathfrak{m}_\xi < \prod_{\xi<a} \mathfrak{m}_\xi$.

6 (AC) Prove that $\aleph_a < \aleph_a^{cf(\aleph_a)}$. (Distinguish between $\alpha = 0$, $\alpha = \beta + 1$ and Lim α and use 10.18, 10.22.) (Cf. 10.26 (ii)).

7 There are arbitrarily large $\mathfrak{m}$ for which $\mathfrak{m}^{\aleph_0} = \mathfrak{m}$ - just as there are (*AC*) arbitrarily large α for which $\aleph_\alpha^{\aleph_0} > \aleph_\alpha$.

8 $\aleph_\alpha$ is called a *strong limit* if $\beta < \alpha$ implies $2^{\aleph_\beta} < \aleph_\alpha$. Prove (*AC*) that arbitrarily large strong limits exist. (Hint: use 10.13 to show that if $\mathfrak{m}_{n+1} = 2^{\mathfrak{m}_n}$, $\sum_{n\in\omega} \mathfrak{m}_n$ is a strong limit.)

9 If $\aleph_0 \leqslant \mathfrak{m} \leqslant \aleph_\alpha$ then $\mathfrak{m}$ is an aleph.

10 *CC* implies that $\omega < \mathrm{cf}(\omega_{\alpha+1})$ (hence that ω_1 is regular.)

11 MODELS

In this section something is said about the functioning of a number of **ZF**-axioms using so-called models. To explain this notion, we shall make a short excursion to algebra, where models have been fruitfully used already for a long time.

Since most people are familiar with groups, let us consider group theory. A group is a structure of the form $\langle A, \cdot, ^{-1}, e\rangle$ where A is a non-empty set, $\cdot$ a binary operation on A, $^{-1}$ a unary operation on A and e an element of A such that certain conditions, the group-*axioms* are satisfied; one also expresses this by saying that $\langle A, \cdot, ^{-1}, e\rangle$ is a *model* of the group axioms.

In exactly the same way a model of the **ZF**-axioms (a **ZF**-*model*, for short) is a structure of the form $\langle A,E\rangle$ where $A \neq \emptyset$, $E \subseteq A \times A$ such that "the **ZF**-axioms are satisfied". By this is meant the following. If Φ is some assertion expressed in the **ZF**-formalism (for instance some **ZF**-axiom) we read '$\forall x$' ('$\exists x$') not any longer as: 'for every set x' ('there exists a set x for which ') but: 'for every element x of A' ('there exists an element x of A for which ') and instead of reading '$x \epsilon y$' as 'x is an element of y', we now read 'x is related by E to y' (i.e., $\langle x,y\rangle \in E$). In this way, Φ is translated in an assertion about the structure $\langle A,E\rangle$ which either is true or false; in the first case we say that $\langle A,E\rangle$ is a *model* of Φ. It often happens that E is the "real" ϵ-relation, i.e., $E = \epsilon_A = \{\langle a,b\rangle \in A \times A \mid a \;\epsilon\; b\}$; in that case we only need to re-interpret the quantifiers and hence $\langle A,\epsilon_A\rangle$ is a model of Φ just in case Φ^A holds (cf. §7.) We noticed for example in §2 that if A is transitive, $\langle A,\epsilon_A\rangle$ is a model of the axiom of extensionality.

Often people are surprised or even worried by the analogy between set theory on the one hand and, say, group theory on the other as far as models are concerned. Probably this is for the following reasons. In the first place the axiom systems of group theory and set theory respectively have completely different purposes: the first is meant to describe a great variety of groups; on the other hand the **ZF**-axioms are meant to describe only one thing namely, 'the set theoretic universe', 'Cantor's absolutum', 'the cumulative hierarchy' or whatever one wants to call this phantasy world of mathematicians; the fact that these axioms also admit models that we did not intend can be regarded merely as an unhappy coincidence, implying that the axioms do not fit their purpose completely (cf. non-standard models for elementary arithmetic.) In the second place there seems to be something funny in talking about models for set-theoretic axioms since on the one hand these axioms are meant to say something about what sets are while on the other hand models are defined in terms of sets. So if models are used to prove freedom of contradiction of the set theoretic axioms one has to be aware of the possibility of circular reasoning. In most cases, however, we shall take freedom of contradiction (consistency) of the basis-theory **ZF** for granted and use models to show that certain axioms cannot be derived from certain other axioms. This is the case if they do not hold in situations (models) where the other axioms are satisfied. To make our position precise: statements are interpreted in models by interpreting the basic relations and operations; and ϵ is a relation symbol as any other.

In the sequel the words 'model' and 'structure' are regarded as synonymous (as in customary in considerations of this kind); a structure in which a certain axiom is satisfied (in which this axiom *holds* or *is valid*) is called a *model of* that axiom.

Structures of type $\langle A, \epsilon_A \rangle$ are called *epsilon-structures* or *standard-models*; as a rule they are identified with their universe A as this determines the structure as a whole. In §2 we have seen that standard models of the extensionality-axiom are isomorphic to transitive epsilon structures since ϵ_A always is well-founded.

The model-theory of transitive epsilon-structures, at least where the interpretation of a number of important formulas is concerned, is very neat as 11.1 shows.

We use the following abbreviations:

$Z(u) := \forall x(x \notin u)$;

$S(u,v) := \forall x(x \in v \leftrightarrow \exists y \in u\ (x \in y))$;

$P(u,v,w) := \forall x(x \in w \leftrightarrow x = u \vee x = v)$;

$PS(u,v) := \forall x(x \in v \leftrightarrow \forall y \in x\ (y \in u))$;

$I(u)$ is the formula stating that $\emptyset \in u$ and if $a \in u$ then $a \cup \{a\} \in u$ also; $O(u)$ finally is the formula expressing that u is an ordinal.

Hence the axioms of the empty set, sum-set, pairing, power-set and infinity are the formulas $\exists u\ Z(u)$, $\forall u\ \exists v\ S(u,v)$, $\forall u\ \forall v\ \exists w\ P(u,v,w)$, $\forall u\ \exists v\ PS(u,v)$ and $\exists u\ I(u)$ respectively.

Lemma 11.1: If A is a non-empty transitive set and $a,b,c \in A$ then

(*i*) $Z^A(a) \leftrightarrow a = \emptyset$;

(*ii*) $S^A(a,b) \leftrightarrow b = \bigcup a$;

(*iii*) $P^A(a,b,c) \leftrightarrow c = \{a,b\}$;

(*iv*) $PS^A(a,b) \leftrightarrow b = A \cap \mathcal{P}(a)$;

(*v*) $I^A(a) \rightarrow \omega \subseteq a$;

(*vi*) $O^A(a) \leftrightarrow a \in OR$.

Proof: (*i*) If $Z^A(a)$, i.e., $\forall x \in A\ (x \notin a)$, then $a = \emptyset$: for if $b \in a$ then (since A is transitive) $b \in A$ also. Hence $\neg\forall x \in A\ (x \notin a)$. Conversely, if $a = \emptyset$ then $\forall x(x \notin a)$ and *a fortiori* $\forall x \in A\ (x \notin a)$.
(*iv*) Assume $PS^A(a,b)$, i.e., if $c \in A$ then $c \in b \leftrightarrow \forall y \in c \cap A\ (y \in a)$. Since A is transitive we get $c \in b \leftrightarrow \forall y \in c\ (y \in a)$, in other words, $b \cap A = A \cap \mathcal{P}(a)$. But $b \cap A = b$, as $b \in A$ and A is transitive. (*vi*) $O(u)$ is $\forall x \in u\ \ \forall y \in x\ (y \in u \wedge \forall z \in y(z \in x))$ so $O^A(a)$ is $\forall x \in a \cap A\ \forall y \in x \cap A\ (y \in a \wedge \forall z \in y \cap A\ (z \in x))$. But if $a \in A$, then this amounts to $O(a)$. Parts (*ii*), (*iii*) and (*v*) are left to the reader. Cf. exercice 1. □

The next result supplies a number of conditions under which transitive epsilon structures satisfy **ZF**-axioms. A set A is called *super transitive* if $a \in A$ and $b \subseteq a$ imply $b \in A$, in other words if $\bigcup_{a \in A} \mathcal{P}(a) \subseteq A$.

Lemma 11.2: If A is non-empty and transitive, then $\langle A, \in_A \rangle$ satisfies the

(*i*) regularity-axiom;

(*ii*) axiom of extensionality;

(*iii*) axiom of the empty set;

(*iv*) pairing-axiom iff $\forall a\ b \in A\ (\{a,b\} \in A)$;

(*v*) sumset-axiom iff $\forall a \in A\ (\bigcup a \in A)$;

(*vi*) powerset-axiom iff $\forall a \in A\ (A \cap \mathcal{P}(a) \in A)$;

(*vii*) axiom of infinity if $\omega \in A$;

(*viii*) axiom of separation if A is supertransitive;

(*ix*) replacement-axiom if $\forall a \in A\ \forall f \in A^{a}\ (\mathrm{Ran}(f) \in A)$;

(*x*) (*AC*) *AC*4 if $\bigcup_{a \in A} \mathcal{P}(\bigcup a) \subseteq A$.

Proof: (*iii*) Notice that $\emptyset \in A$ since it is the unique minimal element of A. By 11.1 (*i*) we have $Z^{A}(\emptyset)$ thus $(\exists u\ Z(u))^{A}$ hence A satisfies the axiom of the empty set. (*iv*), (*v*), (*vi*) are immediate from 11.1. (*vii*): If $\omega \in A$ then (11.1(*v*)) $I^{A}(\omega)$ hence $(\exists x I(x))^{A}$.

(*viii*): Note that in general supertransitivity is a much stronger requirement then to be a model for the separation-axiom since the last only requires membership of "describable" subsets of elements. Something similar can be said in case of (*ix*).

(*x*): Note that if R is a set of representatives for the partition $P \in A$ then $R \in \mathcal{P}(\bigcup P)$. Details are left to the reader. □

A *natural model* is a structure of the form $\langle V_{\alpha}, \in_{V_{\alpha}} \rangle$ where $\alpha \in \mathrm{OR}$, $\alpha > 0$. What is natural about these models is perhaps that one often thinks of the universe of all sets somewhat along the following lines: V "would have been a partial universe if OR would have been a set" (by $V = \bigcup_{\alpha \in OR} V_{\alpha}$.)

<u>Theorem 11.3</u>: If $\mathbf{A} = \langle V_{\alpha}, \in_{V_{\alpha}} \rangle$ with Lim α then $\mathbf{A}$ is a model of all **ZF**-axioms with the possible exceptions of the infinity -and replacement- axioms; if $\alpha = \omega$ then replacement does hold but infinity does not and if $\alpha > \omega$ infinity is satisfied but replacement "in general" (for instance if $\alpha = \omega + \omega$) is not.

Proof: Combine lemma 11.2 and the properties of partial universes given in §4. The following observations concern the axioms of infinity and replacement.

If $(\exists x I(x))^{V_{\omega}}$, for instance $a \in V_{\omega}$ and $I^{V_{\omega}}(a)$ then (11.1(*v*)) $\omega \subseteq a$ hence

$\omega \in V_\omega$, contradicting the fact that $\omega \notin V_\omega$. So infinity does not hold in V_ω. Replacement does hold: if $a \in V_\omega$ and (11.2(ix)) $f: a \to V_\omega$ then $\mathrm{Ran}(f) \subseteq V_\omega$ is finite since a is. Now if $b \subseteq V_\omega$ is finite then $\rho(b) = \bigcup\{\rho(c) + 1 \mid c \in b\}$ is a finite sum of finite ordinals hence finite, therefore $b \in V_\omega$. If $\alpha > \omega$, then $\omega \in V_\alpha$, hence V_a (11.2(vii)) satisfies infinity. But if V_a also satisfies replacement, α must be rather large, as the discussion preceding theorem 7.1 (and exercise 3 of §7) demonstrates. In the case $\alpha = \omega + \omega$ one can write down a formula Φ for which, if $n \in \omega$ and $a \in V_a$: $\Phi^{V_a}(n,a) \leftrightarrow a = \omega+n$. Validity of the replacement-axiom in V_a would imply $\{\omega+n \mid n \in \omega\} \in V_a$ contradicting $\rho(\{\omega+n \mid n \in \omega\}) = \omega + \omega$. □

Corollary 11.4: (*i*) The infinity-axiom is not derivable from the other ZF-axioms.

(*ii*) The replacement-axiom is not derivable from the other ZF-axioms.

Proof: Immediate from 11.3: V_ω demonstrates (*i*) and $V_{\omega+\omega}$ demonstrates (*ii*). □

In Chapter I, it was noticed that the replacement-axiom implies the separation-axiom; 11.4(*ii*) demonstrates that the converse fails. Later on we shall improve on 11.4 . 11.3 and 11.2(ix) imply that a natural model $\langle V_\alpha, \epsilon_{V_\alpha} \rangle$ satisfies every ZF-axiom when $\omega < \alpha$, Lim α and $\forall a \in V_\alpha\ \forall f \in V_\alpha^a\ (\mathrm{Ran}(f) \in V_\alpha)$; 11.7 characterizes these ordinals α under the assumption of the axiom of choice.

Definition 11.5: (*i*) A regular cardinal κ is called *strongly inaccessible* if $\aleph_0 < \kappa$ and κ is a *strong limit*, i.e., $\forall m(m < \kappa \to 2^m < \kappa)$;

(*ii*) $\alpha \in \mathrm{OR}$ is called *strongly inaccessible* if α is an initial number of strongly inaccessible cardinality.

Lemma 11.6 (*AC*): If $\alpha = \omega$ or if α is strongly inaccessible then $|V_a| = |\alpha|$.

Proof: It suffices to show that $|V_\xi| < |\alpha|$ if $\xi < \alpha$, since then $|V_a| = |\bigcup_{\xi<a} V_\xi| \leqslant \sum_{\xi<a} |V_\xi| \leqslant \sum_{\xi<a} |\alpha| = |\alpha|^2 = |\alpha|$, while of course also $|\alpha| \leqslant |V_a|$, as $\alpha \subseteq V_a$. We prove this by induction on ξ: assume as an induction-hypothesis that $|V_\delta| < |\alpha|$ for $\delta < \xi < \alpha$. Then $2^{|V_\delta|} < |\alpha|$, by assumption on α, ar therefore $|V_\xi| = |\bigcup_{\delta<\xi} \mathcal{P}(V_\delta)| \leqslant \sum_{\delta<\xi} |\mathcal{P}(V_\delta)| = \sum_{\delta<\xi} 2^{|V_\delta|} < |\alpha|$, since α is regular. □

The characterization promised is the following
Theorem 11.7 (*AC*): The following assertions are equivalent:

(*i*) $\alpha = \omega$ or α is strongly inaccessible;

(*ii*) $\forall a \in V_\alpha\ \forall f \in V_\alpha^a\ (\mathrm{Ran}(f) \in V_\alpha)$.

Proof: (*i*) → (*ii*): Assume $a \in V_\alpha$ and $f: a \to V_\alpha$.
$\mathrm{Ran}(f) = \{f(b) \mid b \in a\}$ therefore $\rho(\mathrm{Ran}(f)) = \bigcup\{\rho(f(b)) + 1 \mid b \in a\}$ and this is an ordinal $< \alpha$, as α is regular and $|a| < |\alpha|$ by 11.6. Thus $\mathrm{Ran}(f) \in V_\alpha$.
(*ii*) → (*i*): α is regular, for if $\beta < \alpha$ and $f: \beta \to \alpha$, then by (*ii*) $\mathrm{Ran}(f) \in V_\alpha$, thus $\mathrm{Ran}(f) \subseteq \rho(\mathrm{Ran}(f)) < \alpha$. To show that $\beta < \alpha$ implies $|\mathcal{P}(\beta)| < |\alpha|$, it suffices to show that the type of any well-ordering in V_α is smaller than α: for take (*AC*) some well-ordering S of $\mathcal{P}(\beta)$ then $\langle \mathcal{P}(\beta), S\rangle \in V_\alpha$ (note that $\beta < \alpha$ and Lim α); and if γ is the type of $\langle \mathcal{P}(\beta), S\rangle$ and $\gamma < \alpha$ then $|\mathcal{P}(\beta)| = |\gamma| < |\alpha|$. So suppose $\langle A, R\rangle \in V_\alpha$ is a well-ordering and h the order-isomorphism between A and its type γ. Using R-induction it follows that $\gamma = \mathrm{Ran}(h) \subseteq V_\alpha$: for if $h(b) \in V_\alpha$ for all b with bRa then $h(a) \in V_\alpha$, since $h(a) = \{h(b) \mid bRa\}$ and $\{b \mid bRa\} \in V_\alpha$. (Apply (*ii*) with $f := h \restriction \{b \mid bRa\}$.) Thus $h: A \to V_\alpha$. But then, using (*ii*) again, it follows that $\gamma = \mathrm{Ran}(h) \in V_\alpha$ and hence $\gamma < \alpha$. □

Using the remark preceding 11.5, we now know that if α is strongly inaccessible, then (*AC*). $\langle V_\alpha, \in_{V_\alpha}\rangle$ is a model for all ZF-axioms.

The connection between strong and weak (8.9) inaccessibility is given by.

Theorem 11.8: If ω_α is strongly inaccessible it is also weakly inaccessible; conversely, if ω_α is weakly inaccessible and GCH holds, it is strongly inaccessible.

Proof: Assume ω_α is strongly inaccessible. We show that Lim α. Let $\beta < \alpha$. As $\aleph_\beta < \aleph_\alpha$, we have $2^{\aleph_\beta} < \aleph_\alpha$ also thus $2^{\aleph_\beta}$ is an aleph. $\aleph_\beta < 2^{\aleph_\beta}$, so we cannot have $2^{\aleph_\beta} < \aleph_{\beta+1}$, therefore $\aleph_{\beta+1} \leq 2^{\aleph_\beta} < \aleph_\alpha$, hence $\beta+1 < \alpha$ and thus Lim α. Conversely, if Lim α and if GCH holds, then if $\beta < \alpha$: $2^{\aleph_\beta} = \aleph_{\beta+1} < \aleph_\alpha$. □

Do strongly inaccessible cardinals exist? Even if this is the case we cannot prove it on the basis of the ZF-axioms. This is a corollary of the next lemma. IN(x) is a ZF-formula expressing that x is a strongly inaccessible ordinal.

Lemma 11.9: If Lim α and $a \in V_\alpha$ then $\mathrm{IN}^{V_\alpha}(a)$ holds just in case a is a strongly inaccessible ordinal.

One can consider this as an "absoluteness"-result in much the same way as lemma 11.1. An exact proof requires a great deal of verification (depending on the actual choice of IN) which in principle is trivial, but rather tedious. Therefore we omit it, the reader may consult [Drake, 1974].

Corollary 11.10 (AC): If α is the smallest strongly inaccessible ordinal then $(\neg \exists x\ \mathrm{IN}(x))^{V_\alpha}$ and therefore $\exists x\ \mathrm{IN}(x)$ is not derivable from the ZF-axioms (together with AC).

Proof: If not $(\neg \exists x\ \mathrm{IN}(x))^{V_\alpha}$ as for instance, $a \in V_\alpha$ and $\mathrm{IN}^{V_\alpha}(a)$ holds, then by 11.9 a would be a strongly inaccessible ordinal smaller than α contradicting the assumption on α. Furthermore, $\langle V_\alpha, \epsilon_{V_\alpha} \rangle$ satisfies every ZF-axiom (and AC, if assumed from the beginning) (11.7, 11.2). Finally, if there happens to be no strongly inaccessible cardinal at all we certainly cannot prove the existence of one. □

On the other hand, it can be argued to be intuitively plausible that partial universes satisfying the conditions of 11.2 do exist, in other words, that strongly inaccessible cardinals (assuming AC) are a reality: cf. paragraph 12.

Proving 11.2, we already remarked that the condition $\forall a \in A\ \forall f \in A^a$ $(\mathrm{Ran}(f) \in A)$ (or strong inaccessibility of α when $A = V_\alpha$) is far too crude to characterize ZF-models, since the replacement-axiom only requires the condition for functions f describable by means of ZF-formulas. In fact we have the rather surprising

Theorem 11.11 (AC): If α is strongly inaccessible there exists a set $A \subseteq \alpha$ closed and cofinal in α such that $\langle V_\xi, \epsilon_{V_\xi} \rangle$ is a ZF-model for every $\xi \in A$.

Proof: Since $\langle V_\alpha\ \epsilon_{V_\alpha} \rangle$ is a ZF-model, it also satisfies the reflection-principle; in other words, for every ZF-assertion Φ there is a closed cofinal. $A_\Phi \subseteq \alpha$ included in $\{\xi < \alpha \mid \Phi^{V_\xi} \leftrightarrow \Phi^{V_\alpha}\}$. Take A to be the intersection of all (countably many) A_Φ, then (6.12(ii)) A is closed and cofinal in α since

$\omega < \mathrm{cf}(\alpha) = \alpha$. And if $\xi \in A$ and Φ is some ZF-axiom then Φ^{V_ξ} holds since Φ^{V_α} does. (Strictly speaking, this proof contains an incorrect argument. We return to this detail in a moment.) □

Notice that if ξ is the smallest ordinal for which $\langle V_\xi, \epsilon_{V_\xi} \rangle$ is a ZF-model then $\mathrm{cf}(\xi) = \omega$ because otherwise we could repeat the reasoning used in the proof of 11.11 with $\alpha = \xi$.

Hence as a matter of fact we can imagine ourselves to be in a universe, the ordinals of which are given by a countable process; and this, in a striking way, illustrates the weakness of the replacement-axiom.

(N.B. Some readers might remark that a minor modification in the proof of 11.11 yields the result that there exists a closed collection A cofinal in OR such that $\langle V_\xi, \epsilon_{V_\xi} \rangle$ is a ZF-model for every $\xi \in A$ (namely, replace α by OR everywhere and apply 6.12(*iii*) instead of 6.12(*ii*)). This is a fallacy resulting from the fact that the proof of 11.11 is not entirely precise. Without further explanation it is clear that every individual A_Φ exists, but the existence of their intersection A might be questioned. The reason is that in the definition of A a quantification over ZF-assertions is made: $\xi \in A$ iff for all Φ, $\xi \in A_\Phi$. And this is impossible as our language admits quantification over sets only. The solution to this dilemma illustrates the slogan that everything (hence a ZF-assertion also) is a set, but this is to long-winded to explain here. We only mention that in any case it is possible to define a cofinal, closed subset A of α, such that for every Φ, $A \subset A_\Phi$ can be proved; this clearly suffices.)

The replacement-axiom really is not a single statement but produces infinitely many statements, in technical terms: it is an axiom schema. One might wonder whether all those axioms are actually necessary: it might be the case that all ZF-axioms already follow from a *finite* number of ZF-theorems (and hence from *one*: the conjunction of this number.)
The next theorem shows that this is impossible.

<u>Theorem 11.12:</u>

There is no ZF-statement Φ, consistent with Zermelo's axioms, such that every replacement-axiom can be derived from Zermelo's axioms plus Φ.

Proof: Suppose that every replacement-axiom can be derived from Zermelo's axioms Z plus a certain statement Φ. Take Ψ to be the conjunction of Φ and all Zermelo-axioms with the exclusion of the (infinitely many) separation-axioms. As the reflection principle is derivable from Z plus Φ we can also derive the statement $\exists A\,[A \neq \emptyset \wedge A \text{ is transitive} \wedge A \text{ is supertransitive} \wedge \Psi^A]$. (1)

From 11.2(*viii*) it follows that if A is a set whose existence is claimed in (1), then every Z-plus-Φ-axiom holds in it, and hence again the reflection-principle for Ψ holds in A. The axiom of regularity can also be derived in Z plus Φ, therefore we can also derive the statement $\exists A\,[A \neq \emptyset \wedge A \text{ is transitive} \wedge A \text{ is supertransitive} \wedge \Psi^A \wedge (\exists B\,[B \neq \emptyset \wedge B \text{ is transitive} \wedge B \text{ is supertransitive} \wedge \Psi^B])^A \wedge \neg \exists B \in A\,[B \neq \emptyset \wedge B \text{ is transitive} \wedge B \text{ is supertransitive} \wedge \Psi^B]]$. Carrying out the relativization of the middle part of this statement, we easily see that it directly contradicts the last part. Therefore Φ is not consistent with Zermelo's axioms. □

Though no finite set of instances of the replacement axiom allows us to derive all the instances, the number of concrete applications we have made of the axiom is very limited (we do not mean for instance the recursion lemma and the reflection-principle, as they constitute again general theorem-schemas). Practical applications mainly concern the existence of the transitive closure, the representation-theorem for well-founded extensional structures, Hartogs' function and the cumulative hierarchy.

We now discuss a type of models, illustrating the role of the axioms of power-set and sum-set.

Suppose that K is some collection of sets, then $\mathrm{Her}(K) := \{a \in K \mid \mathrm{TC}(a) \subseteq K\}$ is the collection of *hereditarily-K*-sets. The *powerclass* of K is the collection $\mathcal{P}(K) := \{a \mid a \subseteq K\}$ of all *subsets* of K.

<u>Lemma 11.13:</u> $\mathrm{Her}(K)$ is the only solution of the equation $\mathcal{P}(X) \cap K = X$.

Proof: (*i*) $\mathrm{Her}(K)$ satisfies the equation: if $a \subset \mathrm{Her}(K)$ then $\mathrm{TC}(a) \subset \mathrm{Her}(K)$, as $\mathrm{Her}(K)$ is transitive, and $\mathrm{TC}(a) \subseteq K$, as $\mathrm{Her}(K) \subseteq K$. If, moreover, $a \in K$, then $a \in \mathrm{Her}(K)$. Conversely, $\mathrm{Her}(K) \subseteq \mathcal{P}(\mathrm{Her}(K)) \cap K$, since $\mathrm{Her}(K)$ is a transitive part of K.

(ii) Suppose $\mathcal{P}(X) \cap K = X$. We apply epsilon-induction. First, $X \subseteq \mathrm{Her}(K)$: take $a \in X$, then $a \subseteq X$ by assumption and $a \subseteq \mathrm{Her}(K)$ by induction-hypothesis. Thus, $\mathrm{TC}(a) \subseteq \mathrm{Her}(K) \subseteq K$ and as $a \in K$, by assumption, we obtain $a \in \mathrm{Her}(K)$. Also, $\mathrm{Her}(K) \subseteq X$: if $a \in \mathrm{Her}(K)$, then $a \subseteq \mathrm{Her}(K)$, hence $a \subseteq X$ by induction-hypothesis; furthermore $a \in K$, as $\mathrm{Her}(K) \subseteq K$ and therefore $a \in \mathcal{P}(X) \cap K \subseteq X$. □

The next examples illustrate the operation Her. In ZF, without the regularity-axiom, the regular sets (§1) are hereditarily-regular too. The ordinals are precisely the hereditarily-transitive sets.

For our next discussion, let κ be some cardinal $\geq \aleph_0$. We shall be occupied by the sets hereditarily of cardinality $< \kappa$, that is to say with $\mathrm{HC}_\kappa :=$ $\mathrm{Her}\ \{a \mid |a| < \kappa\}$. The elements of HC_κ can be split up in a hierarchy, similar to the cumulative hierarchy (cf. exercise 7):

<u>Definition 11.14</u>: $V^\kappa_\alpha = \bigcup_{\xi<\alpha} \{a \subseteq V^\kappa_\xi \mid |a| < \kappa\}$

<u>Lemma 11.15</u>: *(i)* V^κ_α is transitive;
(ii) V^κ_α is supertransitive;
(iii) $\xi < \alpha \rightarrow V^\kappa_\xi \subseteq V^\kappa_\alpha$;
(iv) $\bigcup_{\alpha \in OR} V^\kappa_\alpha = \mathrm{HC}_\kappa$.

Proof: Induction w.r.t. α.

(i) If $a \in b$, $b \in V^\kappa_\alpha$ then $\xi < \alpha$ exists such that $b \subseteq V^\kappa_\xi$ and hence $a \in V^\kappa_\xi$. Therefore, $|a| < \kappa$ and $a \subseteq V^\kappa_\xi$ by induction-hypothesis. Thus, $a \in V^\kappa_\xi$.

(ii) is left to the reader.

(iii) is immediate from *(i)* and 11.14.

(iv) We show that $V^\kappa_\alpha \subseteq \mathrm{HC}_\kappa$ using induction on α. If $a \in V^\kappa_\alpha$, then $|a| < \kappa$ and for some $\xi < \alpha$, $a \subseteq V^\kappa_\xi$. As $V^\kappa_\xi \subseteq \mathrm{HC}_\kappa$ by induction-hypothesis, we have a $a \subseteq \mathrm{HC}_\kappa$ and hence $a \in \mathrm{HC}_\kappa$ by 11.13. We leave the proof of $\mathrm{HC}_\kappa \subseteq \bigcup_{\alpha \in OR} V^\kappa_\alpha$ to the reader. □

An alternative for 11.14 is the

<u>Lemma 11.16</u>: *(i)* $V^\kappa_0 = \emptyset$;
(ii) $V^\kappa_{\alpha+1} = \{a \subseteq V^\kappa_\alpha \mid |a| < \kappa\}$; *(iii)* $\mathrm{Lim}\ \alpha \rightarrow V^\kappa_\alpha = \bigcup_{\xi<\alpha} V^\kappa_\xi$.

Proof: using 11.14 and 11.15(*iii*); details are left to the reader. □

Theorem 11.17: If β is a regular ordinal then $HC_{|\beta|} = \bigcup_{\alpha<\beta} V_{\alpha}^{|\beta|}$.

Proof: 11.15(*iv*) gives one inclusion. Now suppose that $a \in HC_{|\beta|}$, then $a \subseteq HC_{|\beta|}$ and (hypothesis for epsilon-induction) $a \subseteq \bigcup_{\alpha<\beta} V_{\alpha}^{|\beta|}$. Define $h: a \to \beta$ by $h(b) := \bigcap \{\alpha | b \in V_{\alpha}^{|\beta|}\}$, then $\alpha := \bigcup Ran(h) < \beta$, as $|a| < |\beta|$ and β is regular. Now $a \subseteq V_{\alpha}^{|\beta|}$ (11.15(*iii*)), thus $a \in V_{\alpha+1}^{|\beta|}$. This completes the proof, as $\alpha + 1 < \beta$. □

Corollary 11.18 (*AC*): $HC_{\kappa} \in V$.

Proof: If (AC) $\kappa = \aleph_{\alpha}$ then $HC_{\kappa} \subseteq HC_{\aleph_{\alpha+1}}$; $\omega_{\alpha+1}$ is regular (10.18), hence (11.17) $HC_{\aleph_{\alpha+1}} = \bigcup_{\xi<\omega_{\alpha+1}} V^{\aleph_{\alpha+1}} \in V$. □

The cardinal $\mathfrak{m}$ is called a *strong limit* if $\mathfrak{m} \geq \aleph_0$ and $\mathfrak{n} < \mathfrak{m} \to 2^{\mathfrak{n}} < \mathfrak{m}$. $\aleph_0$ is a strong limit and the strongly inaccessible cardinals are exactly the regular strong limits greater than $\aleph_0$.

Lemma 11.19(AC): If α = 0 or if $\aleph_{\alpha}$ is strongly inaccessible then $V_{\xi}^{\aleph_{\alpha}} = V_{\xi}$ $(\xi \leq \omega_{\alpha})$.

Proof: Induction w.r.t. ξ and 11.16. If $\xi < \omega_{\alpha}$ then

$$\begin{aligned} V_{\xi+1}^{\aleph_{\alpha}} &= \{a \subseteq V_{\xi}^{\aleph_{\alpha}} \mid |a| < \aleph_{\alpha}\} \\ &= \{a \subseteq V_{\xi} \mid |a| < \aleph_{\alpha}\} \\ &= V_{\xi+1} \quad \text{(by 11.6)}. \end{aligned}$$ □

Theorem 11.20 (AC): If $\kappa \geq \aleph_0$ then HC_{κ} is model of the

- axiom of infinity exactly if $\kappa > \aleph_0$;
- sum-set axiom exactly if κ is regular;
- power-set axiom exactly if κ is a strong limit.

Moreover, HC_{κ} is model of every other ZF-axiom.

Proof: If $\kappa > \aleph_0$ then $\omega \in HC_{\kappa}$ and hence HC_{κ} is a model for the axiom of infinity according to 11.2(*vii*). If $\kappa = \aleph_0$ then $HC_{\kappa} = V_{\omega}$ (11.19) and the infinity-axiom does not hold in V_{ω} (11.3). If κ is regular and $a \in HC_{\kappa}$ then $\bigcup a \subseteq HC_{\kappa}$ and (10.11) $|\bigcup a| \leq \sum_{b\in a} |b| < \kappa$, hence $\bigcup a \in HC_{\kappa}$. But then HC_{κ} is a

model of the sum-set axiom, according to 11.2(*v*). If κ is not regular and, say, $\kappa = \aleph_a$, then a partition P of ω_a exists in less then $\aleph_a$ subsets of power $< \aleph_a$. Then $P \in HC_\kappa$ (why ?), but $\omega_a = \bigcup P \notin HC_\kappa$ and therefore (11.2(*v*)) HC_κ is not a model of the sum-set axiom.
If κ is a strong limit and $a \in HC_\kappa$, then $\mathcal{P}(a) \in HC_\kappa$ as $\mathcal{P}(a) \subseteq HC_\kappa$ and $|\mathcal{P}(a)| < \kappa$; now apply 11.2(*vi*). But if $\aleph_a < \kappa \leqslant 2^{\aleph_a}$ then $\omega_a \in HC_\kappa$ while $HC_\kappa \cap \mathcal{P}(\omega_a) = \mathcal{P}(\omega_a) \notin HC_\kappa$ and hence (11.2(*vi*)) HC_κ is not a model of the powerset axiom. Validity of the other axioms in HC_κ follows from the corresponding parts of 11.2. □

<u>Corollary 11.21</u> (AC): The sum-set axiom is not derivable from the other ZF-axiom; the same holds for the powerset axiom.

Proof: There are non-regular strong limits: let $\mathfrak{m}_0$ be any cardinal $\geqslant \aleph_0$ and define $\mathfrak{m}_{n+1} = 2^{\mathfrak{m}_n}$ for all $n \in \omega$; put $\mathfrak{m} := \sum_{n \in \omega} \mathfrak{m}_n$. Then $\mathfrak{m}$ has cofinality $\aleph_0$ and if $\mathfrak{n} < \mathfrak{m}$ then for some n, $\mathfrak{n} \leqslant \mathfrak{m}_n$ (use 10.13) and hence $2^{\mathfrak{n}} \leqslant 2^{\mathfrak{m}_n} = \mathfrak{m}_{n+1} < \mathfrak{m}$. Therefore $HC_{\mathfrak{m}}$ satisfies all ZF-axioms, except the one for sum-sets (according to 11.20).
Also, there are regular cardinals which are not strong limits: any successor-cardinal will do. Application of 11.20 yields the result for the powerset axiom. □

<u>Theorem 11.22</u> (AC): $HC_{\aleph_a} = V_{\omega_a}$ just in case $a = 0$ or $\aleph_a$ is strongly inaccessible.

Proof: One direction is immediate from 11.7; the other from both 11.19 and 11.17. □

We conclude this paragraph with some remarks about the set $HC_{\aleph_0} = V_\omega$ of hereditarily finite sets. We have the following simple characterization:

<u>Theorem 11.23</u>: V_ω is the intersection of all sets A for which (*i*) $\emptyset \in A$ and (*ii*) if $a,b \in A$ then $a \cup \{b\} \in A$.

Proof: V_ω satisfies both (*i*) and (*ii*), so the intersection mentioned is part of V_ω. Now suppose A satisfies (*i*) and (*ii*); we deduce that $V_\omega \subseteq A$. Let $a \in V_\omega$, then $a \subseteq V_\omega$ and we may assume, as an hypothesis for epsilon-induction, that $a \subseteq A$. As a is finite we can carry out an induction on the number of elements of a. If $a = \emptyset$ then $a \in A$ by hypothesis. So assume $a = a' \cup \{b\}$, $b \notin a'$. Then by the second induction-hypothesis, $a' \in A$. Furthermore $b \in a \subseteq A$, hence $b \in A$ and therefore $a' \cup \{b\} \in A$ by assumption (ii). □

There is an amusing way to code, so to speak, V_ω in ω. Define the relation ε between natural numbers by putting $n \mathrel{\varepsilon} m$ iff for some finite $a \subseteq \omega$, $m = \sum_{k \in a} 2^k$ and $n \in a$; in other words: if in the binary notation for m the n-th place (counting from the right, starting with zero) is occupied by the numeral 1. For instance, $41 = 2^5 + 2^3 + 2^0$; the binary notation for 41 is 101001. Thus, $3 \mathrel{\varepsilon} 41$ but $1 \not\mathrel{\varepsilon} 41$.

Theorem 11.24: $\langle \omega, \varepsilon \rangle \cong \langle V_\omega, \in_{V_\omega} \rangle$.

Proof: ε is well-founded on ω as $n \mathrel{\varepsilon} m$ implies $n < m$. Also, ε is extensional on ω for if $a := \{n | n \mathrel{\varepsilon} m\}$ then $m = \sum_{n \in a} 2^n$ hence if $\{n|n \mathrel{\varepsilon} m\} = \{n|n \mathrel{\varepsilon} m'\}$ then $m = m'$. By 2.9 $\langle \omega, \varepsilon \rangle$ is isomorphic to the epsilon-structure on some transitive set A by an isomorphism h. Using induction on n we see that it follows that $h(n) \in HC_{\aleph_0}$, therefore, $A \subseteq V_\omega$. Furthermore, $h(0) = \emptyset \in A$ and if $h(n) = a$, $h(m) = b$ and $b \notin a$, then $h(n + 2^m) = a \cup \{b\} \in A$, so we also get $V_\omega \subseteq A$ by 11.23. □

We now return to our remarks concerning 11.4(i). V_ω satisfies all ZF-axioms with the exception of the axiom of infinity. 11.24 states that an isomorphic copy of V_ω can be defined on ω. Therefore whoever accepts the natural numbers (or the consistency of arithmetic) must also accept the hereditarily finite sets (or the consistency of ZF minus infinity).
The 'coding' of V_ω in ω has a parallel on the syntactical level: there is a translation of the language of ZF into the language of arithmetic and it turns out that the translation of the axioms of ZF, with the exception of infinity, are provable in Peano's arithmetic. As a corollary one obtains the following result: if Peano's arithmetic is consistent, then so is ZF minus infinity.

Exercises

1. Prove 11.1 (*ii*), (*iii*) and (*v*).

2. Complete the proof of 11.2.

3. (AC) there are arbitrarily large α for which $|V_a| = |\alpha|$.

4. (AC) There are strong limits $\aleph_a$ such that ω_a does not satisfy the condition from exercise 3.

5. Compute some values of the function h in 11.24.

6. $V_\omega = \bigcap \{X | \emptyset \in X \wedge \forall a,b \in X\ (a \cup b \in X \wedge \{a,b\} \in X)\}$.

7. $V_a^\kappa = HC_\kappa \cap V_a$.

8. Define $H_\kappa := \{a |\ |TC(a)| < \kappa\}$ and prove using AC (*i*) $H_\kappa \in V$ (hint:use 4.1) and (*ii*) $H_\kappa = HC_\kappa$ just in case κ is regular.

9. (*i*) The well-founded extensional structure $\langle A,R\rangle$ has a type of the form $TC(\{a_0\})$ iff for some $a \in A$, $\forall b \in A\ (b \neq a \rightarrow b\bar{R}a)$. (Call $\langle A,R\rangle$ *topped* in that case.)
   (*ii*) $TC(\{a\}) = TC(\{b\}) \rightarrow a = b$.
   (*iii*) (AC) $a \in H_{\aleph_a}$ iff $TC(\{a\})$ is type of a topped well-founded extensional structure $\langle \beta,R\rangle$ where $\beta < \omega_a$.
   (*iv*) (AC) $|H_{\aleph_a}| = \sum_{\beta<a} 2^{\aleph_\beta}$.

12 MEASURABLE CARDINALS

This section presupposes the material from §1 of chapter III. The axiom of choice will be employed so frequently that from now on we shall not mention its use.

The introduction of the concept of *strong inaccessibility* is motivated by 11.2(*ix*), where the natural, though excessively strong, condition $\forall a \in A\ \forall f \in A^a\ (\text{Ran}(f) \in A)$ is given, which suffices for the non-empty transitive set A to satisfy the replacement-axiom, and by the characterization 11.7. *Mutatis mutandis*, the usual way to think about the universe V of all sets is as if it were a natural model V_a with a strongly inaccessible index α; and this is the reason the existence of strong inaccessibles is so plausible though we cannot prove this (11.10). We even know that on account of a general result by Gödel, the so-called *second incompleteness theorem* it is impossible to prove the mere consistency of this existence-assertion, even if one takes the consistency of ZF for granted.

If one is sensitive to this plausibility argument by means of "reflection" (: the "index of our universe" behaves like a strong inaccessible, therefore strong inaccessibles probably exist), one can argue in the same line that arbitrarily large strong inaccessibles should exist as the index of the universe is an inaccessible greater than any given real one. In terms of notations introduced in §11, we should have

$$\forall x(\mathbb{O}(x) \rightarrow \exists y(x \in y \wedge \mathrm{IN}(y))). \qquad [*]$$

In other words, if θ_ξ is the ξ-th strongly inaccessible ordinal: θ_ξ exists for every $\xi \in$ OR. Of course, this is a rather strong addition to the ZF-axioms: the method of 11.10 will show that, even if θ_γ ($\gamma \in$ OR) exists, this cannot be proved on the basis of the ZF-axioms together with the existence of the θ_ξ ($\xi < \gamma$).
But once [*] is accepted as true, it can be used to motivate yet stronger assertions using reflection on our imaginary universe-index: [*] states that θ constitutes a one-one-correspondence between OR and the collection of strong inaccessibles. As the index of V is strongly inaccessible and comes next to both all ordinals and all inaccessibles, it would be a fixed point of θ if θ had been defined on it. Therefore it is plausible that θ really has fixed points, i.e., that ordinals α with $\theta(\alpha) = \alpha$ do exist.
Again, this is not derivable in ZF using [*] (even if it were true): this is proved by the method of 11.10 using the model V_α where α is the least fixed point of θ.

It is now clear how to proceed further: the next step is to require θ to have arbitrarily large fixed points (call these inaccessible of degree 2), i.e., require the existence of the derivative of θ. Iterating this into the transfinite, we may require the existence of inaccessibles of every degree $\xi \in$ OR. As our imaginary index of the universe will certainly be inaccessible of every degree we see that by our reflexive reasoning ordinals α should exist which are inaccessible of every degree $< \alpha$.

Clearly, there is no end to this kind of arguments: they motivate stronger and stronger requirements of infinity, which cannot be proved from earlier ones. One way to make this series of infinity-statements provable for quite a large initial segment is to postulate the existence of a so-called *measurable cardinal*. For if some initial number κ has measurable cardinality, then it is possible to transform the plausibility-arguments presented above in real proofs in which properties of a universe U are reflected to $V_\kappa \in U$. As a result κ will satisfy all requirements of infinity discussed before (and many more). We first present the necessary definitions:

Definition 12.1: Let $\mathfrak{m}$ be some cardinal. A filter F is called

(*i*) $\mathfrak{m}$-*complete* if for all $G \subseteq F$ such that $|G| \leqslant \mathfrak{m}$, $\cap G \in F$; and

(*ii*) $<\mathfrak{m}$-*complete* if for all $G \subseteq F$ such that $|G| < \mathfrak{m}$, $\cap G \in F$.

Every filter is $<\aleph_0$-complete and every principal filter is $\mathfrak{m}$-complete for all $\mathfrak{m}$. If $\omega < \mathrm{cf}(\alpha)$ then the intersection of less than $|\mathrm{cf}(\alpha)|$ closed cofinal subsets of α is closed and cofinal in α; hence these subsets generate a $<|\mathrm{cf}(\alpha)|$ - complete filter over α.

There is a second, (but, in our opinion, simple-minded) way to produce properties of infinity, namely, to generalize properties of ω. For instance, $\aleph_0$ is a regular strong limit; the other regular strong limits are the strongly inaccessible cardinals.

In this way, the next definition is motivated by the observation (III.1.9) that free $<\aleph_0$-complete ultrafilters over sets of power $\geqslant \aleph_0$ exist:

Definition 12.2: $\aleph_\alpha$ is called *measurable* if $\alpha > 0$ and there exists a free $<\aleph_\alpha$-complete ultrafilter over ω_α.

Though 12.2 looks rather innocent, we shall see, however, that the existence of measurable cardinals has impressive consequences. Motivating this assumption is rather difficult (but cf. W.N. Reinhardt [1974]), though the reader may try to split up the (definable) properties of ordinals according to their applicability to the index of the universe (if only this would have been a real ordinal) in order to get a class of properties resembling an ultrafilter over OR which is free and complete in some sense.
Therefore, it cannot be excluded that some day somebody may prove that measurables just do not exist.

But the work over the last 60 years (and in particular research into measurability has been quite intensive since 1960) make this highly improbable. In particular some work on measurability was done by people (as Scott and Jensen) wishing to prove the non-existence of measurables- but instead of a contradiction derived from the existence-assumption, they just obtained rather surprising theorems. Nowadays even properties stronger than measurability are considered.

We shall now occupy ourselves with the hard facts. The existence of free $< \aleph_0$-complete ultrafilters follows from the axiom of choice. To begin with, the existence of measurable cardinals follows already from the existence of free $\aleph_0$-complete ultrafilters according to Ulam's

Theorem 12.3: If (1) $\mathfrak{m}$ is the smallest cardinal such that $\aleph_0$-complete free ultrafilters exist over sets of power $\mathfrak{m}$ and (2) $|A| = \mathfrak{m}$ and F is a $\aleph_0$-complete free ultrafilter over A, then F is $< \mathfrak{m}$-complete and hence $\mathfrak{m}$ is measurable.

Proof: Suppose that $G \subseteq F$, $|G| < \mathfrak{m}$ but $\cap G \notin F$. We may assume without loss of generality, that if $X, Y \in G$ and $X \neq Y$ then $X \cup Y = A$. (*)

For if $G = \{X_\xi \mid \xi < \alpha\}$ then define $Y_\xi := X_\xi \cup \bigcup_{\delta < \xi} (A - X_\delta)$ $(\xi < \alpha)$ and $G' := \{Y_\xi \mid \xi < \alpha\}$; and we have $G' \subseteq F$, $|G'| < \mathfrak{m}$ and $\cap G' = \cap G$ (trivially, $\cap G \subseteq \cap G'$; and if $a \notin \cap G$ take $\xi < \alpha$ minimal such that $a \notin X_\xi$ - then $a \notin Y_\xi$ and hence $a \notin \cap G'$) hence G' has property (*). Define $F' := \{H \subseteq G \mid \cap H \notin F\}$, then it suffices to show that F' is a $\aleph_0$-complete free ultrafilter over G for this would contradict assumption 1.

(*i*) If, in the definition of F', we consider $\cap H$ as short for $\{a \in A \mid \forall X \in H(a \in X)\}$ then $\cap \emptyset = A \in F$ hence $\emptyset \notin F'$. On the other hand $\cap G \notin F$ so $G \in F'$ and $F' \neq \emptyset$.

(*ii*) If $H \subseteq G$, then $\cap G = \cap H \cap \cap(G-H) \notin F$. Hence at least one of $\cap H$ and $\cap(G-H)$ is outside F, i.e., $H \in F'$ or $(G-H) \in F'$.

(*iii*) If $H \subseteq G$ then by (*) $\cap H \cup \cap(G-H) = A \in F$ and since F is an ultrafilter at least one of $\cap H$ and $\cap(G-H)$ is in F, i.e., $H \notin F'$ or $(G-H) \notin F'$. Suppose that $H_n \in F'$ whenever $n \in \omega$, then $\cap H_n \notin F$ hence $\cap(G-H_n) \in F$ by the previous argument, and as F is $\aleph_0$-complete, $\bigcap_{n\in\omega} \cap(G-H_n) \in F$. But $\bigcap_{n\in\omega} \cap(G-H_n) = \cap(G - \bigcap_{n\in\omega} H_n)$ (show this!), hence $G - \bigcap_{n\in\omega} H_n \notin F'$ and, by (*ii*), $\bigcap_{n\in\omega} H_n \in F'$.

(*iv*) If $H' \supseteq H \in F'$, then $\cap H' \subseteq \cap H \notin F$, hence $\cap H' \notin F$ and $H' \in F'$.

(v) F' is free: if $X \in G$ then $\bigcap\{X\} = X \in F$ thus $\{X\} \notin F'$. □

We shall have occasion to use the next characterisation of $\mathfrak{m}$-completeness:

Lemma 12.4: The ultrafilter F over A is $\mathfrak{m}$-complete iff every $G \subseteq \mathcal{P}(A)$, for which $\bigcup G \in F$ and $|G| \leq \mathfrak{m}$, intersects F.

Proof: Suppose that F is $\mathfrak{m}$-complete and G as described. If $G \cap F = \emptyset$, then $\{A-X \mid X \in G\} \subseteq F$, since F is an ultrafilter. So $\bigcap_{X\in G} (A-X) \in F$. But $\bigcap_{X\in G} (A-X) = A - \bigcup G$, therefore $\bigcup G \notin F$. Conversely, suppose that

$G \cap F \neq \emptyset$ for every $G \subseteq \mathcal{P}(A)$ such that $\bigcup G \in F$ and $|G| \leq \mathfrak{m}$; and take $H \subseteq F$ such that $|H| \leq \mathfrak{m}$. We show that $\bigcap H \in F$: consider $G := \{A-X \mid X \in H\} \cup \{\bigcap H\}$. Now $\bigcup G = \bigcup_{X\in H} (A-X) \cup \bigcap H = A \in F$ and as $|G| = |H| \leq \mathfrak{m}$ we obtain $G \cap F \neq \emptyset$. But $G \cap F \subseteq \{\bigcap H\}$. Therefore $\bigcap H \in F$. □

The first result indicating that measurables are large is the

Theorem 12.5: Measurable cardinals are strongly inaccessible.

Proof: Let $\mathfrak{m}$ be measurable, $|A| = \mathfrak{m}$ and F a $<\mathfrak{m}$-complete free ultrafilter over A.

(i) $\mathfrak{m}$ is regular. For otherwise a partition $G \subseteq \mathcal{P}(A)$ of A in less than $\mathfrak{m}$ subsets of power $< \mathfrak{m}$ exists. By 12.4, as $\bigcup G = A \in F$, we obtain some $X \in F \cap G$. As $|X| < \mathfrak{m}$ and $X = \bigcup_{a\in X} \{a\} \in F$, again by 12.4 we get $a \in X$ such that $\{a\} \in F$, contradicting the fact that F is free.

(ii) $\mathfrak{m}$ is a strong limit. For take $B \underset{1}{<} A$ and suppose that, nevertheless, $A \underset{1}{\leq} 2^B$, say $h: A \to 2^B$ is an injection. If $b \in B$ then $\{a \in A \mid (h(a))(b) = i\}$ ($i = 0,1$) split A, hence exactly one of these sets is in F. Define $f \in 2^B$ by the condition $\{a \in A \mid (h(a))(b) = f(b)\} \in F$. As $|B| < \mathfrak{m}$ and F is $<\mathfrak{m}$-complete, $\bigcap_{b\in B} \{a \in A \mid (h(a))(b) = f(b)\} \in F$. But $\bigcap_{b\in B} \{a \in A \mid (h(a))(b) = f(b)\} = \{a \in A \mid h(a) = f\}$ contains at most one element as h is injective. □

We shall now have a closer look at the situation indicated just before 12.1, in which it is possible to use a reflection argument. There we have a transitive set W and a universe U, for which $W \subseteq U \neq W$, such that $\Phi^W \leftrightarrow \Phi^U$ for

arbitrary **ZF**-formulas Φ with parameters in W (in such a case we call W *elementary substructure* of U and write $W \prec U$). Suppose that some ordinal $\kappa \in U - W$ exists. Then for every formula Ψ we shall have
If $\Psi^U(\kappa)$ then $\{\beta \in W | \Psi^W(\beta)\}$ is cofinal in $W \cap$ OR. [**]. For suppose that $\xi \in W$. Then $\xi < \kappa$ as W is transitive and $\kappa \notin W$. By hypothesis of [**], $\Psi^U(\kappa)$. From 11.1(*vi*) we obtain $O^U(\kappa)$, therefore (take $x = \kappa$!) $[(\exists x)(\xi \in x \wedge O(x) \wedge \Psi(x))]^U$. As $W \prec U$ and $\xi \in W$ we obtain $[(\exists x)(\xi \in x \wedge O(x) \wedge \Psi(x))]^W$, say, $\xi \in b \in W$, $O^W(b)$ and $\Psi^W(b)$. According to 11.1(*vi*), b is an ordinal and we are through: $\forall \xi \in W\, \exists \beta \in W\, [\xi < \beta \wedge \Psi^W(\beta)]$.
Now let κ be an initial number of measurable cardinality and take in the previous discussion $W = V_\kappa$. 12.6 states that V_κ is elementary substructure of a transitive set U for which $\kappa \in U$.

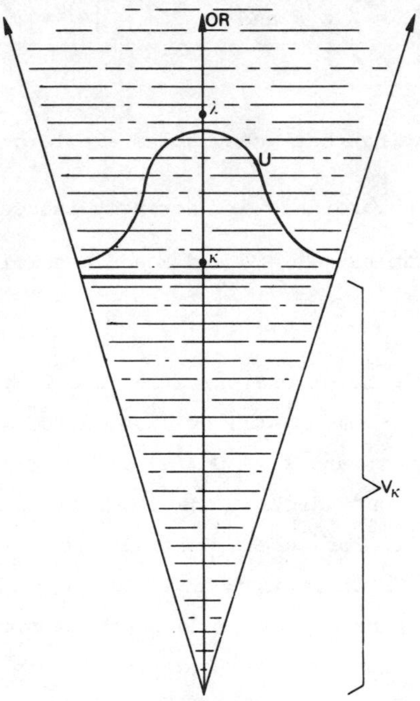

Fig. 3

11.9 states that (*i*) $\alpha \in V_\kappa$ and $\mathrm{IN}^{V_\kappa}(\alpha)$ imply $\mathrm{IN}(\alpha)$; and one may show analogously that (*ii*) if A is transitive and $\alpha \in A$ is strongly inaccessible, then $\mathrm{IN}^A(\alpha)$. Now U is transitive, $\kappa \in U$ and hence by 12.5 and (*ii*) $\mathrm{IN}^U(\kappa)$. From [**] it follows that $\{\beta < \kappa \,|\, \mathrm{IN}^{V_\kappa}(\beta)\}$ is cofinal in κ, and by 11.9 this is exactly the set of strongly inaccessible ordinals below κ. Since κ is regular, this set has type κ, in the terminology of the beginning of this paragraph: $\theta_\kappa = \kappa - \kappa$ is fixed point of the enumeration of inaccessibles. That κ has the other properties, discussed above, follows more or less in the same fashion, using absoluteness-results analogous to (*i*) and (*ii*). For instance, consider $\Psi(\alpha) := \mathrm{IN}(\alpha) \wedge \forall\xi < \alpha\ \exists\delta < \alpha\ (\xi < \delta \wedge \mathrm{IN}(\delta))$. As $\theta_\kappa = \kappa$ we have $\Psi(\kappa)$. Again we can show that $\Psi^U(\kappa)$. [**] says that $\{\beta < \kappa \mid \Psi^{V_\kappa}(\beta)\}$ is cofinal in κ and therefore $\{\beta < \kappa \mid \Psi(\beta)\}$ too is cofinal in κ, which means that κ is the κ-th fixed-point of θ, etc.: κ is inaccessible of every degree $\xi < \kappa$.

Finally we show how to construct the elementary extension U of V_κ in the next

Theorem 12.6: If κ is an initial number of measurable power then V_κ is an elementary substructure of a transitive set U such that $\kappa \in U$.

Proof: Take F to be a $<|\kappa|$ - complete free ultrafilter over κ; the required U is isomorphic with the structure known as the *ultrapower* of V_κ with respect to κ and F which is constructed as follows.
Define the relations $\sim$ and ε on $(V_\kappa)^\kappa$ by
$f \sim g := \{\xi < \kappa \mid f(\xi) = g(\xi)\} \in F$ and
$f \ \varepsilon\ g := \{\xi < \kappa \mid f(\xi) \in g(\xi)\} \in F$.

It is easy to show that $\sim$ is a *congruence* on $(V_\kappa)^\kappa$ with respect to ε, i.e., (*i*) $\sim$ is an equivalence-relation and (*ii*) if $f \sim f'$, $g \sim g'$ and $f\ \varepsilon\ g$ then also $f'\ \varepsilon\ g'$. This means that, in the quotient-structure $(V_\kappa)^\kappa/F$ of equivalence-classes modulo $\sim$ the relation $\bar{\varepsilon}$ can be defined by $|f|\ \bar{\varepsilon}\ |g| := f\ \varepsilon\ g$. ($|f| := \{f' \mid f' \sim f\}$).

With every **ZF**-formula Θ we now associate the formula $\bar{\Theta}$ obtained from Θ by (1) restricting quantifiers to membership in $(V_\kappa)^\kappa/F$ and (2) replacing $\in$ by $\bar{\varepsilon}$. We then get the following result.

Lemma A: $\bar{\Theta}(|f_1|,\ldots,|f_n|) \leftrightarrow \{\xi < \kappa \mid \Theta^{V_\kappa}(f_1(\xi),\ldots,f_n(\xi))\} \in F$.

This is proved by induction on the number of logical symbols in Θ. If there are none then Θ is an identity $x = y$ or has the form $x \in y$; and the result holds by definition. If it is correct for Φ it holds for $\neg\Phi$ as F is an ultrafilter; and if it is correct for both Φ and Ψ it works for $\Phi \wedge \Psi$ too by the filter-properties. Finally, assume that it holds for $\overline{\Phi(y, x_1, \ldots, x_n)}$; we shall show it to hold for $\exists y\ \Phi(y, x_1, \ldots, x_n)$ as well. Suppose that $\overline{\exists y\ \Phi}(y, |f_1|, \ldots, |f_n|)$, that is, for some $g \in (V_\kappa)^\kappa$, $\overline{\Phi}(|g|, |f_1|, \ldots, |f_n|)$. By induction-hypothesis $\{\xi < \kappa \mid \Phi^{V_\kappa}(g(\xi), f_1(\xi), \ldots, f_n(\xi))\} \in F$. But this set is part of $\{\xi < \kappa \mid (\exists y\ \Phi)^{V_\kappa}(y, f_1(\xi), \ldots, f_n(\xi))\}$, which as a result also is in F. Conversely, there certainly exists (AC) a $g: \kappa \to V_\kappa$ for which $(\exists y\ \Phi)^{V_\kappa}(y, f_1(\xi), \ldots, f_n(\xi)) \to \Phi^{V_\kappa}(g(\xi), f_1(\xi), \ldots, f_n(\xi))$, hence if $\{\xi < \kappa \mid (\exists y\ \Phi)^{V_\kappa}(y, f_1(\xi), \ldots, f_n(\xi))\}$ is in F, then so is $\{\xi < \kappa \mid \Phi^{V_\kappa}(g(\xi), f_1(\xi), \ldots, f_n(\xi)\}$. By induction-hypothesis, $\overline{\Phi}(|g|, |f_1|, \ldots, |f_n|)$, and this implies $\overline{\exists y\ \Phi}(y, |f_1|, \ldots, |f_n|)$. □

Now define $\overline{a} \in (V_\kappa)^\kappa$ (when $a \in V_\kappa$) by $\overline{a}(\xi) := a$. Then Lemma A transforms into

<u>Lemma B</u>: $\overline{\Phi}(|\overline{a}_1|, \ldots, |\overline{a}_n|) \leftrightarrow \Phi^{V_\kappa}(a_1, \ldots, a_n)$,

since $\{\xi < \kappa \mid \Phi^{V_\kappa}(a_1, \ldots, a_n)\}$ is either $\emptyset$ or κ; and $\emptyset \notin F$ but $\kappa \in F$. In lemma B one can take the extensionality-axiom for Φ. As this holds in V_κ we get that $\overline{\epsilon}$ is extensional on $(V_\kappa)^\kappa/F$. In order to be able to apply the representation-theorem for well-founded extensional structures we show next that $\overline{\epsilon}$ is well-founded. From theorem 9.8 we see that it suffices to show that no function $f: \omega \to (V_\kappa)^\kappa$ such that $|f(i+1)|\overline{\epsilon}|f(i)|$ $(i \in \omega)$ exists. So suppose such an f does exist. Then for all $i \in \omega$, $\{\xi < \kappa \mid f(i+1)(\xi) \in f(i)(\xi)\} \in F$ and since F certainly is $\aleph_0$-complete, $\bigcap_{i \in \omega}\{\xi < \kappa \mid f(i+1)(\xi) \in f(i)(\xi)\} \in F$. As $\emptyset \notin F$, this set has some member ξ_0. Define $g: \omega \to V_\kappa$ by $g(i) := f(i)(\xi_0)$, then $g(i+1) \in g(i)$ $(i \in \omega)$ contradicting the axiom of regularity. In this way, using 2.9 , the structure $\langle (V_\kappa)^\kappa/F, \overline{\epsilon} \rangle$ is isomorphic to a transitive epsilon-structure $\langle U, \epsilon_u \rangle$. Let h be the isomorphism between the two then, if we define $i: V_\kappa \to U$ by $i(a) := h(|\overline{a}|)$, we get from Lemma B the

<u>Lemma C</u>: $\Phi^U(i(a_1), \ldots, i(a_n)) \leftrightarrow \Phi^{V_\kappa}(a_1, \ldots, a_n)$.

Thus, 12.6 has been proved if we show that i is the identity on V_κ and $\kappa \in U$. The first statement is shown using epsilon induction. By 12.5, κ is strongly inaccessible, thus (11.6) $|a| < |\kappa|$ for all $a \in V_\kappa$. Let $a \in V_\kappa$ and assume as induction hypothesis for epsilon induction that $i(b) = b$ if $b \in a$. Lemma C already implies that $a \subseteq i(a)$. Conversely, suppose that $p \in i(a)$, e.g. $p = h(|f|)$.

$i(a) = h(|\overline{a}|)$, so $|f|\overline{\varepsilon}|\overline{a}|$ as h is an isomorphism. This means that $\{\xi < \kappa \mid f(\xi) \in a\} \in F$ by definition of $\overline{a}$. But $\{\xi < \kappa \mid f(\xi) \in a\} = \bigcup_{b \in a} \{\xi < \kappa \mid f(\xi) = b\}$ and as F certainly is $|a|$-complete, by 12.4 there exists $b \in a$ such that $\{\xi < \kappa \mid f(\xi) = b\} \in F$. Hence $|f| = |\overline{b}|$ and $p = h(|f|) = h(|\overline{b}|) = i(b) = b$ by induction hypothesis and therefore $p \in a$.

Finally we show that $\kappa \in U$. To this end, consider the identity-function $e \in (V_\kappa)^\kappa$ defined by $e(\xi) := \xi$. If $\delta < \kappa$, then $\delta + 1 < \kappa$ and $\delta + 1 \lesssim_1 \kappa$, therefore $\delta + 1 \notin F$ (note that $\delta + 1 = \bigcup_{\xi \in \delta+1} \{\xi\}$; apply 12.4 and remember that F is free). But then $\kappa - (\delta+1) = \{\xi < \kappa \mid \delta \in e(\xi)\} \in F$ and hence $|\overline{\delta}| \,\overline{\varepsilon}\, |e|$; $\delta = i(\delta) = h(|\overline{\delta}|) \in h(|e|)$, i.e., $\kappa \subseteq h(|e|) \in U$. For κ to be in U it now suffices to show that $h(|e|) \in$ OR as U is transitive. From 11.1(vi) we obtain $\forall \xi < \kappa\ O^{V_\kappa}(\xi)$; therefore $\{\xi < \kappa \mid O^{V_\kappa}(\xi)\} = \{\xi < \kappa \mid O^{V_\kappa}(e(\xi))\} = \kappa \in F$. From Lemma A it follows that $\overline{O}(|e|)$ and hence $O^U(h(|e|))$. Applying 11.1(vi) once more, we obtain $h(|e|) \in$ OR. □

<u>Exercises</u>

1 Assume κ to be the *least* measurable initial number and F some $<|\kappa|$-complete free ultrafilter over κ. If W is transitive we can construct, as in the proof of 12.6, a transitive U and an injection $i: W \to U$ such that $\Phi^U(i(a_1),\ldots,i(a_n)) \leftrightarrow \Phi^W(a_1,\ldots,a_n)$ for all Φ and $a_1,\ldots,a_n \in W$, which is the identity on $W \cap V_\kappa$ and such that $\kappa < i(\kappa)$ if $\kappa \in W$.

   (i) Assume that $V_{\kappa+1} \subseteq W$. Show that $G := \{X \subseteq \kappa \mid \kappa \in i(X)\}$ is a $<|\kappa|$-complete free ultrafilter over κ.

   (ii) If $V_{\kappa+1} \subseteq W$ then $V_{\kappa+1} \subseteq U$.

   (iii) $F \notin U$ (thus even if $V_{\kappa+2} \subseteq W$, then, nevertheless, $V_{\kappa+2} \nsubseteq U$ as $F \in V_{\kappa+2} - U$).

   (iv) If $\Phi^{V_{\kappa+1}}(a)$, then $\alpha < \kappa$ exists such that $\Phi^{V_{\alpha+1}}(a \cap V_\alpha)$.

   (v) If G is the ultrafilter defined in (i) and $X_\delta \in G$ where $\delta < \kappa$, then $\{\xi < \kappa \mid \xi \in \bigcap_{\delta<\xi} X_\delta\} \in G$; and if $f: \kappa \to \kappa$ is such that $\{\xi < \kappa \mid f(\xi) < \xi\} \in G$ then $\delta < \kappa$ exists such that $\{\xi < \kappa \mid f(\xi) = \delta\} \in G$; every closed cofinal subset of κ is member of G and finally every normal $f: \kappa \to \kappa$ has a regular fixed point.

2 The collection L of *constructible* sets defined by Gödel in 1938 has the properties (i) to be transitive; (ii) to contain OR as a subclass;

(*iii*) to model all **ZF**-axioms and (*iii*) to be the least collection satisfying (*i*), (*ii*) and (*iii*). Gödel's *constructibility axiom* is the assertion that $V = L$ and it therefore implies that every transitive collection containing OR and modelling all **ZF**-axioms equals V. Prove (Scott, 1961) that if measurable cardinals exist, $V \neq L$. (Hint: take $W = V$ in exercise 1 and apply part (*iii*) of that exercise.)

CHAPTER III
APPLICATIONS

APPLICATIONS

For a number of subjects treated in this chapter, it is assumed that the reader has a basic knowledge of mathematics. In particular we presuppose some knowledge of topology in section 6, which, however, does not exceed a knowledge of the main definitions and some experience. In section 5 we will apply the axiom of choice to concrete situations in certain parts of mathematics. The reader without mathematical experience can safely skip the parts requiring specialised mathematical techniques.
In this chapter we shall, in accordance with mathematical practice, use terms like 'class', 'operation', etc. more loosely.
However, the reader who wishes to do so can easily provide formulations in terms of **ZF**.

1 FILTERS

Filters form very common and convenient tools in many parts of mathematics, for example in topology, analysis and modeltheory. One can define, quite generally, filters on partially ordered sets, e.g. lattices or Boolean algebras (cf. section 2). Here we will study filters of sets, without reference to topological or algebraic structure.

<u>Definition 1.1:</u> $F \subseteq \mathcal{P}(X)$ is a *filter* on X if
(*i*) $\emptyset \notin F$ and $F \neq \emptyset$
(*ii*) $A, B \in F \rightarrow A \cap B \in F$
(*iii*) $A \in F$ and $A \subseteq B \rightarrow B \in F$

Examples

1 $X = \omega$ and $F = \{A \subseteq \omega \mid \{0,1,\ldots,9\} \subseteq A\}$

2 $X = [0,1] \subseteq \underset{\sim}{R}$ and $F = \{A \subseteq [0,1] \mid \mu(A) = 1\}$,
where μ is the Lebesque measure.

3 X is an infinite set and F is the family of all cofinite subsets of X.
(A is cofinite if $|A^c| < \aleph_0$).

Definition 1.2:

a) F is a *principal filter* on X if there is an $a \in X$ such that $F = \{A \subseteq X | a \in A\}$. *b*) F is a *free filter* if $\cap F = \emptyset$.

c) F is an *ultrafilter* (*UF*) if for each filter F' with $F \subseteq F'$ we have $F = F'$

Remark. Filters such as the one in example 1, and free filters, are in a sense at opposite extremes. An ultrafilter is just a maximal filter with respect to inclusion. A principal filter is, trivially, an ultrafilter. The filters in examples 2 and 3 are free filters.
If F is free and $F \subset F'$, then F' is also free.

Definition 1.3: $E \subset \mathcal{P}(X)$ is *centered*, or has the *finite intersection property*, if each finite subset $D \subseteq E$ has a non-empty intersection.

Families with the finite intersection property can be used to construct filters.

Lemma 1.4: If E has the finite intersection property, then there is a unique minimal filter F containing E.

Proof: F is the collection of all sets containing finite intersections of elements of E. It is easy to see that F is a filter and that each filter containing E also contains F. □

We say that the F of lemma 1.4 *is generated by* E.

Theorem 1.5: The following statements are equivalent:

(*i*) F is an ultrafilter

(*ii*) $\forall A \subseteq X \ (A \in F \vee A^c \in F)$

(*iii*) $\forall A,B \subseteq X \ (A \cup B \in F \leftrightarrow A \in F \vee B \in F)$

Proof: (*iii*) → (*ii*) $A \cup A^c = X \in F$ (for F is not empty), so $A \in F$ or $A^c \in F$.

(*ii*) → (*i*) Suppose there is a proper extension F' of F, then there is a set $A \in F' - F$. From $A \notin F$ it follows that $A^c \in F \subseteq F'$, hence $\emptyset = A \cap A^c \in F'$. Contradiction.

i) → *iii*) Since F is a filter,we only have to show $A \cup B \in F \rightarrow A \in F$ or $B \in F$ Suppose $A \cup B \in F$ and $A \notin F$, $B \notin F$. Because F is an ultrafilter $F \cup \{A\}$ nor $F \cup \{B\}$ is centered, hence there are $C,D \in F$ such that $C \cap A = \emptyset$ and $D \cap B = \emptyset$. From the filter properties it follows that $C \cap D \in F$. But then $(C \cap D) \cap (A \cup B) = \emptyset$ holds. Contradiction.

Lemma 1.6: A free filter does not contain finite sets.

Proof: Let $\{a_1,\ldots,a_n\} \in F$, where F is a free filter. Since F is free there exists an $A_i \in F$ for each a_i, such that $a_i \notin A_i$. Therefore $\{a_1,\ldots,a_n\} \cap \bigcap_{i=1}^{n} A_i = \emptyset \in F$. Contradiction. □

Corollary 1.7: *a*. Every free ultrafilter contains all cofinite sets.
b. Every ultrafilter on a finite set is a principal filter.

Theorem 1.8: Every filter is contained in an ultrafilter.

Proof. Consider the collection of all filters containing F, partially ordered by inclusion. A simple application of Zorn's lemma yields the desired result. □

Corollary 1.9: On each infinite set there exist free ultrafilters.

Proof: The filter F of cofinite sets is contained in an ultrafilter F'. Since F is free, so is F'. □

Literature

C.C. Chang and H.J. Keisler. *Model theory*. Amsterdam 1973.
W.W. Comfort and S. Negrepontis. *The theory of ultrafilters*. Berlin 1974.

Exercises

1 Let F be a free ultrafilter. If $A \in F$ and B is finite, then $A - B \in F$.

2 Let E be a countable subset of $\mathcal{P}(\omega)$. Show that E does not generate a free ultrafilter.

3 X is a set with cardinality κ. A subset is called *co-κ* if $|A^c| < \kappa$. Show that the *co-κ* subsets of X form a free filter.

4 If F is a(n) (ultra)filter on X and $A \in F$, then $F \cap \mathcal{P}(A)$ is a(n) (ultra-)filter on A.

5 The set of filters on X is closed under (arbitrary) intersection and union of chains.

2 BOOLEAN ALGEBRAS

In chapter I we have observed that the subsets of a fixed set V behave under the operations of intersection, union and complementation in a way, strongly reminiscent of elementary algebra. In this section we will have a closer look at these properties, without, however, using the specific properties of the set-theoretical operations. We will present an axiomatic theory which formulates the usual properties of $\cup$, $\cap$, ^{c}. In the literature one meets various notations; we list the most familiar ones below.

intersection - infimum - product

$a \cap b$ $\quad a \wedge b$ $\quad a \cdot b$

join - supremum - sum

$a \cup b$ $\quad a \vee b$ $\quad a + b$

complement

a^c $\quad \bar{a}$ $\quad -a$

partial ordening

$a \subseteq b$ $\quad a \leqslant b$

In I. 13.14 we considered structures with *relations* only; in the present section we consider structures with operations as well. In mathematical practice (e.g. algebra) one usually deals with structures with operations - think of groups, rings, fields and vector spaces. It is therefore convenient to extend the concept of 'structure', so as to cover structures with operations.

Remark. Of course, since an operation is a special kind of relation, structures with operators could, in principle, be treated as relational structures. However, it often happens that theories, which are very elegant, when formulated with operations, become extremely cumbersome when formulated with relations.

As a deterrent one simple example: suppose we want to treat addition of natural numbers as a relation, writing $P(x,y,z)$ for $x + y = z$.
A simple statement like $1+(1+2) = 4$ now becomes $\exists u(P(1,2,u) \wedge P(1,u,4))$.

In algebra one needs not only operations and relations, but also *constants*, such as neutral elements, end-points, etc. It is possible to treat a constant as a special case of operation. We shall however introduce constants separately.

Definition 2.1: A *structure* of type $(k_1,\ldots,k_m,l_1,\ldots,l_n,p)$ is an ordered set $\mathbf{A} = \langle A,R_1,\ldots,R_m,F_1,\ldots,F_n,c_1,\ldots,c_p\rangle$, where $A \neq \emptyset$, $R_i \subset A^{k_i}$, $F_j : A^{l_j} \to A$, $c_i \in A$.

A is called the *universe* of $\mathbf{A}$. Observe that each structure is provided with an identity relation, which is *not* listed among the R_i 's. Structures without identity relation are not of interest for us.

Examples

1. $\mathbf{A}$ is a group: $\mathbf{A} = \langle A, \cdot, {}^{-1}, e\rangle$
2. $\mathbf{A}$ is an ordered field: $\mathbf{A} = \langle A, <, +, -, \cdot, {}^{-1}, 0, 1\rangle$
3. $\mathbf{A}$ is a partially ordered set: $\mathbf{A} = \langle A,R\rangle$ with $R \subseteq A^2$.

Definition 2.2: If $\mathbf{A} = \langle A,R_0,\ldots,R_n,F_0,\ldots,F_m,c_0,\ldots,c_p\rangle$ and
$\mathbf{B} = \langle B,S_0,\ldots,S_n,G_0,\ldots,G_m,d_0,\ldots,d_p\rangle$ are of the same type,
then a bijection $f:A \to B$ is called an *isomorphism* if

(*i*) $R_i(a_1,\ldots,a_k) \leftrightarrow S_i(f(a_1),\ldots,f(a_k))$

(*ii*) $f(F(a_1,\ldots,a_l)) = G(f(a_1),\ldots,f(a_l))$

(*iii*) $f(c_i) = d_i$

Notation: $f: \mathbf{A} \simeq \mathbf{B}$

If there exists an isomorphism from $\mathbf{A}$ on $\mathbf{B}$, then we say that $\mathbf{A}$ and $\mathbf{B}$ are *isomorphic* (notation: $\mathbf{A} \simeq \mathbf{B}$).

These general preparations being out of the way, we introduce a new structure, the so-called Boolean algebra.

Definition 2.3: A *Boolean algebra* is a structure

$\mathbf{A} = \langle A, \cup, \cap, ^c, \mathbf{0}, \mathbf{1} \rangle$, satisfying the following axioms:

1 *Commutativity*

$a \cup b = b \cup a$ $\qquad$ $a \cap b = b \cap a$

2 *Associativity*

$(a \cup b) \cup c = a \cup (b \cup c)$ $\qquad$ $(a \cap b) \cap c = a \cap (b \cap c)$

3 *Distributivity*

$a \cup (b \cap c) = (a \cup b) \cap (a \cup c)$ $\qquad$ $a \cap (b \cup c) = (a \cap b) \cup (a \cap c)$

4 *Idempotence*

$a \cup a = a$ $\qquad$ $a \cap a = a$

5 *Reciprocity*

$(a \cup b)^c = a^c \cap b^c$ $\qquad$ $(a \cap b)^c = a^c \cup b^c$

6 $\qquad$ $(a^c)^c = a$

7 $a \cup a^c = \mathbf{1}$ $\qquad$ $a \cap a^c = \mathbf{0}$

8 $a \cup \mathbf{0} = a$ $\qquad$ $a \cap \mathbf{1} = a$

9 $a \cup \mathbf{1} = \mathbf{1}$ $\qquad$ $a \cap \mathbf{0} = \mathbf{0}$

The axioms 1 to 9 are certainly not the most economical ones; for our purpose that is not important.

As usual, in the case of associative operations we dispense with superfluous parentheses, e.g. we write $a \cup b \cup c$ instead of $a \cup (b \cup c)$.
Also we introduce arbitrary finite joins and meets (intersections) by

$$\begin{cases} \bigcup_{i \leqslant 1} a_i := a_1 \\ \bigcup_{i \leqslant n+1} a_i := \bigcup_{i \leqslant n} a_i \cup a_{n+1} \end{cases} \qquad \begin{cases} \bigcap_{i \leqslant 1} a_i := a_1 \\ \bigcap_{i \leqslant n+1} a_i := \bigcap_{i \leqslant n} a_i \cap a_{n+1} \end{cases}$$

When no confusion arises we will just write $\bigcup_i$, $\bigcap_i$ instead of $\bigcup_{i \leqslant n}$, $\bigcap_{i \leqslant n}$.

By means of mathematical induction one easily shows

Lemma 2.4: (*i*) $a \cap \bigcup_i b_i = \bigcup_i (a \cap b_i)$

(*ii*) $a \cup \bigcap_i b_i = \bigcap_i (a \cup b_i)$

(*iii*) $(\bigcap_i a_i)^c = \bigcup_i a_i^c$

(*iv*) $(\bigcup_i a_i)^c = \bigcap_i a_i^c$

From the axioms 6 and 7 we immediately conclude

Lemma 2.5: $\mathbf{1}^c = \mathbf{0}$ and $\mathbf{0}^c = \mathbf{1}$

Examples of Boolean algebras

1 In chapter I we already observed that $\mathcal{P}(V)$, with the usual operations and $\mathbf{0} = \emptyset$, $\mathbf{1} = V$, is a Boolean algebra.

2 Consider in ordinary 2-valued propositional logic the relation $\sim$, defined by $P \sim Q$:= $P \leftrightarrow Q$ is a tautology.

It is a simple exercise in logic to show that

(1) $\sim$ is an equivalence relation

(2) the equivalence classes form a Boolean algebra under the operations

$[P] \cup [Q] := [P \vee Q]$

$[P] \cap [Q] := [P \wedge Q]$

$[P]^c := [\neg P]$

The constants are $\mathbf{1} := [P \vee \neg P]$

$\mathbf{0} := [P \wedge \neg P]$.

The connection between Boolean algebra and logic is exploited in algebraic logic. See for example, H. Rasiowa and R. Sikorski, *The mathematics of metamathematics* (Warsaw, 1963)

3 The set of subsets of a fixed set V, which are either finite or cofinite, is a Boolean algebra under the usual operations.

If V is infinite, then this is a proper subalgebra of $\mathcal{P}(V)$ (example 1.)

4 The Lebesque-measurable subsets of $\mathbb{R}$ form a Boolean-algebra.

5 Consider the special case of example 1, where V contains exactly one element. The Boolean algebra contains only the elements $\mathbf{0}$ and $\mathbf{1}$. The tables for $\cup$, $\cap$ and ^{c} are

$\cup$	$\mathbf{0}$	$\mathbf{1}$
$\mathbf{0}$	$\mathbf{0}$	$\mathbf{1}$
$\mathbf{1}$	$\mathbf{1}$	$\mathbf{1}$

$\cap$	$\mathbf{0}$	$\mathbf{1}$
$\mathbf{0}$	$\mathbf{0}$	$\mathbf{0}$
$\mathbf{1}$	$\mathbf{0}$	$\mathbf{1}$

^{c}	
$\mathbf{0}$	$\mathbf{1}$
$\mathbf{1}$	$\mathbf{0}$

Note that these are precisely the truth tables for $\vee$, $\wedge$, $\neg$. Example 2 suggests that this is not purely accidental.
This algebra is the smallest non-degenerate Boolean algebra. If, in example 1, one takes $V = \emptyset$, then the resulting algebra has only one element (so $\mathbf{0} = \mathbf{1}$). Since this algebra is rather uninteresting, we will always assume that $\mathbf{0} \neq \mathbf{1}$.
In a Boolean algebra one can define a partial ordering; I. 5.4. and I. 5.9. already gave a clue how to proceed.

<u>Definition 2.6</u>: $a \leqslant b := a \cap b = a$

$a < b := a \leqslant b \wedge a \neq b$

<u>Lemma 2.7</u>: (*i*) $a \leqslant a$

(*ii*) $a \leqslant b \wedge b \leqslant a \rightarrow a = b$

(*iii*) $a \leqslant b \wedge b \leqslant c \rightarrow a \leqslant c$

Proof. (*i*) trivial

(*ii*) trivial

(*iii*) $a \cap b = a$ and $b \cap c = b$ are given.

We now evaluate $a \cap c$.

$a \cap c = (a \cap b) \cap c = a \cap (b \cap c) = a \cap b = a$, hence $a \leqslant c$. □

Lemma 2.7 shows that $\leqslant$ is a partial ordering.

We give some alternative characterizations of the relation $\leqslant$.

Lemma 2.8: (*i*) $a \leqslant b \leftrightarrow a^c \cup b = \mathbf{1}$

(*ii*) $a \leqslant b \leftrightarrow b^c \leqslant a^c$

(*iii*) $a \leqslant b \leftrightarrow a \cup b = b$

Proof: (*i*) $a \cap b = a \rightarrow a^c \cup b = (a \cap b)^c \cup b = a^c \cup b^c \cup b = a^c \cup \mathbf{1} = \mathbf{1}$.
Conversely $a^c \cup b = \mathbf{1} \rightarrow a = a \cap \mathbf{1} = a \cap (a^c \cup b) = \mathbf{0} \cup (a \cap b) = a \cap b$.

(*ii*) $a \leqslant b \leftrightarrow a^c \cup b = \mathbf{1} \leftrightarrow a^c \cup (b^c)^c = \mathbf{1} \leftrightarrow b^c \leqslant a^c$.

(*iii*) $a \leqslant b \leftrightarrow b^c \leqslant a^c \leftrightarrow b^c \cap a^c = b^c \leftrightarrow a \cup b = b$.

Lemma 2.9: (absorption laws)

(*i*) $a \cap (a \cup b) = a$

(*ii*) $a \cup (a \cap b) = a$

Proof. To show (*i*) it suffices to prove $a \leqslant a \cup b$.
Apply 2.8 (*i*): $a^c \cup (a \cup b) = \mathbf{1} \cup b = \mathbf{1}$.
(*ii*) follows from (*i*) by reciprocity. □

Lemma 2.10: (*i*) $a \cup b$ is the supremum of a and b.

(*ii*) $a \cap b$ is the infimum of a and b.

Proof. (*i*) By 2.9. we know $a, b \leqslant a \cup b$.
Now let $a, b \leqslant c$, then, by 2.8 (*i*) $a^c \cup c = \mathbf{1}$ and $b^c \cup c = \mathbf{1}$, hence $(a^c \cup c) \cap (b^c \cup c) = \mathbf{1}$, or $(a^c \cap b^c) \cup c = (a \cup b)^c \cup c = \mathbf{1}$.
By 2.8 (*i*) this shows $a \cup b \leqslant c$, i.e. $a \cup b$ is the least element greater or equal than a and b. (*ii*) Analoguous, or apply reciprocity. □

Lemma 2.11: (*i*) $a^c = \sup \{x \mid a \cap x = \mathbf{0}\}$

(*ii*) $a^c = \inf \{x \mid a \cup x = \mathbf{1}\}$

Proof. (*i*) We already know that $a \cap a^c = \mathbf{0}$, so we only have to show that $a \cap x = \mathbf{0} \rightarrow x \leqslant a^c$. From $a \cap x = \mathbf{0}$ it follows that $a^c \cup x^c = \mathbf{1}$ or $x \leqslant a^c$.
(*ii*) similar □

From lemmas 2.10 and 2.11 we see that the boolean operations are definable from the partial order. This technique is used in lattice theory.

Another possible way to define the complement from $\cup$ and $\cap$ is given by

Corollary 2.12: a^c is the unique x satisfying $x \cap a = \mathbf{0}$ and $x \cup a = \mathbf{1}$.

Proof. Apply lemma 2.11. □

We began our list of examples of Boolean algebras with the powerset of a set V and in 3 and 4 we considered subalgebras of such an algebra. It was shown by Stone that $\mathcal{P}(V)$ is a prototype for Boolean algebras, in the sense that each Boolean algebra is a subalgebra of some $\mathcal{P}(V)$, up to isomorphism.

We will prove a special case of Stone's theorem here and indicate a proof of the general theorem in an exercise.

Definition 2.13: An element a of a Boolean algebra is an *atom* if $a \neq \mathbf{0}$ and $\forall b (b < a \rightarrow b = \mathbf{0})$.

Example: In $\mathcal{P}(V)$ the singletons are atoms.

Definition 2.14: A Boolean algebra is called *atomic* if for each $a \neq \mathbf{0}$ there is an atom b such that $b \leqslant a$.

$\mathcal{P}(V)$ is atomic, but the algebra of open and closed sets of Cantorspace is not (cf. exercise 7).

The following fact concerning atoms will come in handy .

Lemma 2.15: If p is an atom and $p \leqslant a \cup b$, then $p \leqslant a$ or $p \leqslant b$.

Proof. Suppose $\neg\, p \leqslant a$ and $\neg\, p \leqslant b$, then $p \cap a \neq p$ and $p \cap b \neq p$. Since p is an atom this implies $p \cap a = p \cap b = \mathbf{0}$.
By $p \leqslant a \cup b$, we have $p \cap (a \cup b) = p$.
So $p = p \cap (a \cup b) = (p \cap a) \cup (p \cap b) = \mathbf{0}$.
Contradiction. Hence $p \leqslant a$ or $p \leqslant b$. □

For atomic Boolean algebras we have the following representation theorem:

Theorem 2.16: Every atomic Boolean algebra is isomorphic to a subalgebra of some power set.

Proof. Let V be the set of all atoms of the atomic Boolean algebra.
Define for each element $x, F(x) = \{a \in V | a \leqslant x\}$.
We claim that F is the desired isomorphism.

1. F is injective
   If $a \neq b$, then $\neg\, a \leqslant b$ or $\neg\, b \leqslant a$
   so $a \cap b^c \neq \mathbf{0}$ or $b \cap a^c \neq \mathbf{0}$.
   Let $a \cap b^c \neq \mathbf{0}$, then there is an atom p such that $p \leqslant a \cap b^c$, or $p \leqslant a$ and $p \leqslant b^c$. Suppose $p \leqslant b$, then $p \leqslant b \cap b^c = \mathbf{0}$. Contradiction ;
   so $p \in F(a) - F(b)$, or $F(a) \neq F(b)$.
2. $F(\mathbf{0}) = \emptyset$ and $F(\mathbf{1}) = V$. Immediate
3. $F(a \cup b) = F(a) \cup F(b)$.
   $F(a) \cup F(b) \subseteq F(a \cup b)$ is immediate.
   For $F(a \cup b) \subseteq F(a) \cup F(b)$ apply lemma 2.15.
4. $F(a \cap b) = F(a) \cap F(b)$. Immediate.
5. $F(a^c) = (F(a))^c$.
   It suffices to show $F(a^c) \cup F(a) = V$, since trivially $F(a^c) \cap F(a) = \emptyset$.
   For any $p \in V$, we have $p \leqslant \mathbf{1} = a^c \cup a$.
   Now apply 3. □

A subalgebra of a powerset is called a *field of sets*. So we can reformulate the above representation theorem:
An atomic Boolean algebra is isomorphic to a field of sets.

Corollary 2.17: Each finite Boolean algebra is isomorphic to a Boolean algebra $\mathcal{P}(V)$.

Proof. A finite Boolean algebra is trivially atomic, so we can apply the previous theorem. We only have to show now that F, defined above, is surjective.
Let $\{p_1, \ldots, p_n\} \subseteq V$. We show that $\{p_1, \ldots, p_n\} = F(p_1 \cup \ldots \cup p_n)$.
For convenience we take $n = 2$.

(*i*) From $p_i \leqslant p_1 \cup p_2$, it follows that $\{p_1, p_2\} \subseteq F(p_1 \cup p_2)$.

(*ii*) Let $p \leqslant p_1 \cup p_2$, then by lemma 2.15 $p \leqslant p_1$ or $p \leqslant p_2$. Since p_1, p_2 are atoms we have $p = p_1$ or $p = p_2$. Hence $p \in \{p_1, p_2\}$.

Each subset of V, therefore, is in the range of F. □

From 2.17 it follows that the number of elements of a finite Boolean algebra is always a power of 2, and that the number of elements determines the finite Boolean algebra up to isomorphism (see ex. 3).

The generalization of theorem 2.16 is the

Representation theorem of Stone :

Each Boolean algebra is isomorphic to a field of sets.

For a proof see ex. 1.

Literature

P.R. Halmos. Lectures on Boolean Algebras Princeton 1963.

R. Sikorski. Boolean Algebras. Berlin 1960.

Exercises.

1. A filter ∇ on a Boolean algebra $\mathbf{A}$ is a subset with the properties

   (*i*) $a, b \in \nabla \rightarrow a \cap b \in \nabla$

   (*ii*) $a \in \nabla$ and $a \leqslant b \rightarrow b \in \nabla$

   ∇ is a *proper filter* if $\nabla \neq A$ (equivalently $\mathbf{0} \notin \nabla$). An *ultrafilter* is a maximal, proper filter.

   a. Let ∇ be a filter with $a \notin \nabla$. Show that ∇ is contained in an ultrafilter ∇' with $a \notin \nabla'$ (apply Zorn's lemma).

   b. Show that for an ultrafilter ∇, $a \cup b \in \nabla \leftrightarrow a \in \nabla$ or $b \in \nabla$, and $a^c \in \nabla \leftrightarrow a \notin \nabla$.

   c. Define $F(a) = \{\nabla \mid \nabla$ is an ultrafilter and $a \in \nabla\}$

      Show that: F is injective (apply a.) ; $F(a^c) = (F(a))^c$

      $F(a \cap b) = F(a) \cap F(b)$; $F(\mathbf{0}) = \emptyset$

      $F(a \cup b) = F(a) \cup F(b)$; $F(\mathbf{1}) = \{\nabla \mid \nabla$ is an ultrafilter$\}$

   d. Prove Stone's representation theorem.

2. Let V be an infinite set.
Define $A + B := (A - B) \cup (B - A)$
$A \sim B := A + B$ is finite
Show that: a. $\sim$ is an equivalence relation
b. $\mathcal{P}(V)/_{\sim}$ is a Boolean algebra with operations induced by $\cup, \cap, {}^c$.
c. $\mathcal{P}(V)/_{\sim}$ does not contain atoms.

3. a. Show that the number of elements of a finite Boolean algebra is a power of 2 .
b. Show that finite Boolean algebras with the same number of elements are isomorphic.

4. Finitely generated Boolean algebras are finite .

5. Represent graphically, as a partially ordered set, the Boolean algebra generated by the distinct atoms a,b,c.
What are the subalgebras?

6. In an atomic Boolean algebra the following holds:
$b = \sup \{a \mid a \leqslant b$ and a is an atom$\}$.

7. The clopen (i.e. closed and open) subsets of Cantor space form a non-atomic Boolean algebra.

8. $a \leqslant b \leftrightarrow a \cap b^c = \mathbf{0}$

9. Show that the axioms 5 and 6 in definition 2.3 can be deduced from the remaining ones .

10. Show that one of the distributive axioms in definition 2.3 can be deduced from the remaining axioms.

11. Show $a = \mathbf{0} \leftrightarrow$ for all b $b = (a \cap b^c) \cup (a^c \cap b)$.

12. Define the operation $\rightarrow$ by $a \rightarrow b := a^c \cup b$ (observe the analogy with logic). Show: $a \rightarrow a = \mathbf{1}$,
$(a \rightarrow b) \cap (b \rightarrow c) \leqslant (a \rightarrow c)$.

13. Define $a \leftrightarrow b := (a \rightarrow b) \cap (b \rightarrow a)$ and
$a \sim b := a \leftrightarrow b \in \nabla$, for a filter ∇.
Show that $\sim$ is an equivalence relation.
Show $a \sim b \leftrightarrow (\exists c \in \nabla)(a \cap c = b \cap c)$.

If we modify the definition of isomorphism (2.2) by leaving out the bijectivity condition, the resulting mapping is called a *homomorphism*.

14. Show that the kernel of a homomorphism f (i.e. $\{a|f(a) = \mathbf{1}\}$) is a filter.

15. Let ∇ be a filter on $\mathbf{A}$ and $\sim$ the corresponding equivalence relation (ex. 13). Denote the equivalence class of a by $[a]$.
Define $[a] \cup [b] := [a \cup b]$
$[a] \cap [b] := [a \cap b]$
$[a]^c := [a^c]$
Show that these operations are well-defined (i.e. independent of the choice of representatives).
Show that the resulting structure is a Boolean algebra. What are the $\mathbf{0}$ and $\mathbf{1}$ of this algebra?
The resulting algebra is called the quotient algebra. Notation: $\mathbf{A}/_\nabla$.

16. Show that $a \to [a]$ is a homomorphism of $\mathbf{A}$ onto $\mathbf{A}/_\nabla$, with kernel ∇.

17. Show that ∇ is an ultrafilter iff $\mathbf{A}/_\nabla$ is (isomorphic to) the two-element algebra.

18. Let A_2 be the two element algebra and let a be a non-zero element of $\mathbf{A}$. Show that there exists a homomorphism $h: \mathbf{A} \to A_2$ such that $h(a) = \mathbf{1}$.
(Hint: apply ex. 1.a)

Definition: (*i*) a polynomial is an expression constructed from variables, $\mathbf{0}$ and $\mathbf{1}$, by means of the operations $\cup, \cap$ and ^{c}.

(*ii*) An *equation* is of the form $p = q$, where p and q are polynomials.

(*iii*) The equation $p = q$ holds in the Boolean algebra $\mathbf{A}$ if all substitutions of elements of $\mathbf{A}$ for the variables in p and q yield identical elements.

(*iv*) $p = q$ is true if it holds in all Boolean algebras.

19. If $\mathbf{A} = \langle A, \cup, \cap, ^c, \mathbf{0}, \mathbf{1} \rangle$ is a Boolean algebra, then so is $\mathbf{A}^d = \langle A, \cap, \cup, ^c, \mathbf{1}, \mathbf{0} \rangle$ (the *dual* algebra).

20. Define the dual of a polynomial p by $x^d := x$; $(p \cup q)^d := p^d \cap q^d$; $(p \cap q)^d := p^d \cup q^d$; $\mathbf{0}^d = \mathbf{1}$, $\mathbf{1}^d = \mathbf{0}$, $(p^c)^d = (p^d)^c$.
Show: $(p^d)^d = p$ is true.

Show that $p^d = q^d$ is true if $p = q$ is true (the *duality principle*).

21. Show that the mapping $a \to a^c$ is an isomorphism. What is the image of **A** ?

22. Define the following mapping on polynomials:

$$x^* := x^c \ ; \ (p \cup q)^* = p^* \cap q^* ; \ (p \cap q)^* = p^* \cup q^* ;$$
$$(p^c)^* = (p^*)^c ; \ \mathbf{0}^* = \mathbf{1} \ ; \ \mathbf{1}^* = \mathbf{0}$$

Show: $(p^*)^* = p$ is true

$p^* = p^c$ is true

23. Show that $p = q$ is true iff it holds in the two-element Boolean algebra. (This property of Boolean algebras is called *equational completeness*). Find a decision method for truth of equations.

3 ORDER TYPES

Adopting the algebraic practice of identifying isomorphic structures, one can, in the theory of order, identify isomorphic ordered sets. From the viewpoint of ordering, for example, ω and the set of primes are indistinguishable. When we consider ordered sets, we call the isomorphism types *order types*. We use here the theory and notation from I.14. If there is no confusion we will write A instead of $\langle A, R \rangle$. Furthermore we will try to denote, whenever possible, the order relation by $<$ or $\prec$.

For convenience, we briefly recapitulate some facts.

$f : \langle A, < \rangle \to \langle B, \prec \rangle$ is an isomorphism if (*i*) f is bijective

(*ii*) $a < a' \leftrightarrow f(a) \prec f(a')$.

If there exists an isomorphism from $\langle A, < \rangle$ to $\langle B, \prec \rangle$, then $\langle A, < \rangle$ and $\langle B, \prec \rangle$ are called isomorphic.

Notation: $\langle A, < \rangle \simeq \langle B, \prec \rangle$ or $A \simeq B$.

As in the case of the cardinality 'relation', $=_1$, $\simeq$ is in general not a relation, for the reason that $\simeq$ is not a set. Likewise $\{A | A \simeq B\}$ is in general not a set. Thus the theory of order types meets with the same difficulties as the theory of cardinals. The solution for the latter theory (that is to say, the intervention of the *bottom operation*), presented in II.10, can easily be adopted here. However, we adopt a naive attitude and assume that the difficulties have somehow been satisfactorily resolved (e.g. by working in a fixed set).

Indeed, one can derive most of the benefits of the theory of order types by working simply with individual ordered sets.

The order type of an ordered set A will be denoted by $[A]$. All we actually need, for operating with order types, is the property $A \simeq B \leftrightarrow [A] = [B]$.

The first question one asks, after the introduction of some specific isomorphism type, is: can the types be classified in some systematic way, such that some nice structure emerges? For order types this task seems rather hopeless. Contrary to the pleasing state of affairs in the case of the cardinals, the order types do not yield an interesting structure.

We will illustrate the situation by showing below a number of non-isomorphic orderings of ω.

1. . . . . natural ordering
   0 1 2 3 4

2. .... inverse ordering
   4 3 2 1 0

3. . . . .
   0 2 4 6 1 3 5 7

   even numbers before odd numbers, each in their natural ordering

4. ....
   5 3 1 0 2 4 6

   odd numbers before even numbers, the first in inverse order, the latter in natural order.

5. . . . .
   2 4 6 8 0 7 5 3 1

   positive even numbers in increasing order with 'limit' 0, followed by odd numbers in decreasing order.

   Fortunately there are some countable orderings with a simple characterization. The best known example was provided by Cantor in 1895.

   He showed

<u>Theorem 3.1</u>: Any two countable sets with a dense ordering without endpoints are isomorphic.

A set is densely ordered if between any two points of the set there is another point of the set. The absence of endpoints is expressed by: for each element there is a smaller and a larger one. We can formulate those two conditions in logical symbols:

$$\forall xy\ (x < y \rightarrow \exists z\ (x < z \wedge z < y))$$

$$\forall x \exists y \exists z\ (x < y \wedge z < x)$$

The fact that the theory of dense order without endpoints has, up to isomorphism, only one countable model, is expressed in model theory by the phrase 'the theory of dense order without endpoints is $\aleph_0$-categorical'.
Since $\underset{\sim}{Q}$ is densely ordered without endpoints (in its natural order), we can reformulate Cantor's theorem as follows:

<u>Theorem 3.1.a</u>. Every countable, densely ordered set without endpoints is isomorphic to $\underset{\sim}{Q}$.

We now prove theorem 3.1 by means of Cantor's *back-and-forth* method.

Let A and B be two countable, densely ordered sets without endpoints. By definition we have bijections $f\colon \underset{\sim}{N} \rightarrow A$ and $g\colon \underset{\sim}{N} \rightarrow B$. For convenience we write $a_i = f(i)$ and $b_i = g(i)$. We want an isomorphism $F\colon A \rightarrow B$; this isomorphism will be constructed from 'finite approximations', so called *partial isomorphisms*.

In our proof partial isomorphisms are isomorphisms from finite subsets to finite subsets. We first prove a lemma about partial isomorphisms:

Let U and V be countable densely ordered sets without endpoints and let H be a partial isomorphism from U to V, with $x \notin \text{Dom } H$, then there is a partial isomorphism H' with domain $\text{Dom } H \cup \{x\}$, which extends H.

Proof: $U = \{u_i \mid i \in \underset{\sim}{N}\}$, $V = \{v_i \mid i \in \underset{\sim}{N}\}$, $\text{Dom } H = \{p_1, \ldots, p_n\} \subseteq U$. Suppose $p_1 < p_2 < \ldots < p_n$ (this is no restriction). The order properties imply that (i) $x < p_1$, or (ii) $p_i < x < p_{i+1}$ for some $i < n$, or (iii) $p_n < x$.
<u>Case (i)</u> There exists a $y \in V$ such that $y < H(p_1)$ (V has no endpoints). Determine the first k such that $v_k < H(p_1)$ and define $H'(x) = v_k$ and $H'(p_i) =$

$H(p_i)$.

Case (*iii*). Idem.

Case (*ii*) . There is a y such that $H(p_i) < y < H(p_i)$. (because of the dense ordering). Determine the first k such that $H(p_i) < v_k < H(p_{i+1})$ and put $H'(x) = v_k$ and $H'(p_i) = H(p_i)$.

From the construction it appears that H' is a partial isomorphism. This proves the lemma. (Note that we have given a specific algorithm for the extension by prescribing the choice of $H'(x)$; this plays a role below).

We now define a sequence of partial isomorphisms from A to B.

1. Define $H_1\ (a_0) = b_0$.
2. Consider the partial isomorphism H_1^{-1} and apply the lemma to H_1^{-1} and b_1. The result is H_1'. Put $H_2 = (H_1')^{-1}$.

. . . .

. . . .

(2*n*+1). Let H_{2n} be defined. If a_i is the first point in the sequence $a_0, a_1, a_2, \ldots$ such that $a_i \notin \text{Dom } H_{2n}$, apply the lemma to H_{2n} and a_i. Result: an extension H_{2n}' with domain $\text{Dom } H_{2n} \cup \{a_i\}$.
Put $H_{2n+1} = H_{2n}'$.

(2*n*+2). Let H_{2n+1} be defined. If b_i is the first point in the sequence $b_0, b_1, b_2, \ldots,$ such that $b_i \notin \text{Ran } (H_{2n+1})$, then apply the lemma to b_i and H_{2n+1}^{-1}. Result: an extension G with domain $\text{Ran } (H_{2n+1}) \cup \{b_i\}$.
Put $H_{2n+2} = G^{-1}$.

Finally we define $H = \bigcup\{H_n \mid n \in \underset{\sim}{N}\}$.

From the construction we immediately see that $\text{Dom } H = A$, $\text{Ran } H = B$. It is also clear that H preserves the order: Let $a_i < a_j$, determine an n such that $a_i, a_j \in \text{Dom } H_n$, then $H_n(a_i) < H_n(a_j)$. Since H is an extension of H_n, we also have $H(a_i) < H(a_j)$. □

The most important and most interesting ordertypes are those belonging to well-ordered sets. Historically, this was how Cantor introduced his ordinals. Later Von Neumann gave an extremely elegant definition of ordinal, avoiding the introduction of non-sets. (cf II, 3.). We will discuss Cantor's method nonetheless because his notion of ordinal (based on well-orderings) has an intuitive and geometrical perspicuity that provides a good motivation for a lot of work in the theory of ordinals. This involves the equivalence class approach (although most of the time we will, in fact, deal with concrete well-orderings). Note, however that in this case we can avoid difficulties by invoking II. 2.13.

Definition 3.2: An *ordinal* is the order type of a well-ordered set.

The reader must be aware that one cannot define something that has already been defined. The above definition is, in a sense local, i.e. it is for this section. The relation between the two kinds of ordinals is seen as follows: Consider the ordinals (II.3) in the sense of Von Neumann, which are well-ordered sets, $\alpha, \beta, \gamma, \ldots$ An application of the recursion lemma (II.2.8) shows that for each well ordered set A there is an ordinal α such that $A \simeq \alpha$. Namely define $f(\beta)$ = the least element of $A - f''\beta$ (i.e. the least element in A that is not yet an image of some $\gamma < \beta$ under f).
Since A is a set, there is an α such that $f\colon \alpha \simeq A$.
By I.14.15 and I.14.16 there is exactly one such ordinal.
In short, each ordinal à la Definition 3.2 contains exactly one ordinal à la Von Neumann. From now on we will talk about ordinals in this section and mean order types. The reader should, however, keep the connection, described above, in mind.

Notation: $\alpha, \beta, \gamma, \ldots$ will be used to denote ordinals.
In I.14. the reader has already seen some well-orderings, so we can introduce already some ordinals. Because ordinals are generalizations of natural numbers we start with presenting the analogues of the natural numbers.

Definition 3.3:
$$\underset{\sim}{0} := [\emptyset]$$
$$\underset{\sim}{1} := [\{0\}]$$
$$\underset{\sim}{2} := [\{0, 1\}]$$
$$\cdots\cdots\cdots$$
$$\underset{\sim}{n} := [\{0, \ldots, n-1\}]$$

Since $\underset{\sim}{N}$ is well ordered, we can make one more step: $\omega := [\underset{\sim}{N}]$

Note that in each case we considered a set with an ordering, namely the natural ordering.

We call $\underset{\sim}{0}$, $\underset{\sim}{1}$, $\underset{\sim}{2}$, ... *finite ordinals.*

There are lots of ordinals, namely as many as well ordered sets. We will try to visualize a few, using the operations of addition and multiplication.

Definition 3.4: Let $\langle A, \underset{A}{\leq} \rangle$ and $\langle B, \underset{B}{\leq} \rangle$ be disjunct, ordered sets, then

$$\langle A, \underset{A}{\leq} \rangle + \langle B, \underset{B}{\leq} \rangle := \langle A \cup B, < \rangle \text{ with } x < y := x \underset{A}{\leq} y \text{ and } x,y \in A \text{ or } x \in A \text{ and } y \in B \text{ or } x \underset{B}{\leq} y \text{ and } x,y \in B.$$

We call this the *ordered sum*, and we write $A + B$ if no confusion arises.

Lemma 3.5: If A and B are well-ordered, so is $A + B$.

Proof. Immediate □

Lemma 3.6. If $A \simeq A'$ and $B \simeq B'$, then $A + B \simeq A' + B'$.

Proof. Immediate. □

Now we can consider the addition of ordinals. First we note that for ordinals α and β we can always find disjunct A and B such that $\alpha = [A]$ and $\beta = [B]$. Let, for example, $\alpha = [C]$ and $\beta = [D]$, then we define $A = C \times \{0\}$ and $B = D \times \{1\}$, with orderings $<c,0> < <c',0> \leftrightarrow c < c'$

$<d,1> < <d',1> \leftrightarrow d < d'$.

Clearly $A \cap B = \emptyset$ and $A \simeq C, B \simeq D$.

Definition 3.7: $\alpha + \beta = [A + B]$ for disjunct A and B with $\alpha = [A]$ and $\beta = [B]$.

Note that this addition of ordinals is well defined by lemma 3.6.

Examples

1. $\underset{\sim}{2} + \underset{\sim}{3} = \underset{\sim}{5}$, for

   $\underset{\sim}{2} = [\{0,1\}]$, $\underset{\sim}{3} = [\{2,3,4\}]$ and $\underset{\sim}{2} + \underset{\sim}{3} = [\{0,1,2,3,4\}] = \underset{\sim}{5}$.

2. $\underset{\sim}{1} + \omega = \omega$, for
   $\underset{\sim}{1} = [\{0\}]$ and $\omega = [\{1,2,3, \ldots\}]$
   hence $\underset{\sim}{1} + \omega = [\{0,1,2,3, \ldots\}] = \omega$.

3. $\omega + \underset{\sim}{1} \neq \omega$, for
   $\omega = [\{0,1,2, \ldots\}]$, $\underset{\sim}{1} = [\{\frac{1}{2}\}]$ and
   $\omega + \underset{\sim}{1} = [\{0,1,2, \ldots, \frac{1}{2}\}]$ with $n < \frac{1}{2}$ for all n.
   However $\{0,1,2, \ldots\}$ and $\{0,1,2, \ldots, \frac{1}{2}\}$ are not isomorphic (the latter set has an end point).

4. $\omega + \omega = [\{0,2,4,6, \ldots\}] + [\{1,3,5, \ldots\}]$

Gradually we get a longer and longer sequence of ordinals:

$\underset{\sim}{0}, \underset{\sim}{1}, \underset{\sim}{2}, \ldots, \underset{\sim}{n}, \ldots, \omega, \omega + \underset{\sim}{1}, \omega + \underset{\sim}{2}, \ldots, \omega + \omega, \omega + \omega + \underset{\sim}{1}, \ldots,$
$\omega + \omega + \omega, \ldots$

We can make even larger ordinals by introducing multiplication. It seems plausible to use the lexicographical ordering of cartesian products for this purpose. However, in order to preserve the analogy to II. 5, the anti-lexicographical ordering is better suited.

<u>Definition 3.8</u>: $\langle A, \underset{A}{\leq} \rangle \otimes \langle B, \underset{B}{\leq} \rangle := \langle A \times B, < \rangle$
where $<a,b> \, < \, <a',b'> \; := b \underset{B}{<} b' \vee (b = b' \wedge a \underset{A}{<} a')$.

$\langle A \times B, < \rangle$ is the *ordered product* of $\langle A, \underset{A}{\leq} \rangle$ and $\langle B, \underset{B}{\leq} \rangle$. For convenience we simply write $A \otimes B$.

<u>Theorem 3.9</u>: If A and B are well-ordered, then so is $A \otimes B$.

Proof: Exc. II, 5 no 4. □

Geometrically one can visualize the ordering on $A \otimes B$ as follows (see fig.1): $<a,b>$ is smaller than $<a',b'>$ if either $<a,b>$ is on a horizontal line below $<a',b'>$, or they are on the same horizontal line, but $<a,b>$ is to the left of $<a',b'>$.

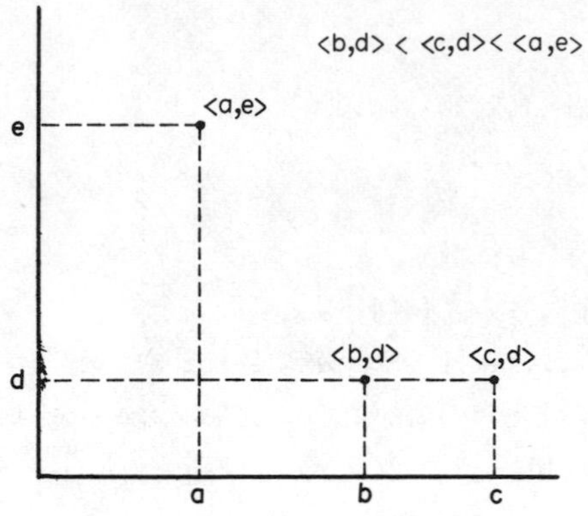

Fig. 1

In fig. 2 we have indicated the least element in a set, with respect to the anti-lexicographical ordering.

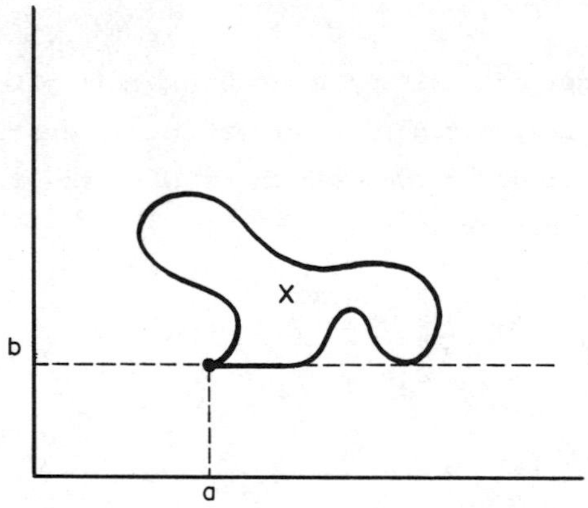

Fig. 2

<u>Lemma 3.10</u>: If $A \simeq A'$ and $B \simeq B'$ then $A \otimes B \simeq A' \otimes B'$.

Proof: left to the reader. □

<u>Definition 3.11</u>: $\alpha.\beta = [A \otimes B]$ if $\alpha = [A], \beta = [B]$.

<u>Examples</u>.

1. $\underset{\sim}{2}.\underset{\sim}{3} = \underset{\sim}{6}$, for
   $\underset{\sim}{2} = [\{0,1\}]$, $\underset{\sim}{3} = [\{0,1,2\}]$ and

$\underset{\sim}{6} = [\{ <0,0> , <1,0> , <0,1> , <1,1> , <0,2> , <1,2> \}]$

2. $\omega.\underset{\sim}{2} = \omega + \omega$, for

   $\underset{\sim}{N} \otimes \{0,1\} = \{<0,0>, <1,0>, <2,0>\ldots, <0,1> , <1,1> , <2,1>\ldots \}$ where we have indicated the ordering from left to right. Now let $\underset{\sim}{N}'$ be a disjoint copy of $\underset{\sim}{N}$, then clearly $\underset{\sim}{N} + \underset{\sim}{N}' \simeq \underset{\sim}{N} \otimes \{0,1\}$.

3. $\underset{\sim}{2}\cdot\omega = \omega$, for

   $\{0,1\} \bigcirc \underset{\sim}{N} = \{<0,0>, <1,0>, <0,1>, <1,1>,\ldots, <0,n>, <1,n>,\ldots \}$

   and clearly $\{0,1\} \bigcirc \underset{\sim}{N} \simeq \underset{\sim}{N}$.

Thus we see that addition and multiplication are not commutative. One can also use the above definitions of + and · to establish positive properties, such as associativity, etc. See II. 5.3 and II. 5.6.

Successor ordinals and limit ordinals can be introduced as in II.3.6. An equivalent, but simpler, definition is

<u>Definition 3.12:</u>

i) α is a *successor* if there is a β such that $\alpha = \beta + 1$

ii) α is a *limit* if $\alpha \neq 0$ and α is not a successor.

Examples of limits are $\omega, \omega + \omega, \omega.\underset{\sim}{n}, \omega^2, \ldots$

<u>Definition 3.13:</u> If $a = [\langle A, <\rangle]$, then the *converse* a^* of a is $[\langle A, >\rangle]$.

Example $[\underset{\sim}{Z}] = \omega^* + \omega$

$[\underset{\sim}{Q}]^* = [\underset{\sim}{Q}]$.

Note that only finite ordinals have ordinals as converses.

<u>Exercises</u>

1. Show $\underset{\sim}{n} + \omega = \omega$
2. If η is the order type of $\underset{\sim}{Q}$, show then $\eta + \eta = \eta$, $\eta.\eta = \eta$, $\eta.\omega = \eta$.
3. Express the order types of densely ordered, countable sets in terms of η.
4. Consider the class C of ordered sets X satisfying: $a)$ X contains a countable ,densely ordered set D without endpoint, which is dense in X

(i.e. between each two elements of X there is an element of D).

b) Every Dedekind cut has a supremum in X (cf. I.12.4.)

Show that each set in C is isomorphic with $\underset{\sim}{R}$.

5. Define an ordering on the ordinals by $[A] < [B]$ iff A is isomorphic to a proper initial segment of B.

   Show that $<$ has the properties of an ordering.

4 INDUCTIVE DEFINITIONS

There are several ways to define the subgroup H of a group G, generated by a set A.

a. H is smallest subgroup (with respect to $\subseteq$) of G containing A.

b. H is the intersection of all subgroups of G containing A.

c. (*i*) $a \in A \rightarrow a \in H$

(*ii*) $a,b \in H \rightarrow ab \in H$

(*iii*) $a \in H \rightarrow a^{-1} \in H$

(*iv*) $a \in H$ by a finite number of applications of (*i*), (*ii*) and (*iii*).

In case *a* we presuppose all subgroups of G containing A (likewise in *b*). Since H itself is one of those subgroups, *a* presents us with a so-called *impredicative definition*. If, in the definition of an object, we refer to a collection of which the object itself is a member, then the definition is called impredicative. Such a definition is, in a certain sense, circular, because that what is to be defined occurs in its own definition. A well-known impredicative definition in mathematics is that of supremum: the supremum of A is the least element of the set of all numbers which are greater or equal than all elements of A (where A is a bounded set of reals).

Poincaré and Russell considered impredicativity to be the source of the paradoxes that were presented at the turn of the century. Russell banned impredicative definitions by his *vicious circle principle*: No totality can contain members definable only in terms of this totality, or members involving or presupposing this totality. In mathematical logic it has extensively been studied which objects (sets) are, or are not, impredicatively definable. It is generally agreed nowadays that impredicativity is not wholly eliminable from mathematics. In the present section we are not so much concerned with an analysis of impredicativity; we will rather present a somewhat more informative alternative to the rough, global definitions *a* and *b*. In case *c* we start out with *A* and successively add more and more elements by taking products and inverses of elements previously obtained. We consider *H* as the *closure* of *A* under product and inverse. The distinction between *a* and *c* can be characterised as follows: *a* defines *H* "from above", *c* defines *H* "from below". *c* has a certain constructive content. Here we will concern ourselves with definitions of the form *c*., which we will call *inductive definitions*. The inductive nature is evident: if certain elements already belong to *H*, we throw in certain specified elements.

First another example of an inductive definition. *The implicational fragment* (*IF*) of propositional logic consists of all propositions, containing only the connective $\rightarrow$.

We now present the inductive definition:

(*i*) If A is an atom, then $A \in IF$

(*ii*) If $A, B \in IF$, then $(A \rightarrow B) \in IF$

The extra clause is left out. As a matter of fact one generally uses a different final clause, i.e. (*iii*) nothing belongs to *IF* except on the ground of (*i*) and (*ii*). This clause is a kind of *economy clause*; it forbids, for example, $\rightarrow A \rightarrow \in IF$. Another form of this economy clause is (*iii'*) *IF* is the smallest set containing the atoms and which is closed under the formation of implication. Clearly this brings us back to impredicativity, as in case *a*. For that reason we used the 'finite number of applications' in *c*. In general, however, a finite number of steps is not sufficient. For this reason we must develop a theory of inductive definitions. From now on we do not formulate the economy clause, it will be assumed tacitly. Apart from the constructive aspect, inductive definitions possess two more virtues: (*i*) one can use inductive proofs for inductively defined classes, (*ii*) one obtains extra insight into the structure of inductively defined classes.

We now return to the subgroup *H* with *A* as its set of generators. Closure under the operations 'product' and 'inverse' can be formulated with an operator Φ, defined by

$$\Phi (X) := \{z \mid \exists x, y \in X\ (z = xy)\} \cup \{z \mid \exists x \in X\ (z = x^{-1})\} \cup X.$$

Define $\Phi^0(X) = X$ and

$$\Phi^n(X) = \Phi(\Phi^{n-1}(X)),$$

then $H = \bigcup_{n\in\omega} \Phi^n(X)$

We generalise this procedure as follows:

Theorem 4.1: Let $\Phi: \mathcal{P}(X) \to \mathcal{P}(X)$ be a monotonic operator. Then there exists a set X_0 such that $\Phi(X_0) = X_0$ (This is a special case of the fixed point theorem of Knaster-Tarski).

Proof: From the monotonicity of Φ with respect to inclusion, we conclude $\Phi(\bigcap_{i\in I} A_i) \subseteq \bigcap_{i\in I} \Phi(A_i)$, for $\bigcap_{i\in I} A_i \subseteq A_j$ (for $j \in I$) so $\Phi(\bigcap_{i\in I} A_i) \subseteq \Phi(A_j)$ for each $j \in I$. Hence $\Phi(\bigcap_{i\in I} A_i) \subseteq \bigcap_{i\in I} \Phi(A_i)$. Now define $X_0 = \bigcap \{Y \subseteq X \mid \Phi(Y) \subseteq Y\}$. Note that $W = \{Y \subseteq X \mid \Phi(Y) \subseteq Y\} \neq \emptyset$, since $\Phi(X) \subseteq X$ holds trivially. We show that X_0 is a fixed-point of Φ.

(*a*) for each $Y \in W$ we have $X_0 \subseteq Y$, so $\Phi(X_0) \subseteq \Phi(Y) \subseteq Y$ for each $Y \in W$. Hence $\Phi(X_0) \subseteq \bigcap W = X_0$.

(*b*) Put $Z = \Phi(X_0)$. From $\Phi(X_0) \subseteq X_0$, it follows that $\Phi(Z) \subseteq Z$, or $Z \in W$.

Now by definition $X_0 \subseteq Z = \Phi(X_0)$.

From (*a*) and (*b*) we conclude $X_0 = \Phi(X_0)$. □

Suppose that Y is some fixed-point of Φ, then $Y = \Phi(Y)$, so $Y \in W$.

Therefore $X_0 \subseteq Y$.

Conclusion: X_0 is the *minimal* fixed-point of Φ.

The minimal fixed-point X_0 of Φ can be considered in a natural way as the union of a well-ordered increasing sequence of sets.

Define $X^0 = \Phi(\emptyset)$

$$X^{\alpha+1} = \Phi(X^\alpha)$$

$$X^\lambda = \bigcup_{\alpha<\lambda} X^\alpha \text{ if } \mathrm{Lim}(\lambda).$$

Claim: there exists a γ such that $X^\gamma = X^{\gamma+1}$. We first note that $\alpha < \beta \to X^\alpha \subseteq X^\beta$ (proof by transfinite induction). Now suppose $\forall\alpha(X^\alpha \neq X^{\alpha+1})$. According to Hartog's lemma (II.8.3, (*i*)) there is a δ such that $\neg\, \delta \leqslant_1 X$. From our supposition we conclude that $\delta \leqslant_1 X^\delta \subseteq X$. Contradiction.

Define $\sigma = \mu\beta\,[X^\beta = X^{\beta+1}]$, i.e. the ordinal at which the sequence first becomes stationary. One easily checks that $X_0 = X^\sigma$.

The ordinal σ obtained above is called the *length* of the minimal fixed-point of Φ.

At this point a few observations should be made. The length of the minimal fixed point of Φ is somewhat dubious. For, by giving a more efficient definition, it might be possible to reduce the length. Since, clearly, the length depends on Φ, a more correct terminology would be the Φ-length of the minimal fixed-point! For convenience, however, we will stick to the first terminology.

The 'transfinite approximation' of X_0, given above, shows clearly that the objects of X_0 are partially ordered by the relation "a is obtained before b". This relation has the following characterisation in terms of the X^α's: there is an α such that $b \in X^\alpha$ and $a \notin X^\alpha$. Now it is immediate that the relation " ... is obtained before ... " is well-founded. Hence we can prove properties over inductively defined sets by induction (II.2.6). This induction is called (somewhat loosely) *induction on the construction* of X_0. Alternatively, one can make use of proof by transfinite induction on α, since the inductive definition yields a sequence $\langle X^\alpha \rangle$ $(\alpha < \sigma)$ such that $X_0 = \cup\{X^\alpha \mid \alpha < \sigma\}$.

<u>Definition 4.2</u>: A monotonic operation Φ is called continuous if $a \in \Phi(A) \leftrightarrow$ there exists a finite $D \subseteq A$ such that $a \in \Phi(D)$.

<u>Theorem 4.3</u>: The length of the minimal fixed-point of a monotonic, continuous Φ is at most ω.

Proof: We show (in the above notation) $X^\omega = X^{\omega+1}$. If $a \in X^{\omega+1} = \Phi(X^\omega)$, then there is a finite set $\{a_1, \ldots, a_n\} \subseteq X^\omega$ such that $a \in \Phi(\{a_1, \ldots, a_n\})$.
$X^\omega = \bigcup_{k \in \omega} X^k$, so $a_i \in X^{k_i}$ for certain k_i $(i \leq \omega)$.
Let $k = \max\{k_1, \ldots, k_n\}$. Since $\{X^k \mid k \in \omega\}$ is a chain $\{a_1, \ldots, a_n\} \subseteq X^k$.
As Φ is monotonic $\Phi(\{a_1, \ldots, a_n\}) \subseteq X^{k+1}$, therefore $a \in X^{k+1} \subseteq X^\omega$.
We now conclude $X^{\omega+1} \subseteq X^\omega$, so $X^\omega = X^{\omega+1}$. □

The construction of the minimal fixed-point X_0, as exhibited above, is a transfinite iteration of a certain operation. This is the constructive aspect, hidden in inductive definitions. The set X_0 is built up "from below". In contrast, the definition using 'the smallest set such that...', is so to speak given at a higher level, where we already have all the sets such that ... and we are required to single out the smallest one.

Continuous operators frequently occur in various branches of mathematics. We shall give a few examples.

The formulas of predicate logic.

(*i*) $A \subseteq$ FORM (A is the set of atoms)

(*ii*) $\Psi_1, \Psi_2 \in \text{FORM} \rightarrow (\Psi_1 \vee \Psi_2) \in \text{FORM}$,

(*iii*) $\Psi \in \text{FORM} \rightarrow (\neg \Psi) \in \text{FORM}$,

(*iv*) $x \in \text{VAR}, \Psi \in \text{FORM} \rightarrow (\exists x)\Psi \in \text{FORM}$ (where VAR is the set of variables).

Consider the set W of all words (i.e. finite strings) over the alphabet of predicate logic. We define Φ on $\mathcal{P}(W)$ by

$$\begin{aligned}\Phi(X) := {} & \{\Psi \mid \exists \Psi_1, \Psi_2 \in X\ (\Psi = (\Psi_1 \vee \Psi_2))\} \cup \\ & \{\Psi \mid \exists \Psi_1 \in X(\Psi = (\neg \Psi_1))\} \cup \\ & \{\Psi \mid \exists\, x \in \text{VAR}, \exists\, \Psi_1 \in X(\Psi = (\exists x)\Psi_1)\} \cup \\ & X\end{aligned}$$

FORM is the minimal fixed-point of Φ. As Φ is continuous FORM has length $\leqslant \omega$. It is immediate that the length of FORM actually is ω. We can now give a simple definition of the *rank of a formula*:

<u>Definition</u>: $r(\Psi) := \mu\,\alpha\,[\Psi \in \text{FORM}^{\alpha}]$

Example. Let Ψ_1 and Ψ_2 be atoms.

$\Psi_1, \Psi_2 \in \Phi(\emptyset) = A$, so $r(\Psi_1) = r(\Psi_2) = 0$

$(\neg \Psi_1) \in \Phi(A) = \Phi^2(\emptyset) = \text{FORM}^1$ and

$(\neg \Psi_1) \notin A = \text{FORM}^0$, so $r(\neg \Psi_1) = 1$

$(((\neg \Psi_1) \vee \Psi_2) \vee \Psi_1) \in \text{FORM}^3 - \text{FORM}^2$, so

$r(((\neg \Psi_1) \vee \Psi_2) \vee \Psi_1) = 3$.

Now we have the rank function, we can prove properties of formulas by ordinary mathematical induction, instead of induction over a well-founded set. Note that in this example we treated a predicate logic with few connectives. A richer language is treated completely analogously.

Theorems of a formal theory

We consider a first-order theory with set of axioms AX. The rules of inference are mappings $F_i : (\text{FORM})^{n_i} \rightarrow \text{FORM}$ ($i \leqslant k$), for example $F(A \vee B, \neg A \vee C) = B \vee C$, cf. [Shoenfield, 1967, p. 21]. Φ is defined by $\Phi(X) := X \cup \bigcup_{i \leqslant k} F_i\,''(X)^{n_i} \cup \text{AX}$.

Again Φ is continuous, so the set of theorems, THM, has length $\leqslant \omega$. In most theories (e.g. predicate logic, arithmetic) the length is exactly ω .

Consider predicate logic. If the length were finite, then one could choose a system with cutfree rules of deduction (e.g. Gentzen's system) and give a decision procedure for provability. This contradicts Church's theorem.

The function classes of Baire, or the analytically representable functions.

The class of Baire, $\underline{\underline{H}}$, is a subset of $\underset{\sim}{R}^{R}$, inductively defined by

(i) f is continuous $\rightarrow f \in \underline{\underline{H}}$

(ii) $\forall n (f_n \in \underline{\underline{H}}) \wedge \lim_{n\to\infty} f_n = f \rightarrow f \in \underline{\underline{H}}$

$\underline{\underline{H}}$ is the smallest set, containing the continuous functions, that is closed under limits of convergent sequences. The operation Φ is defined by $\Phi(X) := \{f | \exists < f_n > (f_n \in X \wedge \lim f_n = f) \cup \mathrm{CONT}$, where CONT is the set of continuous functions. Imitating the proof, one easily proves that the length of $\underline{\underline{H}}$ is at most ω_1 (use the regularity of ω_1). The proof that the length is exactly ω_1 is not trivial, cf. [Natanson, 1961].

The Borel sets

The theory of Borel sets is treated separately in section 6. The analysis of the inductive definition is in that case more refined, as there are two dual hierarchies. There is a close connection between the classes of Baire and the Borel sets, cf. [Kuratowski, 1952, I. p. 299].

The perfect kernel of a closed set

Consider a topological space with a countable base. We want to determine the maximal perfect subset of a closed set A. Define the monotonic operation Φ by $\Phi(X) := X \cup \{x | x$ is an isolated point of $A - X\}$. Let D be the minimal fixed-point of Φ; then $A - D$ is perfect, as follows immediately from the properties of D. Moreover $A - D$ is the maximal perfect subset of A. To determine the cardinality of D, we prove the following:

Lemma: Let $< F_\alpha >_{\alpha < \beta}$ be a decreasing sequence of closed sets. Then there is a $\gamma < \omega_1$ such that $F_\gamma = F_{\gamma+1}$.

Proof: Let $\omega_1 \leqslant \beta$ and suppose $\forall\alpha < \beta \; (F_\alpha \neq F_{\alpha+1})$. Choose for each $\alpha < \beta$ $p_\alpha \in F_\alpha - F_{\alpha+1}$.

Since the space has a countable base each p_α has a neighbourhood $U_{k(\alpha)}$ in the base such that $U_{k(\alpha)} \cap F_{\alpha+1} = \emptyset$. Since $U_{k(\alpha)} \cap F_\alpha \neq \emptyset$, the sets $U_{k(\alpha)}$ are distinct. By our supposition this implies that there are at least $\beta(\geqslant \omega_1)$

basic neighbourhoods. Contradiction. □

Next note that each D^{α} is closed, hence D has a countable length. Moreover each D^{α} is countable, so we have proved the Theorem of Cantor-Bendixson: Each closed set in a space with countable base is the disjoint union of a countable set and a perfect set. Moreover, the partition is unique.

Trees. See section 7.

In (pure) set theory one also applies inductive definitions in cases where Φ is an operation (a class) but not a function. Examples are the cumulative hierarchy (II.4), or Gödel's constructible sets.

Finally some foundational remarks on inductive definitions. There are important notions in mathematics, such as Borel set, that do not possess a predicative definition. For convenience we call these notions impredicative. A number of those do possess an inductive definition, which lends them a certain respectability. To be more precise, inductively definable sets are of a sufficiently constructive nature to be acceptable to some very critical schools in the foundations of mathematics. In particular,Brouwer used an inductive definition for his (intuitionistic) ordinals. One can state that inductive definitions allow us to cross, in an acceptable manner, the boundaries of predicativity. Mathematical means for studying the problems of predicativity are to be found in recursion theory (in particular hierarchy theory).

Literature: Y.N. Moschovakis. Elementary Induction on Abstract Structures. Amsterdam 1974.

5 APPLICATIONS OF THE AXIOM OF CHOICE

In spite of all controversies, the axiom of choice has been a most successful tool in set theory. The first applications were already made right after Zermelo's influential paper 'Beweis dass jede Menge wohlgeordnet werden kann' was published.

Here we present a number of applications which are important for mathematical practice. We will freely use various equivalents of the axiom of choice.

Theorem 5.1: (Hamel, 1905). Every vector space has a basis.

Proof. Consider, in a vector space V, the set X of all independent sets of vectors. Apply Zorn's lemma to X, partially ordered by inclusion. Let K be a chain in X. Then $\cup K \in X$. For suppose that $\cup K$ is dependent, then there are $v_1,\ldots,v_n \in \cup K$, which are dependent. Because K is a chain there are $A_1,\ldots,A_n \in K$ such that $v_i \in A_i$ and $A_i \subseteq A_p$ $(i \leqslant n)$ for some $p \leqslant n$. But then A_p is dependent. Contradiction. Hence $\cup K$ is independent. It is clear that $\cup K$ majorizes the chain K, i.e. $A \subseteq \cup K$ for each $A \in K$. By Zorn's lemma there is a maximal element B in X. We claim that B is a basis for V:

(*i*) B is independent by definition of X.

(*ii*) If $x \in V$, then $B \cup \{x\}$ is dependent on account of the maximality of B. So there are $b_1,\ldots,b_n \in B$ such that $\{b_1,\ldots,b_n,x\}$ is dependent, i.e. $\alpha_1 b_1 +\ldots+ \alpha_n b_n + \alpha x = 0$.
Since $\alpha \neq 0$, we have $x = \alpha^{-1}\alpha_1 b_1 +\ldots+ \alpha^{-1}\alpha_n b_n$.
This shows that B is a basis. □

Theorem 5.2: If R is a ring with unit element and I a proper ideal in R, then I is contained in a maximal ideal.

Proof. Since I is a proper ideal, $1 \notin I$. Consider the set X of all ideals J of R such that $I \subseteq J$ but $1 \notin J$. Clearly $X \neq \emptyset$, X is partially ordered by inclusion. Apply Zorn's lemma to X. Let K be a chain in X, then $\cup K \in X$. Since:
1. $\cup K$ is an ideal. $x,y \in \cup K \rightarrow x \in J \in K, y \in J' \in K$ and $J \subseteq J'$ or $J' \subseteq J$. Take, for example, $J \subseteq J'$ then $x,y \in J'$ so $x - y \in J' \subseteq \cup K$. Closure of $\cup K$ under multiplication is immediate 2. $1 \notin \cup K$, this is immediate.
$\cup K$ majorizes K, so every chain is bounded. By Zorn's lemma there is a maximal element in X and this is the required maximal ideal containing I. □

Theorem 5.2 is a prototype of a class of theorems asserting the existence of maximal so-and-so's. For example, maximal ideals (filters) in Boolean algebras, maximal abelian subgroups.

Theorem 5.3: If $\langle A,\leqslant\rangle$ is a partially ordered set, then $\leqslant$ can be extended to a total order on A.

Proof. Consider the set of all partial orderings of A extending $\leqslant$, ordered by inclusion and apply Zorn's lemma to it. As before each chain is majorised by its union, so there is a maximal partial ordering R extending $\leqslant$. We claim that R is a total ordering. Suppose not, then there are a,b such that neither aRb, nor bRa. We define a new partial ordering R' by squeezing a and b in line.

We put $xR'y$ if xRy, or
if $x = a$, $y = b$, or
if xRa and bRy.

By checking the various cases, one establishes the transitivity of R'. But then R' is a partial ordering properly extending R, contradicting the maximality of R. □

Corollary 5.4: Each set can be ordered.

Proof. Observe that the empty relation is a partial ordering. □

Of course, the corollary also follows from the well-ordering theorem.

Theorem 5.5: (Tychonov's theorem).
The topological product of compact hausdorff spaces is compact.

Proof. Let $T = \underset{i \in I}{\times} T_i$, where each T_i is compact.

A basis for T is formed by all sets $\underset{i \in I}{\times} O_i$, where $O_i = T_i$ for all but finitely many $i \in I$. We show the following: If Z is a family with the finite intersection property, then $\bigcap \{\bar{A} | A \in Z\} \neq \emptyset$.

Consider the set of all Z's with the finite intersection property, ordered by inclusion. A routine application of Zorn's lemma tells us that each Z is contained in a maximal Z'. So we can restrict our attention to maximal Z's for the purpose of showing $\bigcap\{\bar{A} | A \in Z\} \neq \emptyset$. Let Z_0 be some maximal set of sets with the finite intersection property. Z_0 has the following properties:

i) $A \supseteq B \in Z_0 \rightarrow A \in Z_0$

ii) $A, B \in Z_0 \rightarrow A \cap B \in Z_0$

iii) $\forall A \in Z_0 (C \cap A \neq \emptyset) \rightarrow C \in Z_0$

i) and *ii*) are immediate; for *iii*) one has to observe that the premiss states that $Z_0 \cup \{C\}$ has the finite intersection property. By the maximality of Z_0, therefore, $C \in Z_0$.

Now let π_i be the projection of T on the ith coordinate T_i. Put
$\pi_i(Z_0) = \{\pi_i{}''A \mid A \in Z_0\}$. Clearly $\pi_i(Z_0)$ has the finite intersection property.
By the compactness of T_i we have $D_i = \bigcap\{\bar{A}_i \mid A_i \in \pi(Z_0)\} \neq \emptyset$.
Apply the axiom of choice to $\{D_i \mid i \in I\}$: there is a mapping $f\colon I \to \bigcup D_i$ such that $f(i) \in D_i$. That is to say, $f \in \times T_i$.
Claim: $f \in \bigcap\{\bar{A} \mid A \in Z_0\}$.
Consider a neigbourhood U of $\pi_i(f)$.
$\pi_i^{-1}(U) \cap A \neq \emptyset$ for all $A \in Z_0$, so $\pi_i^{-1}(U) \in Z_0$. Now we easily conclude that each basis neighbourhood of f belongs to Z_0. Hence each $A \in Z_0$ meets each neighbourhood of f. So $f \in \bar{A}$ for each $A \in Z_0$, i.e., $f \in \bigcap\{\bar{A} \mid A \in Z_0\}$. This finishes the proof of the theorem, as the property we have established is but an equivalent formulation of compactness. □

Remark: we just showed that in **ZF** $AC \to$ Tychonov. As a matter of fact one can even show Tychonov $\to AC$ in **ZF**.

Theorem 5.6: (Vitali) There is a set of reals which is not Lebesgue measurable.

Proof. We use the fact that the Lebesgue measure on $\mathbb{R}$ is invariant under translations and σ-additive. For $a,b \in [0,1]$, define $a \sim b$ iff $a - b \in \mathbb{Q}$. Clearly $\sim$ is an equivalence relation.
Consider the equivalence classes $[a]$. By AC there is a set A which intersects each $[a]$ in one point (i.e. $|A \cap [a]| = 1$). Suppose A is measurable. Now define for each $r \in \mathbb{Q}$ $A^r = \{x \mid x - r \in A \text{ or } (x-r) + 1 \in A\}$ (A^r is the shift of A over a distance r modulo 1). Note that for $r \neq 0$ $A \cap A^r = \emptyset$ and $\mu(A) = \mu(A^r)$.
Since, by the definition of A, $\bigcup\{A^r \mid r \in \mathbb{Q}\} = [0,1]$ we have
$1 = \mu[0,1] = \sum_{r \in \mathbb{Q}} \mu(A^r)$.

This sum is infinite, hence $\mu(A)$ cannot be positive, so $\mu(A) = 0$. But then $1 = \Sigma\, \mu(A^r) = 0$. Contradiction. Conclusion: A is not Lebesgue measurable. □

5.7. Various continuity definitions for real functions.

In elementary analysis one can define continuity in a number of ways. The best known are

A. f is continuous at a iff

$$\forall \varepsilon > 0 \ \exists \delta > 0 \quad \forall x \ (|x - a| < \delta \rightarrow |f(x) - f(a)| < \varepsilon)$$

B. f is s-continuous (sequence-continuous) at a iff for each sequence $<a_n>$

$$\lim_{k\to\infty} a_k = a \rightarrow \lim_{k\to\infty} f(a_k) = f(a).$$

One easily shows that continuity implies s-continuity (at a point). For the converse *AC* is required. Suppose f is not continuous at a. Then $\exists \varepsilon > 0 \ \forall \delta > 0 \ \exists x(|x - a| < \delta \wedge |f(x) - f(a)| > \varepsilon)$. Let such an ε be given; then we can apply *AC* to the $\forall\exists$ formula: there is a function $s: \omega \rightarrow \underset{\sim}{R}$ such that for each $n \in \omega \quad |s(n) - a| < 2^{-n} \wedge |f(s(n)) - f(a)| > \varepsilon$.
From the first part it follows that $\lim_{n\to\infty} s(n) = a$ and from the second part that $\lim_{n\to\infty} f(s(n))$, if it exists, is distinct from $f(a)$.
This contradicts the s-continuity of f. From this reductio ad absurdum it follows that s-continuity implies continuity.

Remark: of course one can generalize the above facts to abstract spaces, but the striking thing is that *AC* is required almost at the beginning of analysis (Jaegermann, 1965).

5.8. The algebraic closure of a commutative field.
Theorem (Steinitz).Each commutative field has (up to isomorphism) a unique algebraic closure.

The proof consists of two parts, an existence and a uniqueness part. In both parts we employ *AC*.

Existence.
Let K be a commutative field. We will exhibit a field K_{ac} such that (1) K_{ac} is algebraic over K, and (2) K_{ac} is algebraically closed.

Consider $K[X] \times \underset{\sim}{N}$. We embed K in $K[X] \times \underset{\sim}{N}$ via the mapping $\hat{}$ with $\hat{a} := \langle X - a, 0 \rangle$, i.e. the set $\hat{K} = \{<X-a, 0> \mid a \in K\}$ is a field isomorphic with K, with operations induced by $\hat{}$. To be precise $\langle X - a, 0 \rangle \ \square \ \langle X - b, 0 \rangle = \langle X - a \ \square \ b, 0 \rangle$, where $\square$ is one of the operations $+ , - , \cdot , :$. In $K[X] \times \underset{\sim}{N}$ we consider the set V of all fields K', which extend $\hat{K}$ and which satisfy $\hat{p}(<p,n>) = \hat{0}$ for all $<p,n> \in K'$. Here $\hat{p}$ is defined in a natural way: $(a_m X^m + \ldots + a_1 X + a_0)\hat{} = \hat{a}_m X^m + \ldots + \hat{a}_1 X + \hat{a}_0$.

Note that $\hat{K} \in V$, since for $< X - a, 0 > \in \hat{K}$ we have $< X - 1,0 > X - < X - a,0 > = (X - a)\hat{}$ and $< X - 1,0 > < X - a,0 > - < X - a,0>$ $= < X - a,0 > - < X - a,0 > = < X, 0 > = \hat{0}$.

As a consequence V is not empty.

The fields K' are, with respect to the operations, in no way restricted. We just require their domains to be subsets of $K[X] \times \underset{\sim}{N}$, as long as the fields are extensions of $\hat{K}$.

A routine application of Zorn's lemma to V (partially ordered by inclusion) establishes the existence of a maximal K_{ac} in V.

Claim 1. K_{ac} is algebraic over $\hat{K}$.

This follows from the fact that each element $< p,n>$ of K_{ac} satisfies the equation $\hat{p}(< p,n >) = \hat{0}$, where $\hat{p} \in \hat{K}[X]$.

Claim 2. K_{ac} is algebraically closed.

Suppose K' is an algebraic extension of K_{ac}. We will embed K' in $K[X] \times \underset{\sim}{N}$. For a start we map all elements of K_{ac} onto themselves. Now let $u \in K' - K_{ac}$. u is algebraic over K_{ac}, hence also over $\hat{K}$. Let $\hat{p}$ be the minimum polynomial of u (over $\hat{K}$). Since $\hat{q}(< q,n >) = \hat{0}$ for all $q \in K_{ac}[X]$ and for all n and since $\hat{p}(< p,n >) = \hat{0}$ holds for only finitely many $< p,n> \in K_{ac}$, there is an n_p such that $< p,n_p > \notin K_{ac}$.

Map u on $< p,n_p >$. We do this for all $u \in K' - K_{ac}$ (application of AC) and we carry the field operations over to the image K'' of K' in $K[X] \times \underset{\sim}{N}$. Now let $< p,n> \in K''$, then $\hat{p}$ is the minimum polynomial of some u. Since the operations on K'' have been copied from K', we also have $\hat{p}(< p,n >) = \hat{0}$. Therefore $K' \in V$ and by the maximality of K_{ac} we conclude $K' = K_{ac}$.

A field satisfying (1) and (2) is called an *algebraic closure* of K. Note that the algebraic closure K_{ac} was obtained in a rather arbitrary way. The arbitrariness is algebraically speaking, only apparent:

Uniqueness. If K_1 and K_2 are algebraic closures of K then $K_1 \cong K_2$.

Consider the set I of all isomorphisms of subfields of K_2 into K_1, which leave K pointwise fixed (Note: it is no restriction to require K to be contained in both K_1 and K_2). I is partially ordered by inclusion. By another

routine application of Zorn's lemma we find a maximal f.
Suppose f is not defined on all of K_2; then there is a $u \in K_2 - \mathrm{Dom}(f)$, with minimum polynomial p over $K' = \mathrm{Dom}(f)$. Let p' be the corresponding polynomial over $K'' = \mathrm{Ran}(f)$, then p' has a zero v in K_1. Now extend f to $\bar{f}$ by defining $\bar{f}(u) = v$ and extending the mapping algebraically. Clearly f is a proper subfunction of $\bar{f}$, contradicting the maximality of f.
Conclusion: f is an isomorphism of K_2 onto K_1. □

So far we have not encountered any shocking consequences of the axiom of choice. We now cite an example which, in a sense, violates our geometrical intuition. Define two subsets A and B of the three-dimensional euclidean space to be *piecewise* congruent iff they can be partitioned into subsets $A_1, \ldots, A_k$, $B_1, \ldots, B_k$ such that B_i is congruent to A_i.

Theorem 5.9: (Banach, Tarski) A closed ball can be partitioned into two parts each of which is piecewise congruent with the ball itself.
For a proof see [Jech, 1973, p. 3].

Literature

A.A. Fraenkel, Y. Bar-Hillel, A. Levy. D. van Dalen, *Foundations of Set Theory*. Amsterdam 1973.

T.J. Jech. *The axiom of choice*. Amsterdam 1973.

Exercises

1. Show that the following is equivalent to AC: Each set of sets contains a maximal subset consisting of pairwise disjoint sets.

2. Show that there is a discontinuous $f: \mathbb{R} \to \mathbb{R}$ such that $f(x + y) = f(x) + f(y)$ (apply 5.1).

3. Show that all bases of a vector space have the same cardinality.

4. Show that Tychonov's theorem implies AC. Hint: requested a choice function for $\{X_i \mid i \in I\}$. Choose a 'new' element a and consider $Y_i = X_i \cup \{a\}$ with the topology: X_i, Y_i and finite sets are closed. Show that Y_i is compact and apply Tychonov's theorem to $F_i = \{f \in \times_{i \in Z} Y_i \mid f(i) \in X_i\}$.

6 THE BOREL HIERARCHY.

In this section we present a topic which belongs to topology rather than to pure set theory. The reason for including Borel sets here is that (1) the theory contains non-trivial applications of Cantor's diagonalization method, (2) Borel sets are (partly) the fruit of a pre-intuitionistic constructive program in analysis, (3) Borel sets play an important role in certain parts of analysis (measure theory, potential theory etc.) and (4) the Borel sets form a classical example of a hierarchy and as such they are paradigmatic for much that has been going on in recursion theory.

The part of set theory that deals with Borel sets and their generalizations (analytic sets, projective sets, etc.) is called *descriptive set theory*.

The constructive aspect of the theory can roughly be characterized as "approximation of sets by 'nice' sets". The 'nicest' sets of the continuum (i.e. $\underset{\sim}{R}$) are open intervals. Next there are countable unions of intervals (i.e. open sets). One can go on closing this class under the operations of complementation, countable union, countable intersection, thus obtaining ever 'wilder' sets.

For use in this section we list a few topological concepts.

1. A *metric space* is a pair $\langle X,d\rangle$, where d: $X^2 \to \underset{\sim}{R}$ such that

$$d(p,q) = d(q,p)$$
$$d(p,q) = 0 \leftrightarrow p=q$$
$$d(p,r) \leqslant d(p,q) + d(q,r)$$

We call $d(p,q)$ the *distance* from p to q. For all p and q we have $d(p,q) \geqslant 0$. d is called a *metric* on X. For convenience we will call X a metric space.

2. A metric space is *complete* if every fundamental sequence (Cauchy sequence) has a limit.
3. Every metric space is in a natural way a topological space: the open sets are unions of open balls (of the form $\{p|d(p,q)<a\}$).
4. A metric space X is called *separable* if it contains a countable subset which is dense in X.
5. $C = \{0,1\}^\omega$ is the *Cantor space*. Its topology is the product topology.
6. $B = \omega^\omega$ is the *Baire space*. Its topology is the product topology.
7. One can define a metric on B and C:

$$d(f,g) = \begin{cases} 0 \text{ if } f = g \\ (\mu x\,[\,f(x) \neq g(x)] + 1)^{-1} \text{ else.} \end{cases}$$

8. $\mathcal{C}$ and $\mathcal{B}$ have a clopen (i.e. closed and open) basis. They are totally disconnected, complete and separable. Moreover they are o-dimensional. $\mathcal{C}$ is compact.
9. A complete, separable metric space is called a *polish space*.

From now on we will restrict ourselves to polish spaces. The reader who is not after generality can concentrate on the Baire space. For more extensive information on descriptive set theory one should consult the literature. E.g. Kuratowski (1950), Sierpinski (1952), Lyapunow et al. (1955), Bourbaki (1958) and Shoenfield (1967).

Definition 6.1: $\underset{\sim}{B}(X)$ is the smallest class of subsets of X which

i) contains the open sets,

ii) is closed under complementation,

iii) is closed under countable union.

$\underset{\sim}{B}(X)$ is called the *Borel class* of X. For convenience we write $\underset{\sim}{B}$ for $\underset{\sim}{B}(X)$. The elements of $\underset{\sim}{B}$ are called *Borel sets*.

Note that $\underset{\sim}{B}$ has an inductive definition. One can vary the choice of initial class and closure operations. We give an example:

Lemma 6.2: $\underset{\sim}{B}$ is the smallest class of subsets of X which

i) contains the closed sets,

ii) is closed under countable union,

iii) is closed under countable intersection.

Proof: Call the class determined by the above conditions $\underset{\sim}{B}'$. We must show $\underset{\sim}{B} = \underset{\sim}{B}'$.

a) We show that $\underset{\sim}{B}$ satisfies *i*), *ii*) and *iii*) above.

i) Let A be closed. $A^c \in \underset{\sim}{B}$, so $(A^c)^c \in \underset{\sim}{B}$.

ii) Trivial.

iii) Let $A_n \in \underset{\sim}{B}$, then $A_n^c \in \underset{\sim}{B}$ and $\bigcup A_n^c \in \underset{\sim}{B}$.
So $(\bigcup A_n^c)^c = \bigcap A_n \in \underset{\sim}{B}$.

b) We show that $\underset{\sim}{B}'$ satisfies *i*), *ii*), *iii*) of definition 6.1.

i) Let A be open. We show that A is a union of countably many closed sets. Since X is separable, each open set is a countable union of open balls. Therefore it suffices to show that each ball is a countable union of little closed balls.
If $U_\varepsilon(x) = \{y \mid d(x,y) < \varepsilon\}$, then $U_\varepsilon(x) = \bigcup\{B_n \mid n \in \omega\}$, where $B_n = \{y \mid d(x,y) \leqslant \varepsilon\ (1 - 2^{-n})\}$.

iii) trivial.

ii) Let $A \in \underset{\sim}{B}'$. We proceed by induction on the construction of A (cf. section 4). There are three cases:
(*a*)' A is closed, then A^c is open and by (*i*) above $A^c \in \underset{\sim}{B}'$,
(*b*)' $A = \bigcap A_n$, with $A_n \in \underset{\sim}{B}'$. By induction hypothesis $A_n^c \in \underset{\sim}{B}'$. Now $\bigcup A_n^c \in \underset{\sim}{B}'$, so $A^c = (\bigcup A_n^c)^c \in \underset{\sim}{B}'$.
(c)' $A = \bigcup A_n$, with $A_n \in \underset{\sim}{B}'$. Analogous to (*b*)'.
Conclusion: $\underset{\sim}{B} \subset \underset{\sim}{B}'$.
Together with (*a*)' this yields $\underset{\sim}{B} = \underset{\sim}{B}'$. □

There is a traditional notation for Borel sets, based on the following:
If $\underset{\sim}{K} \subseteq \mathcal{P}(X)$ then $\underset{\sim}{K}_\sigma$ is the set of all countable unions of elements of $\underset{\sim}{K}$.
Analogously for $\underset{\sim}{K}_\delta$ and intersections.
The class of open sets is denoted by $\underset{\sim}{G}$ and the class of closed sets by $\underset{\sim}{F}$.

From (*b*), *i*) of lemma 6.2 it follows immediately that we have

Corollary 6.3: $\underset{\sim}{G} \subseteq \underset{\sim}{F}_\sigma$
$\underset{\sim}{F} \subseteq \underset{\sim}{G}_\delta$

In definition 6.1 we have characterized the Borel sets 'from above', i.e. we have considered the class of all sets satisfying the conditions, and taken the least one (which amounts to taking the intersection of all of them). We can also approximate $\underset{\sim}{B}$ 'from below', by starting with $\underset{\sim}{F}$ and $\underset{\sim}{G}$ and iterating the operations σ and δ. In general one has to interate transfinitely often. The process has been discussed *in abstracto* in the previous section.

We now define recursively two sequences of sets $\underset{\sim}{F}_\alpha$, $\underset{\sim}{G}_\alpha$.

<u>Definition 6.4:</u> $\underset{\sim}{F}_0 := \underset{\sim}{F}$ $\qquad$ $\underset{\sim}{G}_0 := \underset{\sim}{G}$

$$\begin{cases} \underset{\sim}{F}_\alpha := (\bigcup_{\mu<\alpha} \underset{\sim}{F}_\mu)_\delta \\ \underset{\sim}{G}_\alpha := (\bigcup_{\mu<\alpha} \underset{\sim}{G}_\mu)_\sigma \end{cases} \quad \text{for even } \alpha \text{ (i.e. } \alpha = \xi.2 \text{ or } \lim\alpha)$$

$$\begin{cases} \underset{\sim}{F}_\alpha := (\bigcup_{\mu<\alpha} \underset{\sim}{F}_\mu)_\sigma \\ \underset{\sim}{G}_\alpha := (\bigcup_{\mu<\alpha} \underset{\sim}{G}_\mu)_\delta \end{cases} \quad \text{for odd } \alpha \text{ (i.e. } \alpha = \xi.2 + 1)$$

<u>Theorem 6.5:</u> $\underset{\sim}{B} = \bigcup_{\alpha<\omega_1} \underset{\sim}{F}_\alpha = \bigcup_{\alpha<\omega_1} \underset{\sim}{G}_\alpha$

Proof: We first establish a few properties of the $\underset{\sim}{F}_\alpha$'s and $\underset{\sim}{G}_\alpha$'s.

(a) $\alpha < \beta \rightarrow \underset{\sim}{F}_\alpha \subseteq \underset{\sim}{F}_\beta$

$\underset{\sim}{G}_\alpha \subseteq \underset{\sim}{G}_\beta$

$\underset{\sim}{F}_\alpha \subseteq \underset{\sim}{G}_\beta$

$\underset{\sim}{G}_\alpha \subseteq \underset{\sim}{F}_\beta$

The first two of these are trivial.

To show $\underset{\sim}{F}_\alpha \subseteq \underset{\sim}{G}_\beta$ we apply transfinite induction on α.

Suppose that for all $\gamma < \alpha$ $\underset{\sim}{F}_\gamma \subseteq \underset{\sim}{G}_\delta$ has been proved for all $\delta > \gamma$.

α *is even.* $A \in \underset{\sim}{F}_\alpha \rightarrow \bigcap A_n = A$ with $A_n \in \underset{\sim}{F}_{\gamma_n}$.

By the induction hypothesis $A_n \in \underset{\sim}{G}_\alpha$, so $A \in \underset{\sim}{G}_{\alpha+1}$.

The desired result follows from $\underset{\sim}{G}_{\alpha+1} \subseteq \underset{\sim}{G}_\beta$.

α *is odd.* $A \in \underset{\sim}{F}_\alpha \rightarrow A = \bigcup A_n$ with $A_n \in \underset{\sim}{F}_{\gamma_n}$.

By the induction hypothesis $A_n \in \underset{\sim}{G}_\alpha$, so $A = \bigcup A_n \in \underset{\sim}{G}_{\alpha+1} \subseteq \underset{\sim}{G}_\beta$.

The case $\underset{\sim}{G}_\alpha \subseteq \underset{\sim}{F}_\beta$ is left to the reader.

(b) $A \in \underset{\sim}{F}_\alpha \leftrightarrow A^c \in \underset{\sim}{G}_\alpha$.

Induction on α. Suppose the equivalence has been established for all $\beta < \alpha$. We distinguish two cases.

α *is even.* $A \in \underset{\sim}{F}_\alpha \leftrightarrow A = \bigcap A_n$, with $A_n \in \underset{\sim}{F}_{\gamma_n}$.

Induction hypothesis: $A_n \in \underset{\sim}{F}_{\gamma_n} \leftrightarrow A_n^c \in \underset{\sim}{G}_{\gamma_n}$.

By definition $\bigcup A_n^c \in \underset{\sim}{G}_\alpha$, i.e. $A^c \in \underset{\sim}{G}_\alpha$.

α *is odd.* Left to the reader.

(c) $\underset{\sim}{F}_\alpha \cup \underset{\sim}{G}_\alpha \subseteq \underset{\sim}{B}$. Induction on α. Use definition 6.1 and lemma 6.2.

(d) $\bigcup \underset{\sim}{F}_\alpha = \bigcup \underset{\sim}{G}_\alpha$. Immediate from (a).

(e) $\bigcup \underset{\sim}{F}_\alpha \subseteq \underset{\sim}{B}$. Immediate from (c).

(f) $\bigcup \underset{\sim}{F}_\alpha$ is closed under countable unions.

Let $A = \bigcup A_n$, with $A_n \in \bigcup\{\underset{\sim}{F}_\alpha \mid \alpha < \omega_1\}$.
For each $n < \omega$ there is an $\alpha < \omega_1$, such that $A_n \in \underset{\sim}{F}_\alpha$.
By the axiom of choice there is a function $f\colon \omega \to \omega_1$ such that $A_n \in \underset{\sim}{F}_{f(n)}$.
Since ω_1 is regular (cf. II.10.18, 10.17) the range of f is bounded, i.e. for some $\beta < \omega_1$ $A_n \in \underset{\sim}{F}_\beta$ for all n. One easily sees that $\bigcup A_n \in \bigcup \underset{\sim}{F}_\alpha$, using this boundedness.

(g) $\underset{\sim}{B} \subseteq \bigcup \underset{\sim}{F}_\alpha$. From (b) and (d) it follows that $\bigcup \underset{\sim}{F}_\alpha$ is closed under complementation. This, combined with (f) and $\underset{\sim}{G} \subseteq \bigcup \underset{\sim}{F}_\alpha$, yields (g)

(h) $\underset{\sim}{B} = \bigcup \underset{\sim}{F}_\alpha$. Immediate from (e) and (g). □

<u>Definition</u> $\underset{\sim}{D}_\alpha = \underset{\sim}{F}_\alpha \cap \underset{\sim}{G}_\alpha$

From part (a) of the proof of 6.5 it follows that we can make the following diagram

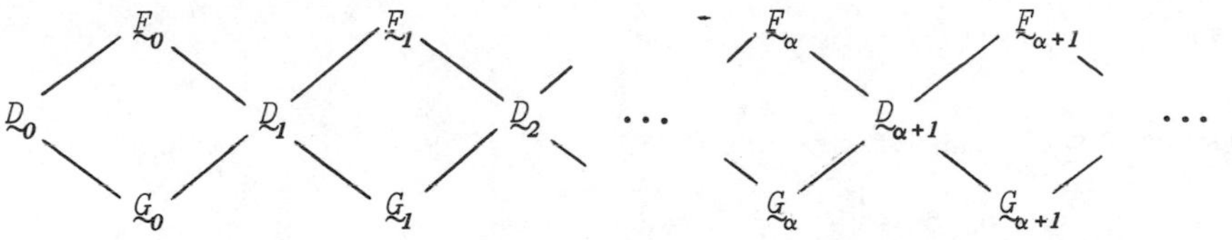

From left to right we have inclusions. The question to be asked is: are the inclusions proper, or, is it possible that the sequence becomes stationary? If one considers a topological space, with discrete topology, then the whole diagram collapses to a point. Moral: the topological space should not be too simple !

We will set ourselves the task of showing the diagram does not stop before ω_1 . For this purpose we need more facts about Borel sets.

<u>Lemma 6.6</u>: Let $f\colon X_1 \to X_2$ be continuous and $A \in \underset{\sim}{F}_\alpha(X_2)$. Then $f^{-1}(A) \in \underset{\sim}{F}_\alpha(X_1)$. Likewise for $\underset{\sim}{G}$.

Proof. By induction on α. Use the following properties:

$A \in \underset{\sim}{F}_0(X_2) \rightarrow \overset{-1}{f}(A) \in \underset{\sim}{F}_0(X_1)$

$\overset{-1}{f}(\cup A_i) = \cup \overset{-1}{f}(A_i)$

$\overset{-1}{f}(\cap A_i) = \cap \overset{-1}{f}(A_i)$ □

Lemma 6.7: $A \in \underset{\sim}{F}_\alpha(X),\ B \in \underset{\sim}{F}_\alpha(Y) \rightarrow A \times B \in \underset{\sim}{F}_\alpha(X \times Y)$.
Likewise for $\underset{\sim}{G}$.

Proof. Induction on α. □

Lemma 6.8: Let f be a homeomorphism from X onto Y. Then $A \in \underset{\sim}{F}_\alpha(X) \leftrightarrow f(A) \in \underset{\sim}{F}_\alpha(Y)$.
Likewise for $\underset{\sim}{G}$.

Proof. Apply lemma 6.6. □

Another way of expressing lemma 6.8 is: The $\underset{\sim}{F}_\alpha$'s and $\underset{\sim}{G}_\alpha$'s are topological invariants.

Lemma 6.9: Let $A \subseteq X^2$ and define

$A_x = \{y \mid \langle x,y\rangle \in A\}$, $A_y = \{x \mid \langle x,y\rangle \in A\}$

$A_\Delta = \{x \mid \langle x,x\rangle \in A\}$.

Then $A \in \underset{\sim}{F}_\alpha \rightarrow A_x, A_y, A_\Delta \in \underset{\sim}{F}_\alpha$.
Likewise for $\underset{\sim}{G}$.

Proof. Note that on relativizing to a subspace the Borel class is preserved.
More precisely: $A \in \underset{\sim}{F}_\alpha(Y) \rightarrow A \cap X \in \underset{\sim}{F}_\alpha(X)$

$A \in \underset{\sim}{G}_\alpha(Y) \rightarrow A \cap X \in \underset{\sim}{G}_\alpha(X)$, where X is a subspace of Y.

Let $Z = \{\langle x,y\rangle \mid x = x_0\}$. Z is homeomorphic to Y. Let $A^*_{x_0} = A \cap Z$ (see fig. 3), so $A^*_{x_0} \in \underset{\sim}{F}_\alpha(Z)$. By 6.8 now $A_{x_0} \in \underset{\sim}{F}(Y)$ since the homeomorphism from Z to Y carries $A^*_{x_0}$ to A_{x_0}.

Likewise for A_y.

For the intersection with the diagonal, A_Δ, we make use of homeomorphism f with $f(x,x) = x$. □

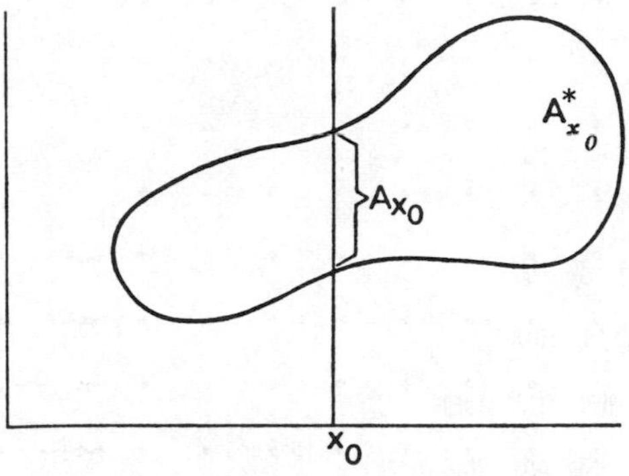

Fig. 3

The natural numbers had the convenient property that $\omega \times \omega$ could be coded onto ω (I. 17.7). The Baire space has an even stronger property: we can code countable sequences of elements of $\mathcal{B}$ onto $\mathcal{B}$.
This is formulated as follows:

Theorem 6.10: There is a bijective homeomorphism $h\colon \mathcal{B} \to \mathcal{B}^{\omega}$

Proof. We use the standard bijection $J\colon \omega^2 \to \omega$, introduced in theorem I, 17.5. J was defined by $J(m,n) = \frac{1}{2}(m+n+1)(m+n) + n$. The 'inverses' of J are K and L, i.e. if $J(m,n) = p$, then $K(p) = m$ and $L(p) = n$.
The precise formula for J is, as a rule, not particularly relevant. For most purposes one only needs the following properties:

$$J(K(z), L(z)) = z$$
$$K(J(x,y)) = x$$
$$L(J(x,y)) = y$$

Let $f \in \mathcal{B} = \omega^{\omega}$, we split f into countably many functions g_n, defined by $g_n(k) = f(J(n,k))$. Now define $h(f) = \lambda n.g_n$ (i.e. the sequence $\langle g_n \rangle_{n \in \omega}$) ,see fig. 4.

1. h is injective.

$$h(f) = h(f') \to \lambda n.g_n = \lambda n.g_n'$$
$$\to \lambda n.(\lambda k.f(J(n,k)) = \lambda n.(\lambda k.f'(J(n,k))$$
$$\to f(J(n,k)) = f'(J(n,k)) \text{ for all } n \text{ and } k$$

Since J is a bijection(i.e. $J(n,k)$ ranges over all natural numbers) we may conclude $f = f'$.

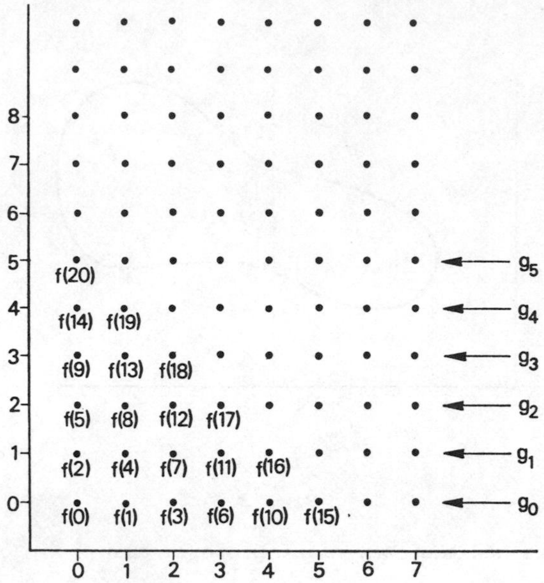

Fig. 4

2. h is surjective.

Let g_n^* be given for all $n \in \omega$. Define f by $f(a) = g_{K(a)}^*(L(a))$, then

$h(f) = \lambda n \cdot (\lambda k . g_n(k)) = \lambda n \cdot (\lambda k \cdot f(J(k,n))) =$

$= \lambda n \cdot \lambda k (g_n^*(k)) = \lambda n \cdot g_n^*$.

3. h is continuous.

Choose a point $\tilde{g} = \langle g_0, g_1, g_2, \ldots \rangle \in \mathcal{B}^\omega$.

Consider an arbitrary, basic neighbourhood of $\tilde{g}$, i.e. the product of basic neighbourhoods $U_0, U_1, U_2, \ldots$ of $\widetilde{\langle g_0, g_1, g_2, \ldots \rangle}$, where only finitely many of the U_i 's are not the whole space $\mathcal{B}$. For each i, U_i consists of all functions in $\mathcal{B}$ which coincide with g_i on some specified finite set $S_i \subseteq \omega$ (i.e. $U_i = \{f \in \mathcal{B} \mid \forall n \in S_i \ (f(n) = g_i(n))\}$, where $S_i = \emptyset$ if $U_i = \mathcal{B}$). In figure 5 the relevant g_i 's (g_1, g_2, g_3, g_6, g_8) with their relevant values have been indicated. We want a neighbourhood of $f = h^{-1}(\lambda n \cdot g_n)$ such that the g_i 's, which occur as images under h, assume the prescribed values in the starred points.

Keeping in mind that f was constructed from the g_i 's by walking over the plane along successive anti-diagonals (cf. I 17.5), we determine a triangle containing all the stars (see the dotted line). Determine the initial segment of f, given the values of the g_i 's in this triangle. Then every f', with this same initial segment is mapped onto a sequence $\langle g'_n \rangle$ which take the prescribed values in the starred points. These f' form a basic neighbourhood of f in $\mathcal{B}$, which is mapped into the given neighbourhood of g. This establishes the continuity of h.

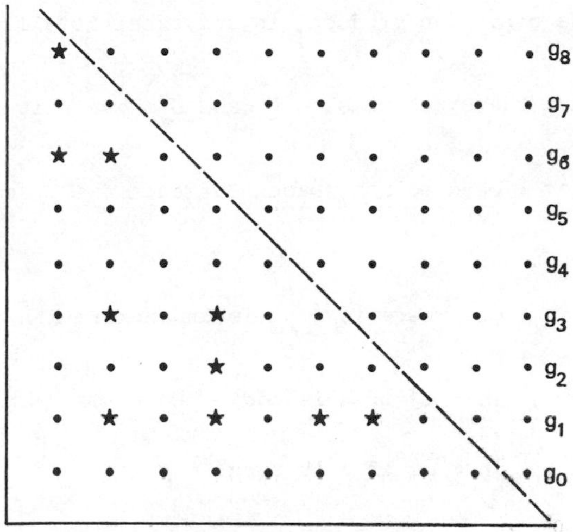

Fig. 5

4. h^{-1} is continuous.

Use the same recipe as in 3. □

Definition 6.11: The projections $pr_n : \mathcal{B}^{\omega} \to \mathcal{B}$ are given by $pr_n(\lambda m.\, g_m) = g_n$

Lemma 6.12: pr_n is continuous.

Definition 6.13: $h_n = pr_n \circ h$.

Lemma 6.14: For each limit ordinal $\lambda < \omega_1$ there is a sequence $\lambda_0 < \lambda_1 < \lambda_2 < \ldots < \lambda_n < \ldots$ such that $\lambda = \bigcup\{\lambda_n \mid n \in \omega\}$.

Proof: $\{\alpha \mid \alpha < \lambda\}$ is countable, so there is a bijection $f: \omega \to \lambda$. Now define g by $g(0) = f(0)$ and $g(n+1) = f(k)$ for the first k such that $f(k) > g(0), \ldots, g(n)$. The sequence $\lambda_n = g(n)$ satisfies the lemma. □

With the help of the axiom of choice we make a fixed assignment of increasing sequences $\langle \lambda_n \rangle_{n \in \omega}$ to countable ordinals.

Definition 6.15: $U \subseteq X_1 \times X_2$ is called *universal* for a class $\underset{\sim}{K} \subseteq \mathcal{P}(X_1)$ if $\underset{\sim}{K} = \{U_y \mid y \in X_2\}$.

In words: U, in the cartesian product, is universal if all elements of $\underset{\sim}{K}$ are sections of U.

We will show that the Borel classes $\underset{\sim}{F}_\alpha$ and $\underset{\sim}{G}_\alpha$ have universal sets.

Definition 6.16: Let X be a polish space. For each $\alpha < \omega_1$ we define $U^\alpha \subseteq X \times B$.

1. Let $f \in \mathcal{B}$

$U^0_f = \bigcup\{C_{f(n)} \mid n \in \omega\}$, where $\{C_i\}_{i \in \omega}$ is an enumeration of a basis of X (including $\emptyset$).

$$U^{\alpha+1}_f = \bigcup\{U^\alpha_{h_n(f)} \mid n \in \omega\} \text{ if } \alpha \text{ is odd}$$

$$\bigcap\{U^\alpha_{h_n(f)} \mid n \in \omega\} \text{ if } \alpha \text{ is even.}$$

$$U^\lambda_f = \bigcup\{U^{\lambda_n}_{h_n(f)} \mid n \in \omega\} \text{ if } \lambda \text{ is a limit.}$$

2. $U^\alpha = \{\langle x, f \rangle \mid x \in U^\alpha_f\}$.

Motivation: We are looking for a universal U^α for $\underset{\sim}{G}_\alpha$, so we mimic the definition of $\underset{\sim}{G}_\alpha$ on a grand scale. For $\alpha = 0$ we get all open sets if f runs through all functions. For $\alpha + 1$ we get all possible countable unions or intersections of the preceding stage. In the limit case we get all unions (λ is even).

Theorem 6.17: *a*) U^α is universal for $\underset{\sim}{G}_\alpha$

b) $U^\alpha \in \underset{\sim}{G}_\alpha(X \times \mathcal{B})$.

Proof. We show $A \in \underset{\sim}{G}_\alpha \rightarrow \exists f(A = U^\alpha_f)$ by induction on α.

$\alpha = 0$. A is open, so $A = \bigcup\{C_{f(n)} \mid n \in \omega\}$ where $C_{f(k)}$ is an enumeration of the basic open sets which yield A as a union. Should A be the union of finitely many open sets then we just keep adding the last one (or the empty set). Hence the required f always exists. Result: $A = U^{\circ}_{f}$.

<u>$\alpha = \delta+1$ and δ even.</u>

$A \in \underset{\sim}{G}_{\delta+1}$, so $A = \bigcap A_n$ with $A_n \in \underset{\sim}{G}_{\delta}$.

By induction hypothesis $\forall n \, \exists f (A_n = U^{\delta}_{f})$. Apply AC: there is an $F\colon \omega \to \mathcal{B}$ such that $A_n = U^{\delta}_{F(n)}$ $\quad (F(n) \in \mathcal{B})$

Now consider the following commuting diagram :

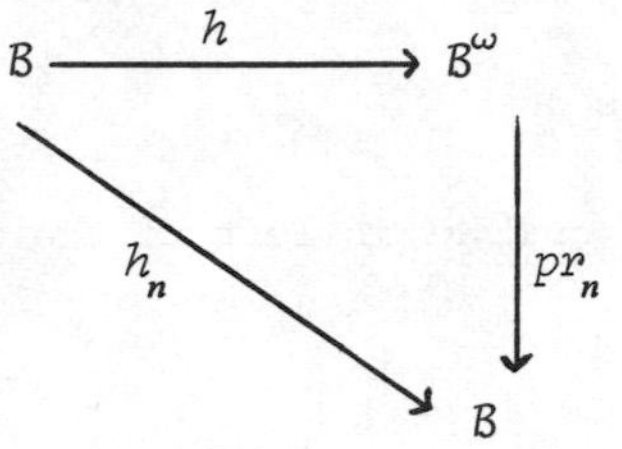

$F \in \mathcal{B}^{\omega}$, h is a bijection so there is a $g \in \mathcal{B}$ such that $h_n(g) = F(n)$.

Now $A = \bigcap_{n \in \omega} A_n = \bigcap_{n \in \omega} U^{\delta}_{h_n(g)} = U^{\delta+1}_{g}$.

<u>$\alpha = \delta+1$ and δ odd.</u> Completely similar.

<u>$\alpha = \lambda$ with $\mathrm{Lim}(\lambda)$.</u>

$A \in \underset{\sim}{G}_{\lambda}$, so $A = \bigcup A_n$. Each A_n is in some class $\underset{\sim}{G}_{\mu}$ with $\mu < \lambda$. We would like to have the A_n 's in $\underset{\sim}{G}_{\lambda_n}$, where $\langle \lambda_n \rangle_{n \in \omega}$ is the standard sequence from lemma 6.14. Here we use the fact that the sequence is strictly increasing, and that $\alpha < \beta \to \underset{\sim}{G}_{\alpha} \subseteq \underset{\sim}{G}_{\beta}$. Determine the first λ_n such that $A_0 \in \underset{\sim}{G}_{\lambda}$, call this λ_{n_0} . Suppose $\lambda_{n_0}, \ldots, \lambda_{n_k}$ have been determined. Take the first $\lambda^{n}_{n} > \lambda_{n_k}$ such that $A_{k+1} \in \underset{\sim}{G}_{\lambda_n}$, call this $\lambda_{n_{k+1}}$.

Now put $B_{n_k} = A_k$ and $B_m = \emptyset$ else. By induction hypothesis $\forall n \exists f (B_n = U^{\lambda_n}_{f})$.

From now on we can mimic the argument of the case of the odd successor.

We have shown that each $\underset{\sim}{G}_{\alpha}$ set is a section of U^{α}.

We now show $U^{\alpha} \in \underset{\sim}{G}_{\alpha}(X \times \mathcal{B})$, by induction on α.

<u>$\alpha = 0$.</u> $U^{\circ} = \{\langle x,f\rangle \mid x \in \bigcup\{C_{f(n)} \mid n \in \omega\}\}$

$= \bigcup_{n \in \omega} \{\langle x,f\rangle \mid x \in C_{f(n)}\}$.

Choose $\langle x_0, f_0\rangle \in U^{\circ}$. We shall exhibit a neighbourhood of this point in U°.

$x_0 \in C_{f_0(n)}$ for some n.

Determine a basic neighbourhood W in $\mathcal{B}$ such that $W = \{f \mid f\lceil(n+1) = f_0 \lceil (n+1)\}$ (remember that $n+1 = \{0,\ldots,n\}$). Then $C_{f_0(n)} \times W \subseteq U^0$, since

$$\begin{aligned}\langle x,f\rangle \in C_{f_0(n)} \times W &\leftrightarrow x \in C_{f_0(n)} \text{ and } f\lceil(n+1) = f_0\lceil(n+1)\\ &\rightarrow x \in C_{f_0(n)} \text{ and } f(n) = f_0(n)\\ &\rightarrow x \in C_{f(n)}\\ &\rightarrow \langle x,f\rangle \in U^0\end{aligned}$$

This establishes that U^0 is open.

The case of the successor ordinal is left to the reader.

<u>$\alpha = \lambda$ and lim (λ)</u>

$$U_f^\lambda = \cup\, U_{h_n(f)}^{\lambda_n} \quad \text{and}$$

$$U^\lambda = \{\langle x,f\rangle \mid x \in U_f^\lambda\} = \{\langle x,f\rangle \mid x \in \cup\, U_{h_n(f)}^{\lambda_n}\}$$

$$= \cup\{\langle x,f\rangle \mid x \in U_{h_n(f)}^{\lambda_n}\}$$

Define $F(\langle x,f\rangle) = \langle x,h_n(f)\rangle$. From the definition of h_n it follows that F is a continuous surjection onto $X \times \mathcal{B}$. Therefore

$$\{\langle x,f\rangle \mid x \in U_{h_n(f)}^{\lambda_n}\} = F^{-1}(\{\langle x,f\rangle \mid x \in U_f^{\lambda_n}\}).$$

Induction hypothesis: $U^{\lambda_n} = \{\langle x,f\rangle \mid x \in U_f^{\lambda_n}\} \in \underset{\sim}{G}_{\lambda_n}$, so, by lemma 6.6., also $\{\langle x,f\rangle \mid x \in U_{h_n(f)}^{\lambda_n}\} \in \underset{\sim}{G}_{\lambda_n}$.

Now it follows immediately that $U^\lambda \in \underset{\sim}{G}_\lambda$.

Finally we conclude that for each f, $U_f^\alpha \in \underset{\sim}{G}_\alpha(X)$ using lemma 6.9. □

Evidently the $\underset{\sim}{F}_\alpha$ must have universal sets too.

<u>Theorem 6.18</u>: For each $\alpha < \omega_1$ there is a $V^\alpha \subseteq X \times \mathcal{B}$ such that

a) V^α is universal for $\underset{\sim}{F}_\alpha$.

b) $V^\alpha \in \underset{\sim}{F}_\alpha(X \times \mathcal{B})$

Proof. Define $V^\alpha := (U^\alpha)^c$.

$A \in \underset{\sim}{F}_\alpha \leftrightarrow A^c \in \underset{\sim}{G}_\alpha \leftrightarrow A^c = U_f^\alpha$ for some $f \leftrightarrow A = (U_f^\alpha)^c$.

$x \in A \leftrightarrow x \notin U_f^\alpha \leftrightarrow \langle x,f\rangle \notin U^\alpha \leftrightarrow \langle x,f\rangle \in V^\alpha \leftrightarrow x \in V_f^\alpha$. So $A = V_f^\alpha$.

By lemma 6.5, (*b*) $V^\alpha \in \underset{\sim}{F}_\alpha(X \times \mathcal{B})$.
From this it follows that each V^α_f belongs to $\underset{\sim}{F}_\alpha\ (X)$. □

We can now show that the classes $\underset{\sim}{F}_\alpha$ and $\underset{\sim}{G}_\alpha$ never coincide on $\mathcal{B}$.

Hierarchy Theorem 6.19. *a*) $\underset{\sim}{G}_\alpha(\mathcal{B}) - \underset{\sim}{F}_\alpha(\mathcal{B}) \neq \emptyset$
b) $\underset{\sim}{F}_\alpha(\mathcal{B}) - \underset{\sim}{G}_\alpha(\mathcal{B}) \neq \emptyset$

Proof. We apply a diagonalization.
Let $Z^\alpha = \{f \mid \langle f,f\rangle \in U^\alpha\}$. By lemma 6.9 $Z^\alpha \in \underset{\sim}{G}_\alpha$. Now suppose $(Z^\alpha)^c \in \underset{\sim}{G}_\alpha$, then $(Z^\alpha)^c = U^\alpha_{f_0}$ for some $f_0 \in \mathcal{B}$, so $f \in (Z^\alpha)^c \leftrightarrow \langle f,f_0\rangle \in U^\alpha$. Now put $f = f_0$, then $f_0 \in (Z^\alpha)^c \leftrightarrow \langle f_0,f_0\rangle \in U^\alpha \leftrightarrow f_0 \in Z^\alpha$. Contradiction.
Conclusion: $(Z^\alpha)^c \notin \underset{\sim}{G}_\alpha$, i.e. $Z^\alpha \notin \underset{\sim}{F}_\alpha$, or $\underset{\sim}{G}_\alpha - \underset{\sim}{F}_\alpha \neq \emptyset$. Similarly $(Z^\alpha)^c \in \underset{\sim}{F}_\alpha$ and $(Z^\alpha)^c \notin \underset{\sim}{G}_\alpha$, so $\underset{\sim}{F}_\alpha - \underset{\sim}{G}_\alpha \neq \emptyset$. □

The trick of the proof is the choice of Z^α, i.e. a projected diagonal. By taking the complement of Z^α we get a set that cannot be in $\underset{\sim}{G}_\alpha$. We can express this as follows: we 'diagonalize' out of $\underset{\sim}{G}_\alpha$.
In order to apply this diagonalization we first had to go through all the labour of constructing universal sets.

<u>Corollary 6.20</u>: The sequences $\underset{\sim}{F}_\alpha$ and $\underset{\sim}{G}_\alpha$ have length ω_1.

Proof. Suppose $\underset{\sim}{F}_\alpha = \underset{\sim}{F}_{\alpha+1}$ for some $\alpha < \omega_1$. Then $\underset{\sim}{D}_{\alpha+1} = \underset{\sim}{F}_{\alpha+1}$, so $\underset{\sim}{F}_{\alpha+1} \subseteq \underset{\sim}{G}_{\alpha+1}$.
Contradiction. □

For applications in analysis, Borel sets on $\underset{\sim}{R}$ are more important than those on $\mathcal{B}$. Could the sequence $\underset{\sim}{F}_\alpha$ become stationary on $\underset{\sim}{R}$? It is known that $\mathcal{B}$ can be mapped homeomorphically onto the set of the irrationals, i.e. we can topologically embed $\mathcal{B}$ into $\underset{\sim}{R}$. Remembering that under relativization Borel classes are preserved, we see that also on $\underset{\sim}{R}$ $\underset{\sim}{F}_\alpha \neq \underset{\sim}{F}_{\alpha+1}$ for $\alpha < \omega_1$.

In the preceding part we have shown that Borel sets can be obtained by starting with simple sets and building more and more complicated sets. The hierarchy theorem shows that the classes increase all the way up to ω_1.

We say that the Borel class $\underset{\sim}{B}$ is split into a hierarchy of classes, the so-called *Borel hierarchy*.

In several parts of mathematics there are similar phenomena. As a rule there is some measure of complexity, which induces a hierarchy.

In particular one finds various hierarchies in logic. The best known are the arithmetical and analytical hierarchies (cf. Shoenfield, 1967, where also the connection with the Borel hierarchy is demonstrated). Further well-known hierarchies are the Grzegorczyk-hierarchy in recursion theory and the Chomsky hierarchy in mathematical linguistics.

We will now make some observations on hierarchies in general. The hierarchies we have in mind are of the same sort as the Borel hierarchy. The fact that, some things, traditionally known as 'hierarchies', are not covered in this approach is the price we have to pay for systematization.

Definition 6.21: A *hierarchy* is a set $\underset{\sim}{H}$ of sets which is well-founded with respect to inclusion.

In the case of the Borel hierarchy we may take
$\{\underset{\sim}{F}_\alpha \mid \alpha < \omega_1\}$ or $\{\underset{\sim}{G}_\alpha \mid \alpha < \omega_1\}$.

The well-foundedness of $\underset{\sim}{H}$ enables as to introduce ordinals as in $\underset{\sim}{B}$.

Lemma 6.22: Let A be a partially ordered set, well-founded with respect to $\leqslant$. Then there is an ordinal α and a function $f:A \to \alpha$ such that
$x < y \to f(x) < f(y)$.

Proof. f is recursively defined by $f(x) := \{f(y) \mid y < x\}$. Clearly each $f(x)$ is an ordinal. By the replacement axiom $f''A$ is a set and is, in fact, the required α. From the definition it follows that $x < y \to f(x) < f(y)$. □

A hierarchical splitting is obtained by putting $M_\beta = \{x \mid f(x) < \beta\}$. Another approach is to start with the sets M_β (cf. exercise 1).

We have already seen an application of this lemma in the case of a transitive set, partially ordered by ϵ (e.g. a standard model of **ZF**). In that case f is the rank function (cf.II,4).

From now on we assume that the sets of $\underset{\sim}{H}$ themselves consist of sets.

<u>Definition 6.23</u>: The domain of $\underset{\sim}{H}$ is $\bigcup\bigcup\underset{\sim}{H}$ (denoted by $\mathcal{D}(\underset{\sim}{H})$).

<u>Definition 6.24</u>: Let $\underset{\sim}{A} \subseteq \mathcal{P}(X)$. Then $\underset{\sim}{A}^c = \{B^c \mid B \in \underset{\sim}{A}\}$.

N.B. A^c is *not* the complement of $\underset{\sim}{A}$ in $\mathcal{P}(X)$.

<u>Definition 6.25</u>: The dual hierarchy $\underset{\sim}{H}^d$ of $\underset{\sim}{H}$ is $\{\underset{\sim}{K}^c \mid \underset{\sim}{K} \in \underset{\sim}{H}\}$ (where complementation is taken in $\mathcal{D}(\underset{\sim}{H})$).

Example: If $\underset{\sim}{F} = \{F_\alpha \mid \alpha < \omega_1\}$ and $\underset{\sim}{G} = \{\underset{\sim}{G}_\alpha \mid \alpha < \omega_1\}$ then $\underset{\sim}{F}^d = \underset{\sim}{G}$ and $\underset{\sim}{G}^d = \underset{\sim}{F}$.

The ordinal α, which by lemma 6.22 belongs to $\underset{\sim}{H}$, is called the *length* of $\underset{\sim}{H}$. Clearly $\underset{\sim}{H}$ and $\underset{\sim}{H}^d$ have the same length. The length of the Borel hierarchy is ω_1.

In topology separation principles play an important role. (Think of Hausdorff spaces, for instance, in which pairs of points are separated by open neighbourhoods). In abstract hierarchies one also considers separation properties, namely for elements of $\underset{\sim}{H}$.

<u>Definition 6.26</u>: Suppose $\underset{\sim}{M} \in \underset{\sim}{H}$. We define

$\mathrm{Sep}_1(\underset{\sim}{M}) := \forall A,B \in \underset{\sim}{M}\,[A \cap B = \emptyset \to \exists C \in \underset{\sim}{M} \cap \underset{\sim}{M}^d\,(A \subseteq C \wedge B \cap C = \emptyset)]$.

Geometrical representation - cf. fig. 6

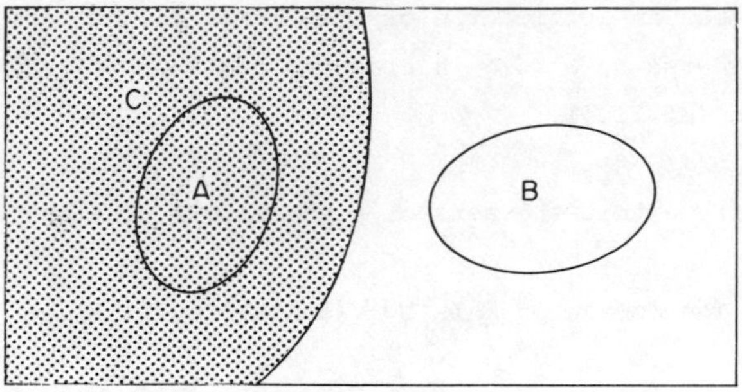

Fig. 6

We say that $\underset{\sim}{M}$ has the *first separation property*.

<u>Definition 6.27</u>: $\mathrm{Sep}_{II}(\underset{\sim}{M}) := \forall A,B \in \underset{\sim}{M}\ \exists C,D\ [A-B \subseteq C \wedge B-A \subseteq D \wedge C \cap D = \emptyset \wedge$
$\wedge\ C^c \in \underset{\sim}{M} \wedge D^c \in \underset{\sim}{M}]$.

Geometrical representation - cf. fig. 7.

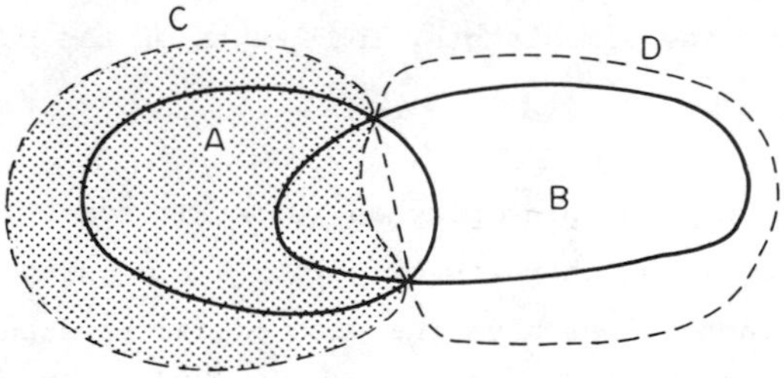

Fig. 7

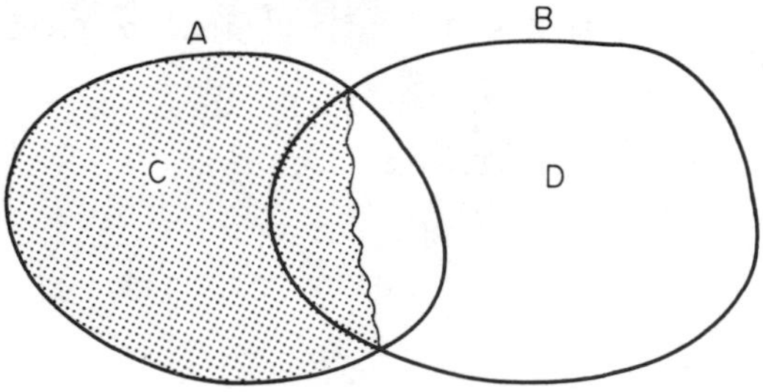

Fig. 8

Finally we define the *reduction property*.

Definition 6.28: $\mathrm{Red}(\underset{\sim}{M}) := \forall A,B \in \underset{\sim}{M} \; \exists C,D \in M \; [A \cup B = C \cup D \wedge C \cap D = \emptyset \wedge$
$\wedge \; C \subseteq A \wedge D \subseteq B]$.

Geometrical representation - cf. fig. 8.

There are a number of simple relations between separation properties and the reduction property.

Theorem 6.29: $\mathrm{Red}(\underset{\sim}{M}) \rightarrow \mathrm{Sep}_I \; (\underset{\sim}{M}^f)$.

Proof. Let $A,B \in M^f$ and $A \cap B = \emptyset$.

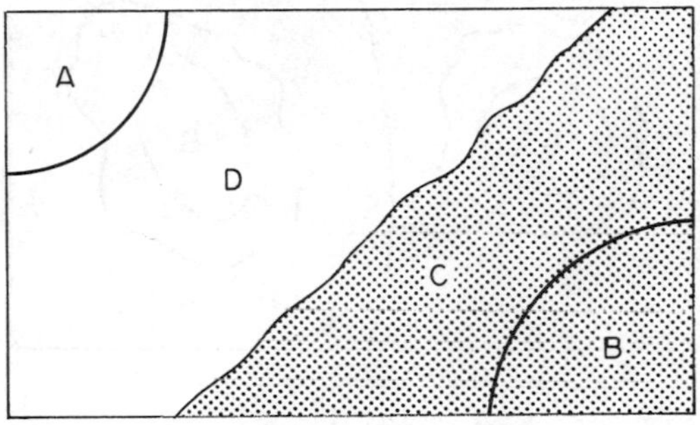

Fig. 9

$A^c, B^c \in \underset{\sim}{M}$. Apply the reduction property: there are $C,D \in \underset{\sim}{M}$ such that $C \cap D = \emptyset$, $C \cup D = A^c \cup B^c = X$, i.c. $D = C^f \in \underset{\sim}{M}^f$. Then $A \subseteq D$, $B \cap D = \emptyset$ and $D \in \underset{\sim}{M} \cap \underset{\sim}{M}^f$. □

Theorem 6.30: $\mathrm{Red}(\underset{\sim}{M}) \rightarrow \mathrm{Sep}_{II} \; (\underset{\sim}{M}^f)$

Proof. Again apply the reduction property to the complements. □

Theorem 6.31: $\mathrm{Red}(\underset{\sim}{M}) \rightarrow (\mathrm{Sep}_I \; (\underset{\sim}{M}) \leftrightarrow \mathrm{Sep}_{II} \; (\underset{\sim}{M}))$

Proof. *a)* Suppose Red $(\underset{\sim}{M})$ and $\mathrm{Sep}_I \; (\underset{\sim}{M})$.
Let A and B be given. Apply the reduction property to A and B. Result: there are disjoint C and D with $C \subseteq A$, $D \subseteq B$, $C \cup D = A \cup B$. Now apply the first separation property to C and D. Result: E, $E^f \in \underset{\sim}{M}$ and $C \subseteq E$, $D \subseteq E^f$. This implies $A - B \subseteq E$ and $B - A \subseteq E^f$.

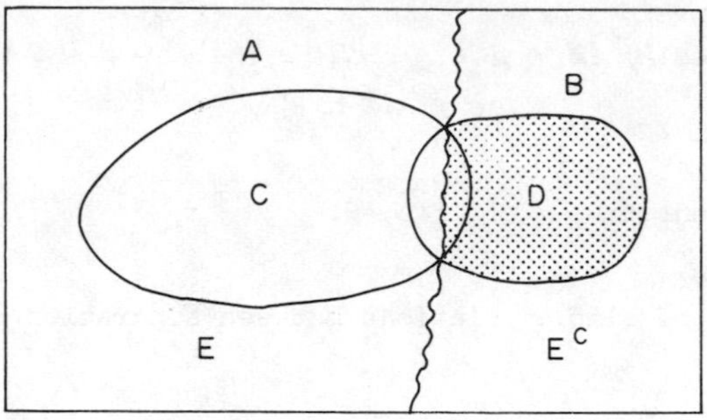

Fig. 10

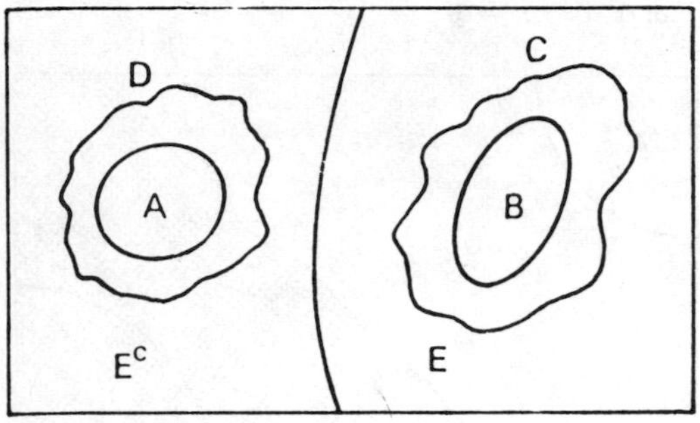

Fig. 11

b) Suppose Red($\underset{\sim}{M}$) and Sep$_{II}$ ($\underset{\sim}{M}$).

Let the disjoint sets A and B be given. A^c, B^c ϵ $\underset{\sim}{M}^c$. By theorem 6.30 we can apply the second separation property: there are $D, C \epsilon \underset{\sim}{M}^c$ such that $B = A^c - B^c \subseteq C$ and $A = B^c - A^c \subseteq D$ and $C \cap D = \emptyset$.

By theorem 6.29 we can apply the first separation property to C and D: there is an $E \epsilon \underset{\sim}{M} \cap \underset{\sim}{M}^c$ such that $C \subseteq E$ and $D \subseteq E^c$. Now we have $A \subseteq E$ and $B \subseteq E^c$. □

In general the converses of the previous theorems do not hold. For further information cf. Kuratowski (1950).

Now we will investigate the reduction and separation properties for the Borel hierarchy. The behaviour of the classes turns out to depend on the closure properties for countable intersections or unions.

Definition 6.32: $\underset{\sim}{Mt}_\alpha = \underset{\sim}{F}_\alpha$ if α is even
$\underset{\sim}{G}_\alpha$ if α is odd

$\underset{\sim}{Ad}_\alpha = \underset{\sim}{G}_\alpha$ if α is even
$\underset{\sim}{F}_\alpha$ if α is odd

$\underset{\sim}{Mt}_\alpha$ and $\underset{\sim}{Ad}_\alpha$ are called the *multiplicative*, resp. *additive* classes.

Lemma 6.33: Let $A \in \underset{\sim}{Ad}_\alpha$ for $\alpha \geqslant 1$. Then A is a countable disjoint union of elements of $\underset{\sim}{D}_\alpha$.

Proof. $\underset{\sim}{D}_\alpha$ is closed under intersection, union and complementation. ($\underset{\sim}{D}_\alpha$ is a field of sets).
$A = \cup\{A_n \mid n \in \omega\}$ with $A_n \in \cup\{\underset{\sim}{Mt}_\beta \mid \beta < \alpha\}$. Since $\underset{\sim}{Mt}_\beta \subseteq \underset{\sim}{D}_\alpha$ if $\beta < \alpha$, $A_n \in \underset{\sim}{D}_\alpha$.

Define $C_0 = A_0$
$C_{n+1} = A_{n+1} - C_n$
Then $A = \cup\{C_n \mid n \in \omega\}$, and the C_n's are mutually disjoint elements of $\underset{\sim}{D}_\alpha$. □

Theorem 6.34: $\mathrm{Red}(\underset{\sim}{Ad}_\alpha)$ for $\alpha \geqslant 1$.

Proof. $A, B \in \underset{\sim}{Ad}_\alpha$ and according to the previous lemma $A = \cup A_n$, $B = \cup B_n$, where $A_n, B_n \in \underset{\sim}{D}_\alpha$ and the A_n's (resp. B_n's) are disjoint.
Define $C_0 = A_0$

$$D_0 = B_0 - A_0$$
$$C_{n+1} = A_{n+1} - \bigcup_{k \leqslant n} B_k$$
$$D_{n+1} = B_{n+1} - \bigcup_{k \leqslant n+1} A_k$$
$$C = \bigcup_{n \in \omega} C_n \qquad D = \bigcup_{n \in \omega} D_n$$

Clearly $C \cup D \subseteq A \cup B$.
Let $x \in A \cup B$. Determine the first set in the sequence $A_0, B_0, A_1, B_1, \ldots\ldots$ containing x. If $x \in A_n$, then $x \in C_n$; if $x \in B_n$ then $x \in D_n$. In both cases $x \in C \cup D$. So $A \cup B \subseteq C \cup D$, and hence $A \cup B = C \cup D$. Moreover $C \cap D = \emptyset$, and $C, D \in \underset{\sim}{Ad}_\alpha$. □

<u>Corollary 6.35:</u> $\mathrm{Sep}_I(\underset{\sim}{Mt}_\alpha)$ and $\mathrm{Sep}_{II}(\underset{\sim}{Mt}_\alpha)$ for $\alpha \geqslant 1$.

Note that theorem 6.34 and cor. 6.35 in general fail for $\alpha = 0$ (cf. exercise 3.) In the Borel hierarchy, Red and Sep propagate in a zig-zag way, as shown in the following diagram.

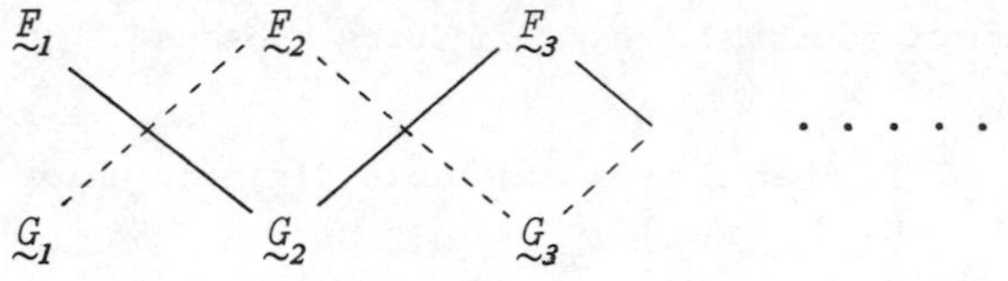

The dotted line connects the classes with the separation property and the continuous line connects those with the reduction property.

It turns out that the Borel hierarchy is not closed under continuous mappings. One can use this fact to extend it to the so-called *projective hierarchy* (see Kuratowski, 1950). Finally we remark that the continuum hypothesis holds for Borel sets: a Borel set is finite, countable or has cardinality $2^{\aleph_0}$ (cf. Sierpinski, 1952, p. 228; Hausdorff, 1927, §32).

If one wants to establish that a certain set is in $\underset{\sim}{B}$ it is very advantageous to make use of the language of logic.
The key properties are: $\{y \mid \exists n\, P(n,y)\} = \bigcup_{n\in\omega} \{y \mid P(n,y)\}$
and $\{y \mid \forall n\, P(n,y)\} = \bigcap_{n\in\omega} \{y \mid P(n,y)\}$.

<u>Examples</u>

1 The set of points of continuity of a real valued function is $\underset{\sim}{G}_1$.
a is a point of continuity of $f \leftrightarrow$

$\forall n\, \exists k\, \forall x(\,|x - a| \leqslant 2^{-k} \rightarrow |f(x) - f(a)| < 2^{-n})$.

Define $A_n = \{a \mid \exists k\, \forall x\, (|x-a| \leqslant 2^{-k} \rightarrow |f(x) - f(a)| < 2^{-n})\}$. Then $A_{n+1} \subseteq A_n$ and the points of continuity of f form the set $A := \bigcap A_n$.
To each $a \in A_n$ we associate a number $k(n)$ such that $|x - a| \leqslant 2^{-k_n(n)} \rightarrow |f(x) - f(a)| < 2^{-n}$. By elementary analysis $a \in A_{n+1} \rightarrow U_{2^{-k(n+1)}}(a) \subseteq A_n$. Put $B_n = \bigcup\{U_{2^{-k(n+1)}}(a) \mid a \in A_{n+1}\}$. Then $A_{n+1} \subseteq B_n \subseteq A_n$ and B_n is open.
Now $A = \bigcap A_n = \bigcap B_n \in \underset{\sim}{G}_\delta$, i.e. $A \in \underset{\sim}{G}_1$.

2. The set of points, in which a given continuous function is differentiable, is $\underset{\sim}{F}_2$.

$$f \text{ is differentiable in } a \leftrightarrow \lim_{x \to 0} \frac{f(a+x) - f(a)}{x} \text{ exists.}$$

Put $F(a,x) = \dfrac{f(a+x) - f(a)}{x}$. Then this condition is equivalent to

$$\underbrace{\forall n\, \underbrace{\exists k\, \underbrace{\forall x\, \forall x'\, \underbrace{((|x| \geqslant 2^{-k} \wedge |x'| \geqslant 2^{-k}) \vee |F(a,x) - F(a,x')| \leqslant 2^{-n})}_{\underset{\sim}{F}_0}}_{\underset{\sim}{F}_0}}_{\underset{\sim}{F}_1}}_{\underset{\sim}{F}_2}$$

3 Let $f_n : \underset{\sim}{R} \to \underset{\sim}{R}$ $(n \in \omega)$, $C = \{x \mid \lim_{n\to\infty} f_n(x) \text{ exists}\}$. $C \in \underset{\sim}{F}_2$

$$x \in C \leftrightarrow \forall k\, \exists n\, \forall m\, (m > n \to |f_n(x) - f_m(x)| \leqslant 2^{-k})$$

$$\leftrightarrow \underbrace{\forall k\, \underbrace{\exists n\, \underbrace{\forall m\, \underbrace{(m \leqslant n \vee |f_n(x) - f_m(x)| \leqslant 2^{-k})}_{\underset{\sim}{F}_0}}_{\underset{\sim}{F}_0}}_{\underset{\sim}{F}_1}}_{\underset{\sim}{F}_2}$$

4 Consider, in the space $\mathcal{B}^{\omega}$, the set D of those sequences f such that Ran(f) is dense in itself. Then $D \in \underset{\sim}{G}_1$, since

$$\text{Ran}(f) \text{ is dense in itself} \leftrightarrow \underbrace{\forall k\, \forall n\, \underbrace{\exists m\, \underbrace{(0 < |f(n) - f(m)| < 2^{-k})}_{\underset{\sim}{G}_0}}_{\underset{\sim}{G}_0}}_{\underset{\sim}{G}_1} .$$

Remark: we are dealing with a set of functions f. It is easy to check that $\{f \mid 0 < |f(n) - f(m)| < 2^{-k}\}$ is open.

The above examples show how one can estimate an upper bound in the Borel hierarchy for certain sets. It is not shown, however, whether this bound is the best possible. In example 4 it is not a priori impossible that $D \in \underset{\sim}{F}_0$. We shall show that $D \in \underset{\sim}{G}_1$ is actually the best possible result by proving that D is not open or closed (i.e. $D \in \underset{\sim}{G}_1 - (\underset{\sim}{F}_0 \cup \underset{\sim}{G}_0)$).

(1) We construct a converging sequence of elements of D such that the limit does not belong to D.
Let $x_i \in \mathcal{B}$ be the function $x_i : \omega \to \omega$ with $x_i(0) = i$ and $x_i(k+1) = 0$. Define $f_i : \omega \to \mathcal{B}$ by $f_i(j) = x_j$ for $j \leqslant i$, while for the remaining arguments $f_i(j)$ is chosen such that $\mathrm{Ran}(f_i)$ is dense in itself (this is possible because $\mathcal{B}$ is separable). Then $\lim_{i\to\infty} f_i = f$ with $f(j) = x_j$ $(j \in \omega)$ and $\mathrm{Ran}(f)$ is not dense in itself, i.e. $f \notin D$.
Conclusion: D is not closed.

(2) Suppose D is open, then $f \in D \to \exists n\, \forall g\, (f \restriction n = g \restriction n \to g \in D)$. In words: being an element of D depends only on an initial segment. Let $f \in D$ and let $f \restriction n$ be the initial segment concerned. Define

$$g(i) = \begin{cases} f(i) & \text{if } i < n \\ f(n-1) & \text{if } i \geqslant n. \end{cases}$$

Then $\mathrm{Ran}(g)$ is not dense in itself, so $g \notin D$. Contradiction. So D is not open.

We will not state general principles for showing that an evaluation of a set in the Borel hierarchy is the best possible - this is a topic for hierarchy theory (cf. Shoenfield, ch. 7).

Literature

C. Kuratowski, Topologie I, II Warszawa 1952
(English translation New York, Warszawa, 1966)
C. Kuratowski and A. Mostowski. Set Theory. Amsterdam 1968.

Exercises

1 Give another proof of lemma 6.22 by defining $M_0 = \{x \in A \mid x \text{ is minimal}\}$, $M_\beta = \{x \in A \mid \forall y \in A(y < x \to y \in \bigcup_{\gamma<\beta} M_\gamma)\}$.
Show that the sequence M_β becomes stationary.

2 If X is zero dimensional (i.e. has a clopen basis), then theorem 6.34 holds for $\underset{\sim}{Ad}_0$.

3 Exhibit a polish space in which $\text{Sep}_I(\underset{\sim}{F})$, $\text{Sep}_{II}(\underset{\sim}{F})$ and $\text{Red}(\underset{\sim}{G})$ all fail.

7 TREES

There are few concepts in mathematics that occur with greater regularity than that of a tree. The mathematician with a well trained eye will discover trees almost everywhere. With the slogan "everything is a set", in mind, one could, with a bit of exaggeration, say that mathematics is the study of trees (readers with a constructive turn will certainly appreciate this view-point).

The definition of 'tree', presented below, is by no means the most general one. The reader will, however, have no difficulty in switching to other kinds of trees.

In this section we deal with partially ordered sets. We will use the terminology of I. 14. In particular we will use the terms '*successor*', '*predecessor*', '*immediate successor*' and '*immediate predecessor*'.

Definition 7.1: A *tree* is an ordered triple $\langle B, \leqslant, t\rangle$ where

i) B is partially ordered by $\leqslant$,

ii) $|B| \leqslant \aleph_0$,

iii) for each $b \in B$ the set of predecessors forms a finite chain

$$b = b_n < b_{n-1} < \ldots < b_1 < b_0 = t .$$

The elements of B are called nodes. The node t is called the top of the tree. It is the maximal element of B. If $b_0, \ldots, b_n$ is the chain mentioned in *iii*), then we call n the *length* of b_n. Note that each node, except for the top has a unique predecessor.

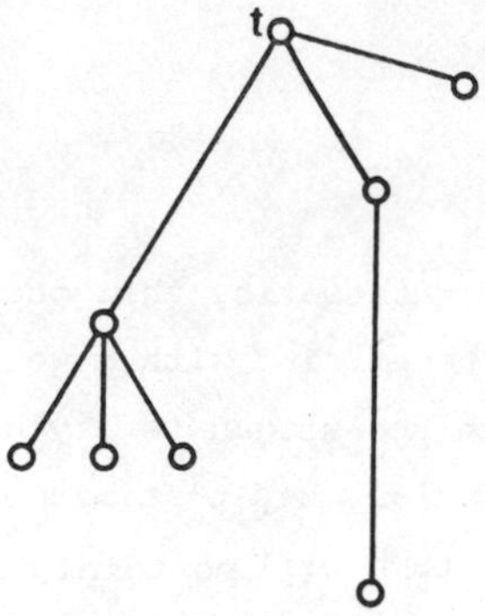

Fig. 12

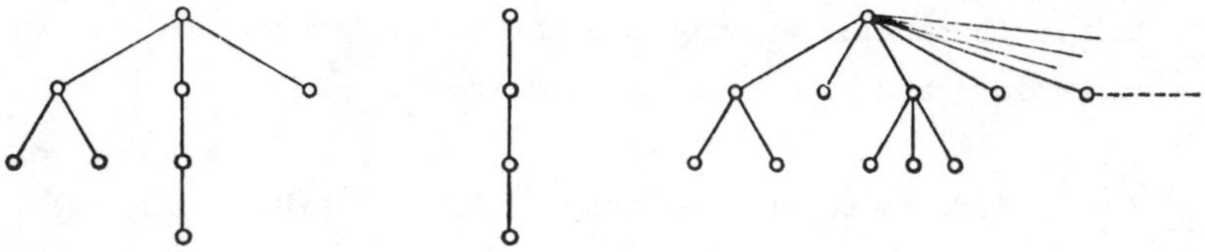

Fig. 13

The next lemma follows immediately from the definition.

Lemma 7.2: Incomparable elements have no common successors.

One, usually, represents trees by diagrams. We have our trees grow downwards (so that the top is on top). See fig. 12, 13.

Warning: some authors have their trees grow upwards or sideways, one should be wary of terms as 'above', 'after', 'successor ', 'top', etc.

Maximal chains are called *branches* (or *paths*). Infinite branches are permitted by the definition.

Note that according to definition 7.1, trees do not form a set. Although this does not really matter we could restrict ourselves to trees consisting of natural numbers (this is possible since $|B| \leq \aleph_0$).

Now we will proceed to exhibit a tree, in which all trees can be isomorphically embedded - a *universal tree*. The nodes of this tree will be finite sequences of natural numbers.

We have already observed that there are two ways of introducing finite sequences - via the ordered pair, or via functions on initial segments of ω (cf. I, 13 exc. 40). Since it is convenient to use the function approach in the context of trees, we will adopt the latter one.

If $a: n+1 \to \omega$ (N.B. $n+1 = \{0,\ldots,n\}$), we write $a = \langle a_0,\ldots,a_n \rangle$ with $a_i = a(i)$. The unique mapping from $\emptyset$ to ω is denoted by $\langle\ \rangle$, and called the *empty sequence*.

Thus the denotation of $\langle a_0,\ldots,a_n \rangle$ here differs from the one in I.8. The point is not very important; either approach could be used.

<u>Definition 7.3:</u> $SEQ := \bigcup\{\omega^n \mid n \in \omega\}$

SEQ is the set of finite sequences of ω (including the empty sequence).

The most important operation on sequences is that of *concatenation*.

<u>Definition 7.4:</u> *i*) If $a : n \to \omega$, then ${}^m a$ is defined by ${}^m a(k) = a(k - m)$ for $m \leq k \leq n + m$, if $n > 0$

${}^m a = \emptyset$ if $n = 0$

ii) If $a: n \to \omega$; $b: m \to \omega$, then $a * b = a \cup {}^n b$.

In *i*) a sequence is shifted over a distance m, in *ii*) two sequences are concatenated.

<u>Example</u>

i) $a = \{\langle 0,1 \rangle , \langle 1,7 \rangle , \langle 2,0 \rangle\}$

${}^3 a = \{\langle 3,1 \rangle , \langle 4,7 \rangle , \langle 5,0 \rangle\}$

ii) $a = \{\langle 0,5 \rangle , \langle 1,1 \rangle\}$; $b = \{\langle 0,0 \rangle, \langle 1,0 \rangle\}$, then

$a * b = \{\langle 0,5 \rangle , \langle 1,1 \rangle , \langle 2,0 \rangle , \langle 3,0 \rangle\}$,

or, in a simplified notation, $a * b = \langle 5,1 \rangle * \langle 0,0 \rangle = \langle 5, 1, 0, 0 \rangle$.

Definition 7.5: $l(a) := \mathrm{Dom}(a)$
(the *length* of the sequence a)

Note that $l\,(<>) = 0$
$l\,(<a_0, \ldots a_n>) = n + 1$
$l\,(a * b) = l(a) + l(b)$.

The partial ordering on SEQ is defined by the inclusion relation:

Definition 7.6: $a \leqslant * \, b := a \subseteq b$ if $a,b \in$ SEQ.

Note that $a \leqslant * \, b \Rightarrow l(a) \leqslant l(b)$.

SEQ is obviously a tree.
One can visualise the diagram of SEQ as a tree in which each node has countably many immediate successors (fig. 14)

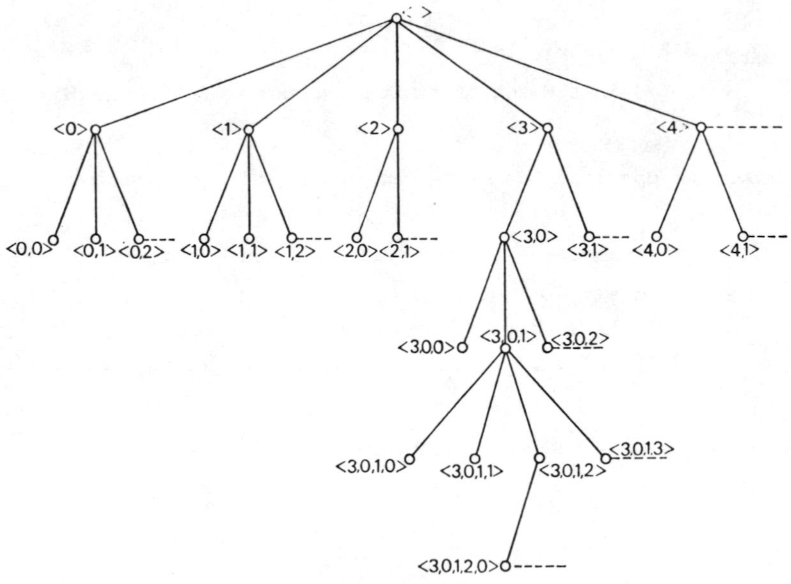

Fig. 14

Theorem 7.7: For each tree B there is an isomorphism from B into SEQ.

Proof. By definition, B is countable, so there is an $\alpha \leqslant \omega$ and an h such that $h: B \underset{1}{=} \alpha$. Let $t, b_1, \ldots, b_n = b$ be the unique maximal chain from t to b, then we define $f(b) = \langle h(b_1), \ldots, h(b_n) \rangle$. At the top we define f by $f(t) = \langle\ \rangle$. Clearly f is an isomorphism from B into SEQ. □

Example: see fig. 15

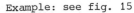

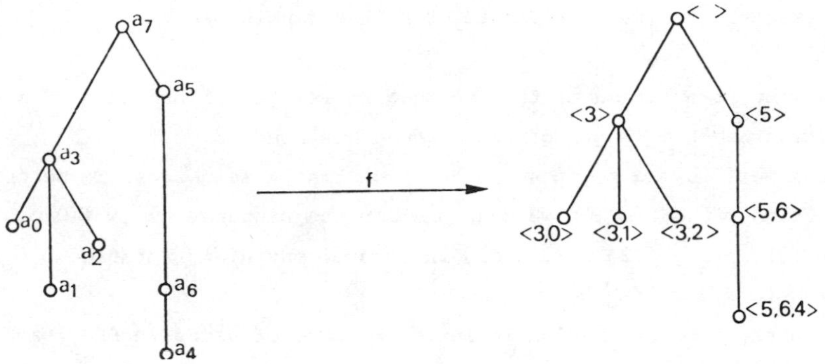

Fig. 15

From theorem 7.7 we see that it suffices to consider subtrees of SEQ.

Definition 7.8: A *finitary tree* or *fan* is a tree in which each node has only finitely many immediate successors.

We now state a well-known theorem from the theory of graphs.

Theorem 7.9. *Konig's infinity lemma* (1926). A fan with infinitely many nodes has an infinite branch.

Proof. Let F be the fan. Consider the subset F' of all nodes with infinitely many successors. Each node in F' has an immediate successor in F'. Apply the axiom of choice: there is a mapping $f: F' \to F'$ such that $f(b)$ is the immediate successor of b. Since F is infinite, $t \in F'$.

Define $a_0 = t$

$a_{n+1} = f(a_n)$

There is not a last node a_k, since $a_k \in F'$ and hence $a_k > f(a_k) = a_{k+1} \in F'$.
Conclusion: $a_0, a_1, a_2, \ldots$ is an infinite branch. □

By contraposition one obtains from the infinity lemma

Theorem 7.10 *Brouwer's fan theorem* (1923):

If all branches of a fan are finite, then the fan is finite.

An equivalent formulation of the fan theorem is: If all branches of a fan are finite, then the length of the branches is bounded.
A function on ω is, so to speak, made up of finite sequences. To be precise, in connection with $f\colon \omega \to \omega$ we can consider the sequence $\langle\ \rangle, \langle f(0)\rangle, \langle f(0), f(1)\rangle, \langle f(0), f(1), f(2)\rangle, \ldots$, or, in a more convenient notation: $f \upharpoonright 0, f \upharpoonright 1, f \upharpoonright 2, f \upharpoonright 3, \ldots$.
That is to say f determines an infinite sequence of nodes in SEQ, each of them being an immediate successor of the previous one. In short, f determines a branch in SEQ. Conversely each branch of SEQ determines a function $f \in \omega^\omega$: let $a^1, a^2, a^3, \ldots$ be a branch, then the corresponding function is $\bigcup\{a^i \mid i \in \omega\}$.

For convenience we will, whenever possible without ambiguity, identify the function and the branch. So, metaphorically, the Baire space is the set of branches of SEQ.
If $a = \langle a_0, \ldots, a_k\rangle \in$ SEQ, then we say that f *passes through* a if $f \upharpoonright k+1 = a$, or $f(0) = a_0, \ldots, f(k) = a_k$.
This is also expressed by '*a is (lies) on f*'. In certain parts of the literature $\langle a_0, \ldots, a_k\rangle$ is replaced by its coding in ω.

Two trees of fundamental importance are

1. The *binary fan*, i.e. the tree of all 0-1-sequences: $\bigcup_n \{0,1\}^n$.

   The corresponding set of branches is $\{0,1\}^\omega$. This can be viewed as the topological product of ω copies of $\{0,1\}$ (with discrete topology), and is known as the *Cantor space*.

2. The *universal tree* SEQ (the name is motivated by theorem 7.7).

The binary fan has, like SEQ, a universal character: Each fan can be embedded in the binary fan (the proof is slightly more involved than that of 7.7, on account of the fact that a node can have a large number of immediate successors). N.B. the relation 'is an immediate successor of' need not be preserved by this embedding.

From now on we will restrict our attention to well-founded trees. In the presence of AC a tree is well-founded if and only if all branches are finite (II. 9.8). The tree itself need of course not be finite, as is shown in the following example.

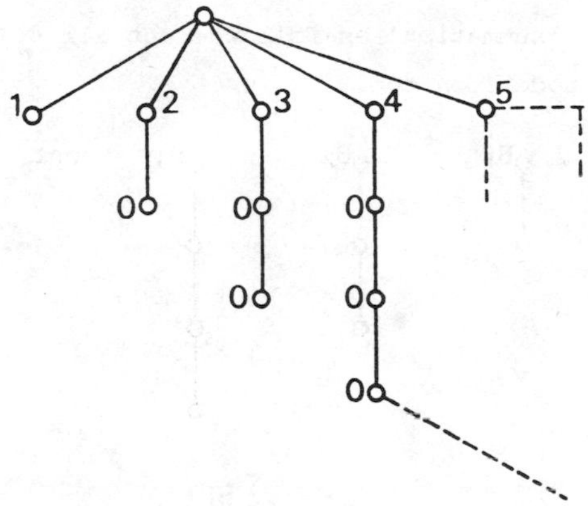

Fig. 16

Consider the functions f_i with $f_i(0) = i$ and $f_i(n) = 0$ $(n > 0)$. The tree in fig. 16 consists of the nodes $f_i \upharpoonright j$ $(j \leq i)$.

To keep the figures manageable we have labeled each node with one number. The corresponding node in SEQ is obtained by going down along the chain belonging to the node, and placing all numbers in a sequence.

There is a remarkable connection between well-founded trees and (countable) ordinals. In order to make the connection transparent we first introduce a few definitions.

<u>Definition 7.11</u>: Let B be a tree. Then $B_b = \{a \in B \mid a \leq b\}$.

Clearly B_b is itself a tree, which we call B_b, the *subtree* of B *under* b. By means of the subtrees we can take apart a tree, and also build trees from smaller ones. This is made explicit in the following inductive definition.

Definition 7.12: *i*) Each singleton is an i-tree.

ii) Suppose, for $\alpha \leqslant \omega$ $\{B_n \mid n < \alpha\}$ is a family of i-trees, and suppose $t \notin \bigcup B_n$. Then $B = \bigcup\{B_n \mid n < \alpha\} \cup \{t\}$ is also an i-tree. The ordering on B is defined by $b < t$ for $b \in \bigcup B_n$ and $b < b' \Leftrightarrow \exists n\, (b, b' \in B_n \wedge b <^n b')$ (where $<^n$ is the ordering on B_n).

Notation: $B = \sum_{n<\alpha} B_n$.

Example. See fig. 17.

One can visualise this 'summation' as follows: put all B_n's in a horizontal row and place the new node t on top.

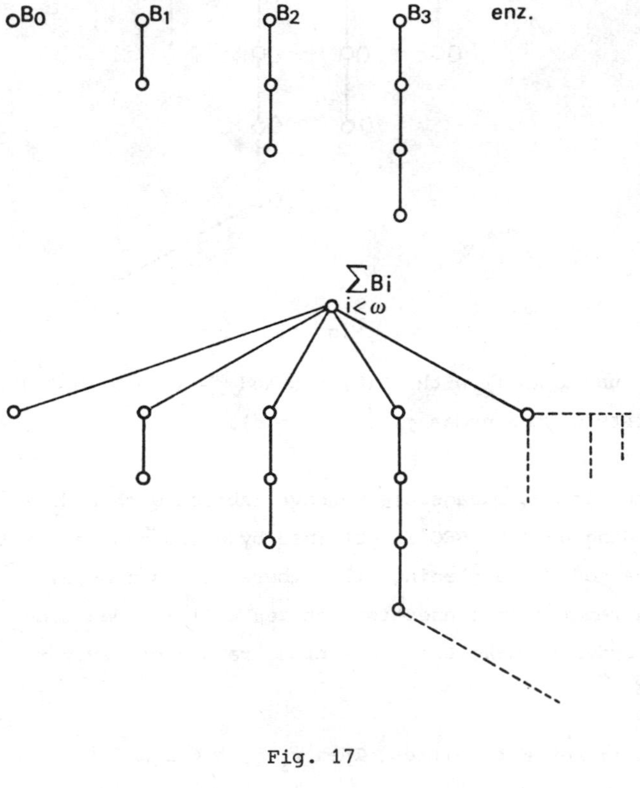

Fig. 17

<u>Theorem 7.13</u>: A tree is well-founded iff it is an i-tree.

Proof. → Since the class of i-trees is inductively defined, we can give a proof by induction.

i) A singleton is well-founded.

ii) Let $t, a_0, a_1, a_2, \ldots$ be an infinite branch in ΣB_n. By definition $a_0 \in B_n$ for some $n < \alpha$. But then $a_0, a_1, a_2, \ldots$ is an infinite branch in B_n. Contradiction.

← Since B is well-founded we may apply the principle of $<^*$-induction (II.2.6) to show that each subtree is an i-tree.
$\forall b <^* a\,[B_b$ is an i-tree$] \to B_a$ is an i-tree (by definition). Hence each subtree B_a is an i-tree, and in particular B is an i-tree. □

We get the following bonus from this theorem: properties of well-founded trees can be established inductively. As the classes of i-trees and well-founded trees coincide, we will only use the term 'well-founded tree'.

Now we assign to each well-founded tree an ordinal.

<u>Definition 7.14</u>: *i*) $\|B\| = 0$ iff B is a singleton

ii) $\|B\| = \bigcup\{\|B_n\| + 1 \mid n < \alpha\}$ iff $B = \sum_{n<\alpha} B_n$.

Example. See fig. 18.

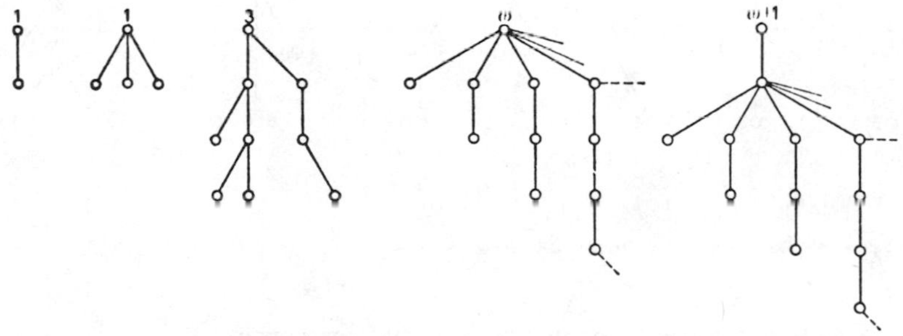

Fig. 18

Well-founded trees are very convenient for giving geometrical representations of (countable) ordinals. Conversely ordinals can be used to establish facts about trees.

Theorem 7.15: $\omega_1 = \{\ \|B\|\ |B$ is a well-founded tree$\}$.

Proof. "$\supseteq$" is trivial. For the converse inequality we can prove: $\alpha \in \omega_1 \to \alpha = \|B\|$ for some B, by transfinite induction on α. □

The above assignment of ordinals is not the only one possible. We will define an ordering in trees which is total and which for well-founded trees, also provides ordinals.

Definition 7.16: Let B be a subtree of SEQ and let $a,b \in B$.

$$a \preccurlyeq b := a <^* b \vee (a = c * \langle m \rangle * d \wedge b = c * \langle n \rangle * e \wedge m < n).$$

In words: $a \preccurlyeq b$ if b is a proper initial segment of a or if, at the first place where a and b differ, a has a smaller value than b (a 'lexicographical' ordering of all sequences).

Lemma 7.17: $\preccurlyeq$ is a total ordering on B.

Proof. The only non-trvial property is transitivity. Let $a \preccurlyeq b$ and $b \preccurlyeq c$. In fig. 19 we have indicated the various possibilities for the relations between a,b and c in B. □

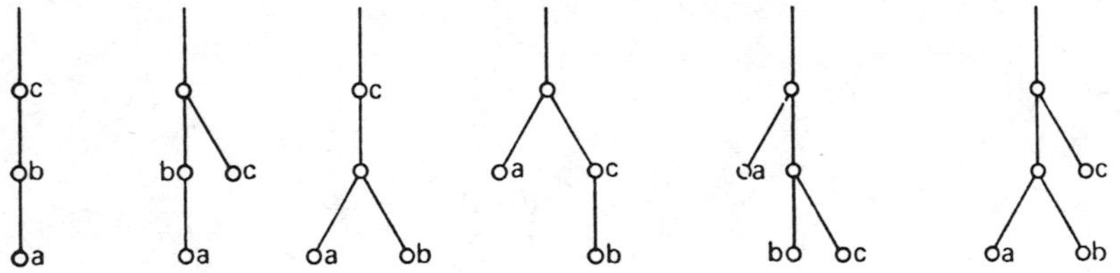

Fig. 19

One easily checks that in each case $a \preccurlyeq c$.

The ordering $\preccurlyeq$ is called the *Brouwer-Kleene* ordering.

Example. See fig. 20.

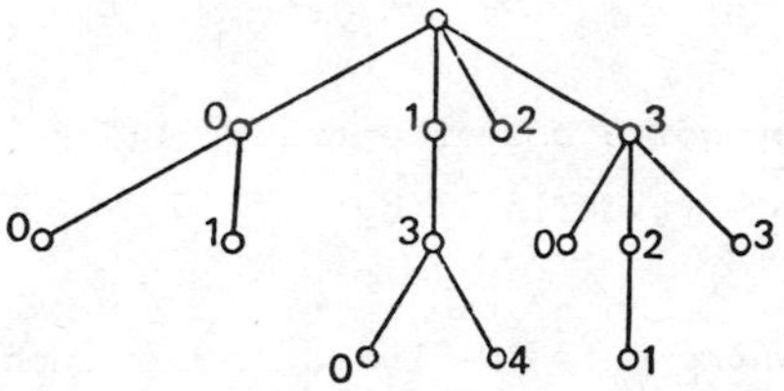

Fig. 20

The order of the nodes in the Brouwer-Kleene ordering is

$\langle 0,0\rangle \preccurlyeq \langle 0,1\rangle \preccurlyeq \langle 0\rangle \preccurlyeq \langle 1,3,0\rangle \preccurlyeq \langle 1,3,4\rangle \preccurlyeq \langle 1,3\rangle \preccurlyeq \langle 1\rangle \preccurlyeq \langle 2\rangle \preccurlyeq \langle 3,0\rangle \preccurlyeq \langle 3,2,1\rangle \preccurlyeq \langle 3,2\rangle \preccurlyeq \langle 3,3\rangle \preccurlyeq \langle 3\rangle \preccurlyeq \langle\ \rangle$.

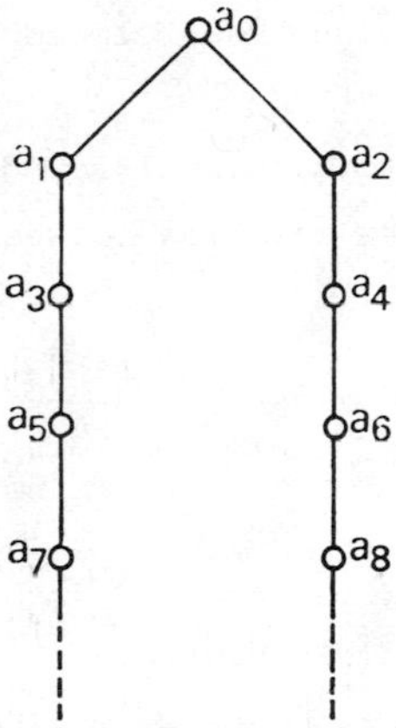

Fig. 21

In fig. 21 we have sketched a tree with exactly two infinite branches. The ordering is $a_0 \succcurlyeq a_2 \succcurlyeq a_4 \succcurlyeq a_6 \succcurlyeq \dots \succcurlyeq a_1 \succcurlyeq a_3 \succcurlyeq a_5 \succcurlyeq \dots$.
The order type of this Brouwer-Kleene ordering is $\omega^* + \omega^*$ (or, that of the set $\{(n+1)^{-1} \mid n \in \omega\} \cup \{1 + (n+1)^{-1} \mid n \in \omega\}$).

Theorem 7.18: The Brouwer-Kleene ordering is a well-ordering iff B is well-founded.

Proof. If B is not well-founded then B has an infinite branch. Along this branch the Brouwer-Kleene ordering has an infinite descending sequence. So $\preccurlyeq$ is not a well-ordering.
Now suppose B is well-founded and let X be a non-empty subset of B. The least

element of X is found by going down the tree, all the time keeping as far left as possible. That is to say, define f by

$f(0) = \langle\ \rangle$

$f(n+1) = f(n) * \langle k \rangle$ where k is the least number such that $f(n) * \langle k \rangle$ is an element of X or has a successor in X.

Since B is well-founded, f has a finite domain and $f(m)$ is the least element of X, where m is the maximum of the domain of f. □

Strictly speaking, any subset of SEQ with a maximal node is a tree, e.g. the set $\{\langle 0 \rangle, \langle 0,0,7 \rangle, \langle 0,2 \rangle\}$. However each of those trees is isomorphic to a subtree of SEQ which is closed under predecessors (by theorem 7.7). So it is no restriction to consider only subtrees closed under predecessors.

By theorem 7.18 we can assign to each well-founded tree B the ordinal $\beta(B)$ of the Brouwer-Kleene ordering. The relation between the ordinals $\|B\|$ and $\beta(B)$ is given by

<u>Theorem 7.19:</u> $\|B\| \leqslant \beta(B)$

Proof. Induction on B. □

The theorems of Brouwer and König established a connection between the set of nodes of a fan and its set of branches. We will exhibit a similar connection for trees in general. The similarity can be expressed in the form of an 'equation': $\frac{\text{fan}}{\text{finite subtree}} = \frac{\text{tree}}{\text{well-founded subtree}}$, where the subtrees are closed under predecessor. We restrict our attention to SEQ.

<u>Definition 7.20:</u> CONT := $\{F \in \omega^{\mathcal{B}} \mid F \text{ is continuous}\}$.

We denote the elements of $\mathcal{B}$ (i.e. ω^{ω}) by $\xi, \eta, \zeta, \ldots$.

For an elegant treatment the continuous mappings from $\mathcal{B}$ into ω we introduce a class of mappings from SEQ into ω.

Let f: SEQ $\to \omega$ be given. Then the *shift of f over a*, f_a , is defined by $f_a(b) = f(a * b)$.

Geometrically speaking, this means that we are only interested in the values of f on SEQ_a. Note that $f_{< >} = f$ and $(f_a)_b = f_{a*b}$. We now define the class K of *Brouwer operations* (after Brouwer and Kreisel).

Definition 7.21: *i*) f is constant and positive $\rightarrow f \in K$

ii) $f(0) = 0 \wedge \forall x \, (f_{<x>} \in K) \rightarrow f \in K$

We have given an inductive definition, as discussed in §4, so we can use the theory of inductive definitions to find a connection between K and ordinals. There is, however, a direct method for associating ordinals with elements of K.

First we will show how to visualise the Brouwer operations. We draw the tree SEQ and label each node with its value under the Brouwer operation f.

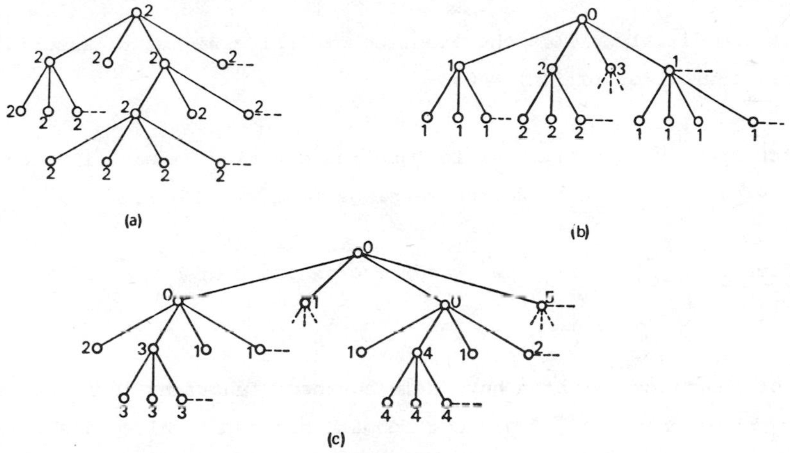

Fig. 22

In figure 22a a constant Brouwer operation is shown. In figure 22b we have a Brouwer operation which is obtained in one step from constant functions, namely $f_{<0>} = \lambda x.1$, $f_{<1>} = \lambda x.2$, $f_{<2>} = \lambda x.3$, $f_{<3>} = \lambda x.1$, etc.

In figure 22c one needs two steps to build f from constant functions. E.g. $f_{<0>} \in K$, since for all x $(f_{<0>})_{<x>} = f_{<0,x>} \in K$ and all $f_{<0,x>}$ are constant and positive.

Notation: if $t(x)$ is a term, then $\lambda x.t(x)$ is the function f with $f(x) = t(x)$ (λ - *abstraction*).

One easily shows by induction on K

Lemma 7.22: For each $f \in K$ i) $f(a) \neq 0 \rightarrow f(a * b) = f(a)$

ii) $\{a \mid f(a) = 0\}$ is well-founded.

Proof. We only consider ii).
If f is constant and positive, then $\{a \mid f(a) = 0\}$ is trivially well-founded. So consider $f \in K$ with $f(\langle\, \rangle) = 0$. Suppose $\{a \mid f(a) = 0\}$ has an infinite descending chain in the tree ordering: $a_0, a_1, a_2, \ldots$ (with $a_0 = \langle\, \rangle$). It is no restriction to assume that a_{i+1} is an immediate successor of a_i ($i \in \omega$). Now $a_1, a_2, a_3, \ldots$ is an infinite descending chain in $\{a \mid f_{a_1}(a) = 0\}$, which contradicts the induction hypothesis. □

The lemma immediately yields the promised ordinal assignment: associate to f the tree ordinal of $\{a \mid f(a) = 0\}$.

Since each Brouwer operation is, in the tree ordering, eventually constant, we can use elements of K to define mappings from $\mathcal{B}$ into ω.

Definition 7.23: $f \in K$ is a *neighbourhood function* of $F \in \omega^{\mathcal{B}}$ if $F(\xi) = k \Longleftrightarrow$ $\Longleftrightarrow \exists m\ f(\xi \lceil m) = k+1$.

The set of functions F, which have neighbourhood functions in K, is called K^*. Evidently the elements of K^* are continuous, since the values are always determined by initial segments of the arguments.
We can do even better:

Theorem 7.24: CONT $= K^*$.

Proof: i) $K^* \subseteq$ CONT is evident (as stated above).

ii) Let $F \in$ CONT. Then (cf exercise 3)

$\forall \xi\ \exists n\ \forall \eta\ (\xi \lceil n = \eta \lceil n \rightarrow F(\xi) = F(\eta))$, i.e.

$X := \{a \mid \exists \xi\ \exists \eta\ \exists n\ (\xi \lceil n = \eta \lceil n = a \wedge F(\xi) \neq F(\eta))\}$ is well-founded.

Now define f: SEQ $\rightarrow \omega$ by $f(a) = \begin{cases} 0 & \text{if } a \in X \\ F(a * \underset{\sim}{0}) + 1 & \text{if } a \notin X \text{ (where } \underset{\sim}{0} = \lambda x.0) \end{cases}$

One easily shows by induction on $\|X\|$ that $f \in K$. From the definition of f it follows that f is a neighbourhood function of F, so $F \in K^*$ (For another proof cf exercise 5). □

The importance of theorem 7.24 for the foundations of analysis is that the *type* of the objects of CONT is lowered. (Notice that the rank of the neighbourhood functions of F is smaller than the rank of F, i.e. the 'complexity' is lowered). We also get for free the insight that CONT is inductively definable (cf. exercise 4).

For real analysis the mappings from $\underset{\sim}{R}$ to $\underset{\sim}{R}$ are far more important than those from $\underset{\sim}{R}$ to ω. Since we are here not specifically interested in the topology of $\underset{\sim}{R}$, we consider the Baire space $\mathcal{B}$ instead of the continuum $\underset{\sim}{R}$. The results from this section can be carried over from $\mathcal{B}$ to $\underset{\sim}{R}$; this is, however, outside the scope of the book.

<u>Definition 7.25:</u>
$\mathrm{CONT}_1 := \{\underset{\sim}{F} \in \mathcal{B}^{\mathcal{B}} \mid \forall \xi\, \forall x\, \exists y\, \forall \eta\, (\xi \restriction y = \eta \restriction y \rightarrow F(\xi) \restriction x = F(\eta) \restriction x)\}$

In words: CONT_1 consists of the functions $\underset{\sim}{F}$ with the property that initial segments of the value are determined by initial segments of the argument; topologically speaking: CONT_1 is the set of continuous mappings from $\mathcal{B}$ to $\mathcal{B}$ (cf exercise 6).

We will now indicate means to simulate continuous mappings from $\mathcal{B}$ to $\mathcal{B}$ by Brouwer operations, and hence by neighbourhood functions.
Let $\underset{\sim}{F}(\xi) = \eta$. We must be able to determine all values $\eta(x)$ for given ξ and x. This is a problem of coding: define $F'$by $F'(\langle x\rangle * \xi) = \underset{\sim}{F}(\xi)(x) = \eta(x)$.

Evidently F' is continuous. Hence F'has a neighbourhood function $f \in K$, and $\eta(x)$ may be computed by f from an initial segment of $\langle x\rangle * \xi$.

<u>Definition 7.26:</u> (*a*) If f is a neighbourhood function of F, write $F := \hat{f}$.
(*b*) For $f \in K$, define $\hat{\hat{f}}: \mathcal{B} \rightarrow \mathcal{B}$ by $\hat{\hat{f}}(\xi) = \eta$ iff $\eta(x) = \hat{f}(\langle x\rangle * \xi)$.
(*c*) $K^{**} = \{\hat{\hat{f}} \mid f \in K\}$.

In complete analogy to theorem 7.24 we have

Theorem 7.27: $\mathrm{CONT}_1 = K^{**}$.

Proof: $K^{**} \subseteq \mathrm{CONT}_1$ follows immediately from definitions 7.25 and 7.26, by induction on K. Now let $\underset{\sim}{F} \in \mathrm{CONT}_1$. Define $G\colon \mathcal{B} \to \omega$ by $G(\langle x \rangle * \xi) = \underset{\sim}{F}(\xi)(x)$. Clearly $G \in \mathrm{CONT}$, so $G = \hat{g}$ for some $g \in K$.
If $\underset{\sim}{F}(\xi) = \eta$ and $\eta(x) = y$, then $G(\langle x \rangle * \xi) = y$, so $\hat{g}(\langle x \rangle * \xi) = y$, i.e. $\hat{\hat{g}} = \underset{\sim}{F}$ and $\underset{\sim}{F} \in K^{**}$. □

As a result of theorem 7.27 even the continuous mappings from Baire space to itself can be reduced to functions from SEQ to ω. Bearing in mind the countability of SEQ we actually have a reduction to functions from ω to ω.

For foundational applications (in an intuitionistic context) the reader is referred to [Kreisel-Troelstra, 1970].

Exercises

1. Prove that the following definition determines the class of trees.
   A tree is a triple $\langle B, v, t \rangle$, where
   i) $|B| \leq \aleph_0$
   ii) $v\colon B - \{t\} \to B$
   iii) for each $b \in B$ there is an n such that $v^n(b) = t$.

2. Let TREE be the set of trees. Since TREE has an inductive definition the sets TREE^α exist (cf 4.1).
   Prove $\|B\| = \alpha \leftrightarrow B \in \mathrm{TREE}^\alpha$ and $B \notin \mathrm{TREE}^\beta$ for $\beta < \alpha$.

   What are the lengths of TREE and K ?

3. Show that $F\colon \mathcal{B} \to \omega$ ($\mathcal{B}$ and ω with their natural topology) is continuous iff $\forall \xi\, \exists n\, \forall \eta\, (\xi \upharpoonright n = \eta \upharpoonright n \to F(\xi) = F(\eta))$.

4. Define $F_a = \lambda\xi \cdot F(a * \xi)$ (the shift of F over a).
   Prove $F \in \mathrm{CONT} \leftrightarrow \forall n\ F_{\langle n \rangle} \in \mathrm{CONT}$.

5. Define a set C of mappings $F\colon \mathcal{B} \to \omega$ by
   i) F is constant $\to F \in C$
   ii) $\forall n\, (F_{\langle n \rangle} \in C) \to F \in C$

Show (*a*) $C = K^*$

(*b*) $\mathrm{CONT} = C$

6. Show that CONT_1 is the set of all mappings from $\mathcal{B}$ to $\mathcal{B}$ which are continuous in the natural topology of $\mathcal{B}$.

7. Determine the length of CONT and CONT_1 (cf. ex. 2).

8. Let A be hereditarily countable. Show that $TC(A)$ with its $\in$-relation is a well-founded tree. What is the relation between its ordinal and its rank?

8 THE AXIOM OF DETERMINATENESS, AD

The axiom system of Zermelo-Fraenkel has shown its usefulness for mathematics by its ability to incorporate existing parts of mathematics. One can get a considerable amount of mathematics without appealing to the more controversial axioms, such as the axiom of choice, the continuum hypothesis, but at certain points one needs a 'richer' universe of sets. E.g. it may be desirable to have available certain choice functions, measurable cardinals, etc. In such cases one usually bluntly postulates the existence of certain functions (AC), or bijections (from $\aleph_1$ on $2^{\aleph_0}$ in the case of CH). For a long time the axiom of choice was the favourite tool of this type used by mathematicians. In the early sixties, however, Mycielski and Steinhaus proposed a new axiom, which was, like the axiom of choice, an existence postulate. The axiom is formulated in terms of games. We will first give some preliminary definitions.

Let A be a subset of X^ω (where X is some non-empty set). Consider two persons, I and II, making alternate choices of elements of X.

I chooses	a_0	a_2	a_4	a_6	...
II chooses	a_1	a_3	a_5	a_7	...

Together I and II determine a sequence $\rho = \langle a_0, a_1, a_2, \ldots \rangle$. We define that I wins if $\rho \in A$ and II wins if $\rho \notin A$.

Examples

1. $X = \omega$ and $A = \{\lambda n.1\}$ (i.e. A only contains the constant function with value 1). Now II can always win, just by choosing $a_1 \neq 1$. So whatever

I does, he cannot beat *II*.

2. $X = \omega$ and $A = \{\rho \mid \rho(2n-1) + \rho(2n)$ is even for all $n > 0\}$.

   We show a number of moves:

I	0		3		40		7		...
II		19		0		7		...	

   Clearly *I* can parry every move of *II* by choosing an odd number after an odd choice of *II* and an even number after an even choice of *II*. *I* has an algorithm for making his choices, given the previous choices of *II*. In short, *I* has a winning *strategy*.

3. $X = \{0,1\}$ and $A = \{\rho \mid \sum_{i=0}^{n} \rho(i)$ is an odd prime $\rightarrow \rho(n+1) = 1\}$.

   I has a simple strategy: at each move he counts the total number of ones that have been previously chosen. If this number is an odd prime, then he chooses a 1, otherwise a 0. *II* is utterly helpless. Even if he stubbornly chooses zeros, *I* wins. Note that in this game *I* wants information on all previous moves.

Definition 8.1: A pair $\langle X,A \rangle$, with $A \subseteq X^\omega$ is a *game*.
If it is clear from the context what X is, we will say 'the game A'.
Now we will present a precise notion of strategy.

In general, as our examples have shown, each move of player *I*, as given by a strategy, depends on his own, as well as *II*'s, previous moves. However, it is easily seen that this can ultimately be determined as a function of *II*'s previous moves only. Hence we suppose that a strategy S for *I* operates on the moves of *II*:

Definition 8.2: A strategy S for a game $\langle X,A \rangle$ is a mapping from $\bigcup \{X^n \mid n \in \omega\}$ to X.

Note that S operates on finite sequences of elements of X (as in definition 7.3, finite sequences are functions). Heuristically, S yields the next move after a number of moves.

If the players I and II make the respective sequences of choices σ and τ, the resulting sequence is in a trivial way composed of σ and τ. We make this precise in

<u>Definition 8.3</u>: $\rho = [\sigma,\tau]$ iff $\rho(2n) = \sigma(n)$
$\rho(2n+1) = \tau(n)$

Example: $[\ \langle 0,0,0,0, \ldots \rangle, \langle 1,1,1,1, \ldots \rangle\ ] = \langle 0,1,0,1,0, \ldots \rangle$.

<u>Definition 8.4</u>: Let S be a strategy, and suppose τ is the (infinite) sequence of choices made by II. Then $S_I(\tau)$ is the corresponding sequence of choices made by I, using S. Similarly for S_{II}.

Whenever the context permits it, we will omit the subscripts.

Hence, if II makes the choices τ and I uses S, then the resulting sequence is $[\ S_I(\tau), \tau]$.

<u>Definition 8.5</u>: S is a winning strategy for I in $\langle X,A \rangle$ if $\forall\tau \in X^\omega\ ([\ S_I(\tau), \tau] \in A)$. S is a winning strategy for II in $\langle X,A \rangle$ if $\forall\sigma \in X^\omega\ ([\sigma, S_{II}(\sigma)] \notin A)$.

In plain words: S is a winning strategy for I if, no matter how II makes his sequence τ of choices, I wins, using S. Likewise for II.

In our examples there always was some plausible winning strategy for one of the players. In general it is not so easy to find winning strategies and it is even problematic whether every game has a winning strategy for some player. We say that a game is *determined* if either I or II has a winning strategy.

Now the *axiom of determinateness* reads

AD Each game is determined.

There is a very convenient way of expressing the determinateness of a game A, using an infinite string of quantifiers. In ordinary logic this is not allowed, but it is a quite plausible extension of the language, which we will adopt for the occasion.

"*II* has a winning strategy" is expressed by
$\forall x_0\ \exists y_0\ \forall x_1\ \exists y_1\ \forall x_2\ \exists y_2 \ldots\ldots (\langle x_0, y_0, x_1, y_1, \ldots \rangle \notin A)$ and "*I* has a winning strategy" is expressed by
$\exists x_0\ \forall y_0\ \exists x_1\ \forall y_1\ \exists x_2\ \forall y_2 \ldots\ldots (\langle x_0, y_0, x_1, y_1, \ldots \rangle \in A)$.
The game is determined if
II has no winning strategy $\Leftrightarrow$ *I* has a winning strategy.
In symbols
$\neg \forall x_0\ \exists y_0\ \forall x_1\ \exists y_1 \ldots (\langle x_0, y_0, x_1, y_1, \ldots \rangle \notin A \leftrightarrow$
$\leftrightarrow \exists x_0\ \forall y_0\ \exists x_1\ \forall y_1 \ldots (\langle x_0, y_0, x_1, y_1, \ldots \rangle \in A)$
In this form the axiom of determinateness appears as a generalization of the laws of De Morgan (concerning the pulling out of negations from prenex forms). Observe, however, that this analogy does not carry much weight if it comes to justifying the axiom.

From now on we will restrict our attention to games in ω^ω. This restriction does not rob us of any significant results.

So far little is known about the status of *AD*. In particular, its consistency relative to **ZF** is, as yet, an open problem (in contrast to our knowledge concerning *AC*). We do know however, from a result of D.A. Martin, that all Borel games are determined (i.e. for all games, with *A* a Borel set).
We shall show something much weaker here.

Theorem 8.6: (Gale and Stewart) *AD* holds for open and for closed games.

Proof: Let $A \subseteq \omega^\omega$ be open. We assume that *I* has no winning strategy and we will show that therefore *II* has one. The underlying idea is: Since *I* has no winning strategy, *II* can, after the first choice, a_0, of *I*, always find a b_0, such that *I* cannot find a strategy to win the continuation of the game. Again: no matter how *I* chooses a_1, *II* can always find a b_1 such that after a_0, b_0, a_1, b_1, *I* has no winning strategy. So if *II* keeps chosing the least b_i such that subsequently *I* has no winning strategy, then he stands a good chance of winning the game.

We make this precise: Let A be a game, and for any finite sequence a, let A^a be the continuation of A after a sequence a of moves: $A^a = \{\sigma | a * \sigma \in A\}$. Now the strategy for II is clear:

$S(a)$ = the least x such that I has no winning strategy for A^b , where

$$b = <a_0 ,\ S(< a_0 >),\ a_1 ,\ S(< a_0 , a_1 >),\ldots a_n ,\ x >.$$

Such an x always exists, for if I has no winning strategy for A^b , then there is a move k such that for no p does I have a winning strategy for $A^{b *<k ,p >}$.

We now show that S is a winning strategy for II. Assume, on the contrary, that for some sequence σ of moves of I, $[\sigma, S(\sigma)] \in A$.

A is open, therefore there is an initial segment $b = < \sigma_0 ,\ \tau_0 ,\ \sigma_1 ,\ \tau_1 ,\ \ldots$ $\sigma_n ,\ \tau_n >$ (with $\tau = S(\sigma)$) such that for all ρ $b * \rho \in A$. This means that I has a winning strategy after move τ_n , which contradicts our hypothesis. So $[\sigma, S(\sigma)] \notin A$ for all σ, in other words, S is a winning strategy for II. There is a similar proof for closed A.   □

The use of the axiom of choice in ordinary real analysis is usually restricted to the case of choice functions for countable families of non-empty subsets of $\underline{R}$. This version is called the axiom of countable choice, CC (cf. II.9.10). Actually one can simplify matters a bit further by replacing $\underline{R}$ by the Baire space. Since for foundational purposes it does not make much difference whether one considers $\underline{R}$ or $\mathcal{B}$ (except that the latter space is easier to handle), logicians in both cases talk of 'reals'.

The form of CC we consider here is

$\forall i \in \omega\ (A_i \subseteq \mathcal{B} \wedge A_i \neq \emptyset) \to \exists F\ \forall i \in \omega\ (F(i) \in A_i)$.

The next theorem shows that the axiom of determinateness implies the axiom of countable choice for the Baire space, so AD lends itself to the development of analysis.

Theorem 8.7: $AD \to CC$.

Proof. Let A_i $(i \in \omega)$ be given with $A_i \subseteq \mathcal{B}$ and $A_i \neq \emptyset$. Consider the game A, where A is defined by $[\sigma, \tau] \in A := \tau \notin A_{\sigma(0)}$. From the definition of A, it follows that only the first move of I is relevant. We see immediately that I has no winning strategy: $A_{\sigma(0)} \neq \emptyset$, so I cannot prevent $\tau \in A_{\sigma(0)}$, i.e.

$[\sigma,\tau] \notin A$. Therefore *AD* implies that *II* has a winning strategy *S*. Now we can use *S* to find a choice function. For, if we define $\sigma_i = \langle i, 0, 0, 0, \ldots \rangle$, then $S(\sigma_i) \in A_i$. Hence the required choice function *F* is $\lambda i.S(\sigma_i)$. □

Note that *CC* carries over from $\mathcal{B}$ to $\underset{\sim}{R}$ (and vice versa), since $\mathcal{B} \underset{1}{=} \underset{\sim}{R}$.

Let us have a closer look at the relation between *AC* and *AD*. We will use some facts about filters (cf. III. 1). We have already shown that *AC* yields the existence of free ultrafilters. We refine this slightly by restricting our attention to filters on ω, and by assuming only a well-ordering on $\mathcal{P}(\omega)$ (instead of *AC* in its full generality).

<u>Theorem 8.8</u>: $\mathcal{P}(\omega)$ is well-ordered $\rightarrow$ there is a free ultrafilter on ω.

Proof. Let F_0 be the (free) filter of cofinite sets on ω. As $\mathcal{P}(\omega)$ is well-ordered, the subsets of ω form a transfinite sequence $\{A_\alpha \mid \alpha < \beta\}$.
Define for $\alpha > 0$, $\alpha < \beta$

$$F_\alpha = \begin{cases} \bigcup_{\gamma<\alpha} F_\gamma \cup \{A_\alpha\} & \text{if this union has the finite intersection property} \\ \bigcup_{\gamma<\alpha} F_\gamma & \text{else} \end{cases}$$

We shall show that F_β is a free ultrafilter.

(1) Note that $\emptyset \notin F_\beta$.

(2) If $A_\alpha, A_{\alpha'} \in F_\beta$, then $A_\alpha \cap A_{\alpha'} = A_\gamma$ for some $\gamma < \beta$.
Suppose $A_\gamma \notin F_\beta$, then $\{A_\gamma\} \cup \bigcup\{F_\delta \mid \delta < \gamma\}$ does not have the finite intersection property. So $A_\gamma \cap A_{\delta_1} \cap \ldots \cap A_{\delta_n} = \emptyset$ for certain $\delta_1, \ldots, \delta_n < \gamma$. Hence $A_\alpha \cap A_{\alpha'} \cap A_{\delta_1} \cap \ldots \cap A_{\delta_n} = \emptyset$, but $A_{\alpha'}, A_\alpha, A_{\delta_1}, \ldots, A_{\delta_n} \in F_\delta$, where $\delta = \max(\alpha, \alpha', \delta_1, \ldots, \delta_n)$. Contradiction.

(3) $A_\alpha \in F_\beta$ and $A_\alpha \subseteq A_\gamma \rightarrow A_\gamma \in F_\beta$. An exercise for the reader.

(4) Since F_β is an extension of a free filter, it is free.

(5) Apply III. 1.5. Suppose $A_\alpha \notin F_\beta$ and $A_\gamma = A_\alpha^c \notin F_\beta$. Then $A_\alpha \cap A_{\alpha_1} \cap \ldots \cap A_{\alpha_n} = \emptyset$ for some $\alpha_1, \ldots, \alpha_n < \alpha$ and $A_\gamma \cap A_{\gamma_1} \cap \ldots \cap A_{\gamma_m} = \emptyset$ for some $\gamma_1, \ldots, \gamma_m < \gamma$. From $A_\gamma = A_\alpha^c$ it follows that
$A_{\alpha_1} \cap \ldots \cap A_{\alpha_n} \cap A_{\gamma_1} \cap \ldots \cap A_{\gamma_m} = \emptyset$. So, by (2) $\emptyset \in F_\beta$. Contradiction.

Hence F_β is an ultrafilter. □

Theorem 8.9: *AD* → There is no free ultrafilter on ω.

Proof. Suppose there is a free ultrafilter F on ω. We now set up a game for *I* and *II*, in which *I* tries to collect an element of F by making suitable choices. *II* tries to prevent this by choosing in such a way that there is not enough left for *I* (remember that elements of free ultrafilters are 'large'). To be precise: if $\rho = [\sigma,\tau]$, then $\rho \in A \leftrightarrow Z := \{\sigma(i) \mid \forall j < i(\sigma(i) \neq \tau(j))\} \in F$. The rules are symmetrical to such an extent that *II* can find a winning strategy if *I* has one, by just copying *I*.
Note that if *I* wins with σ, using S, he can also win with a sequence σ' (using an improved strategy S') in which no move is repeated.
For define

$$\sigma'(a) = \begin{cases} \sigma(a) \text{ if } \sigma(a) \neq \sigma(b) \text{ for all } b <^* a \\ \max\{\sigma(b)+1 \mid b <^* a\} \text{ else.} \end{cases}$$

Observe that if $\forall j < i\ (\sigma(i) \neq \tau(j))$ and if there is a least $i' < i$, such that $\sigma(i') = \sigma(i)$, then *I* chooses $\sigma'(i)$ distinct from $\sigma(i)$. However, $\sigma(i') \neq \tau(j)$ for all $j < i'$, so $\sigma(i) = \sigma(i') = \sigma'(i') \in Z$.
Hence $Z' := \{\sigma'(i) \mid \forall j < i\ (\sigma'(i) \neq \tau(j))\} \supseteq \{\sigma(i) \mid \forall j < i\ (\sigma(i) \neq \tau(j))\} = Z$.
Therefore also $Z' \in F$.

Now let *I* have a strategy S, which never repeats moves. Define a strategy S^* for *II* by $S^*(\langle n\rangle * a) = S(a)$ $(= \tau)$. Now $Y = \{\tau(i) \mid \forall j (0 < j \leqslant i \rightarrow \tau(i) = \sigma(j)\} \in F$. But $Y \cap Z \subseteq \{\sigma(0)\}$, contradicting the assumption that F is a free ultrafilter. Conclusion: *I* has no winning strategy.

Likewise one shows that *II* has no winning strategy. Therefore *AD* fails. The theorem follows by contraposition. □

We now have

Corollary 8.10: *AD* → $\mathcal{P}(\omega)$ cannot be well-ordered.

Corollary 8.11: *AD* → ¬ *AC*.

The axiom of determinateness has another rather curious consequence, namely that all subsets of $\mathbb{R}$ are Lebesque-measurable (which again contradicts *AC*, by Vitali's theorem).

Theorem 8.12: (Mycielski, Swierczkowski).
$AD \Rightarrow$ all sets of reals are Lebesque-measurable.

Proof. We use a *reductio ad absurdum*. The following fact from measure-theory is used without proof. If there is a non-Lebesque-measurable set of reals, then there is also a non-Lebesque-measurable subset of the interval $[0,1]$ with inner measure 0 and outer measure 1. Let W be such a set.

Before we define our game, we will first introduce some auxiliary notions.

$r_0, r_1, r_2, \ldots$ is a sequence of positive rationals such that $\frac{1}{2} > r_0 > r_1 > r_2 > \ldots$ and $\sum_{i=0}^{\infty} r_i < \infty$.

A sequence $A_0, A_1, A_2, \ldots$ of subsets of $[0,1]$ is called *good* if

i) $A_0 \supseteq A_1 \supseteq A_2 \supseteq \ldots$

ii) each A_i is the union of finitely many intervals $[a^i_j, b^i_j]$ with rational endpoints.

iii) the diameter, $d(A_i)$, of A_i is at most 2^{-i}.

iv) the measure of A_i is $r_0 . r_1 . \ldots . r_i$.

The game consists of successive choices of subsets of $[0,1]$, such that a good sequence results. I wins if $p = \cap A_i \in W$, II wins if $p \notin W$. (This game does not have the proper form, as laid down in definition 8.1. In fact, it is not difficult to code the A_i's in ω, such that a proper game is obtained. However, the present formulation is more perspicuous.)

Suppose I has a winning strategy S (the analogous case of a winning strategy for II is left to the reader), then whatever II does, he will lose. Therefore we make II play in such a way that from the course of the game it follows that W has a subset with positive measure.

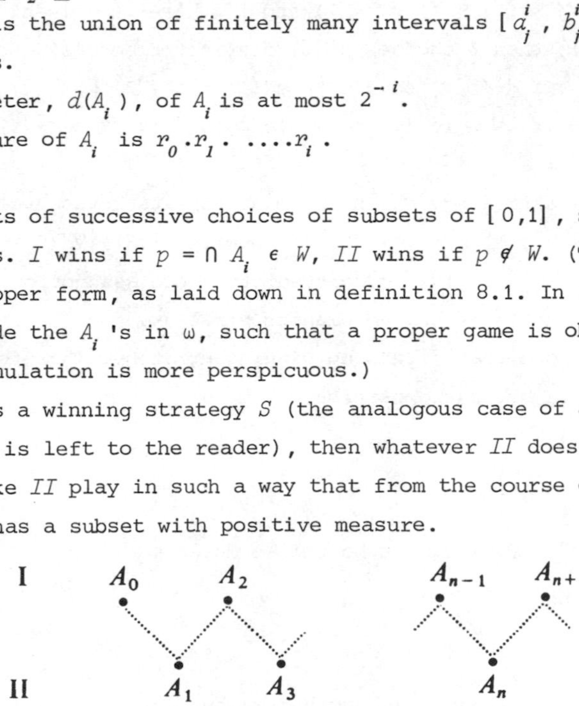

Say I has made the move A_{n-1}. We will select a number of convenient moves A_n for II, such that I(employing strategy S) has a nice store of moves with suitable properties .

(a) There is a finite set $\{B_1, \ldots, B_k\}$ such that $A_0, \ldots, A_{n-1}, B_i$ ($i \leqslant k$) is good and such that for $C_i = S(\langle A_0, \ldots, A_{n-1}, B_i \rangle)$ the following holds

i) $C_i \cap C_j = \emptyset$ if $i \neq j$

ii) $m(C_1 \cup \ldots \cup C_k) \geqslant (1 - 2r_{n+1}).m(A_{n-1})$.

Proof:

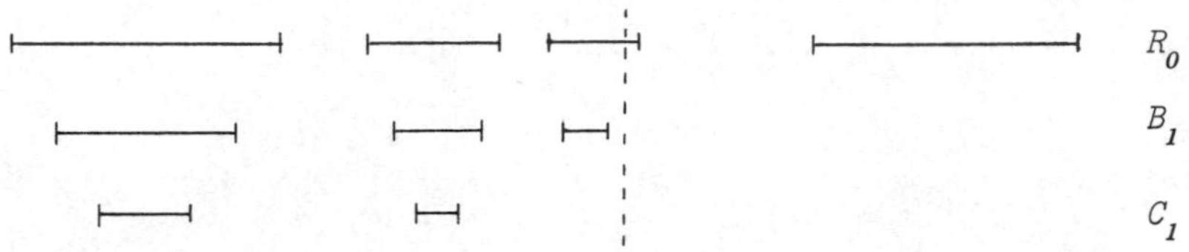

Fig. 23

Consider $R_0 := A_{n-1}$. We know that $m(R_0) > 2.\ m(A_{n-1}).r_{n+1}$ (since $1 > 2\,r_{n+1}$). Partition the smallest segment containing R_0 in two equal parts (see dotted line). One of these parts contains a part P of R_0 with $d(P) \leqslant d(R_0)/2 \leqslant 2^{-n}$ and $m(P) > m\ (A_{n-1}).r_{n+1}$. By making P somewhat smaller we obtain the required B_1. C_1 is obtained by applying S.

Now suppose we have already obtained $B_1, \ldots, B_j$.

Consider $R_j = A_{n-1} - \cup\{C_i \mid i \leqslant j\}$. If $m(R_j) \leqslant 2.\ m(A_{n-1}).r_{n+1}$, then we are done. If not, repeat the above partitioning process, with resulting B_{j+1}. B_{j+1} again forms with $A_0, \ldots, A_{n-1}$ a good sequence and C_{j+1} is determined by an application of S.

By induction hypothesis $C_1, \ldots, C_j$ are disjoint.

$C_{j+1} \subset B_{j+1} \subseteq A_{n-1} - \bigcup_{i \leqslant j} C_i$, so $C_1, \ldots, C_{j+1}$ are also disjoint.

The procedure stops when $m(A_{n-1} - \bigcup_{i \leqslant k} C_i) \leqslant 2.\ m(A_{n-1}).r_{n+1}$, so $m(\bigcup_{i \leqslant k} C_i) \geqslant m(A_{n-1})\ (1-2r_{n+1})$.

(b) We consider all possible good sequences, obtained as under (a), by means of the strategy S. That is to say: if $\{C_1, \ldots, C_k\}$ are constructed as in (a) from A ($=A_{n-1}$), then we put $F(A) := \{C_1, \ldots, C_k\}$.

Define $G_0 := \{A_0\}$, where $A_0 = S(\langle\ \rangle)$

$G_{n+1} := \cup\{F(A) \mid A \in G_n\}$.

and

$$G_n^* := \cup\{A \mid A \in G_n\}$$
$$G^* := \cap\ G_n^*$$

The elements of G_n are the $2n^{\text{th}}$ moves in the game (made by I). From the definition of the game it follows that $a \in G^* \to a \in W$ (by the winning strategy S of I), so $G^* \subseteq W$.

(c) We will estimate the measure of G^*.

Let $G_n = \{A_1^1, \ldots, A_{n_1}^1, \ldots, A_1^k, \ldots, A_{n_k}^k\}$, then

$$G_{n+1} = \{C_{1\ ,1}^1, \ldots, C_{1\ ,m(1\ ,1)}^1, \ldots, C_{n_1\ ,1}^1, \ldots, C_{n_1\ ,m(1\ ,n_1)}^1,$$
$$\ldots,\ C_{1\ ,1}^k, \ldots, C_{1\ ,m(k\ ,1)}^k, \ldots, C_{n_k\ ,1}^k, \ldots C_{n_k\ ,m(k\ ,n_k)}^k\}.$$

Each A_j^i has been transformed into a set of $C_{j\ ,p}^i$'s, which are all disjoint.

The measure of G_n^* satisfies $m(G_n^*) \geqslant r_0 . (1 - 2\ r_2) \ldots (1 - 2\ r_{2n})$.
This is shown by induction on n: $m(G_0^*) = m(A_0) = r_0$;
$$m(G_{n+1}^*) \geqslant r_0 . (1 - 2\ r_2) \ldots (1 - 2\ r_{2n}) . m (\cup C_{j\ ,p}^i)$$
$$\geqslant r_0 . (1 - 2\ r_2) \ldots\ldots\ldots\ldots\ldots (1 - 2\ r_{2n+2}) \text{ by } (a)\ ii).$$
Since $G_{n+1}^* \subseteq G_n^*$, we have $m(G^*) \geqslant r_0 \prod_1^\infty (1 - 2\ r_{2n})$.

The infinite product on the right hand side has a positive value, since $\Sigma\ r_n$ converges (see an analysis text). So we have our contradiction: W contains a set with positive measure.

Conclusion: All subsets of $\underset{\sim}{R}$ are Lebesque-measurable.   □

The axiom of determinateness has further surprising consequences, e.g.

1. $\aleph_1$ is measurable.

2. Every uncountable subset of $\underset{\sim}{R}$ contains a perfect subset.

Literature

J.E. Fenstad. The axiom of determinateness.

   In: Proceedings of the Second Scandinavian Logic Symposium.

      Amsterdam, 1971.

T.J. Jech. The axiom of choice. Amsterdam, 1973.

APPENDIX

In the body of this text we have introduced several terms which may have troubled conscientious readers. The purpose of the present appendix is to bring some clearness to these matters, to comment on, and to justify our somewhat loose handling of the language of set theory.

Our main concern is a number of abstract concepts such as *totality*, *property*, *correspondence*, *functional relationship* etc. and the respective categories of linguistic or formal devices denoting abstractions such as these.

To begin with, it is worth noticing that we could have been more restrictive in our linguistic habits by avoiding those problems altogether. The reason is that Zermelo-Fraenkel set theory, in fact, is concerned with *sets* only; abstractions, as mentioned above, are completely neglected. In practice, however, it happens to be rather cumbersome to content oneself with this state of affairs: a reasonable smooth handling of the theory asks for these more abstract concepts.

The set-theoretic language was rather austere in character: we have *variables* (x, y, a, ...), with which *atomic formulas* ($x = y$, $a \in x$, ...) are formed, which in turn function as the building-blocks for more complex *formulas* together with the logical connectives (conjunction , negation, quantification,.). A variable always denotes a *set*. At some places it was convenient to introduce *constants* for special sets, such as: $\emptyset$, ω, 3, $\aleph_1$ etc. The assertion expressed by a formula often was not properly distinguished from that formula itself. For instance the axiom of the empty set (i.e., the formula $\exists x\, \forall y\, (y \notin x)$) *asserts* the existence of an elementless set.

So formulas may express assertions. But there are also formulas representing *properties* (or *conditions*): for instance, the formula $\forall y (y \notin x)$ expresses the property (of x) of emptiness. There are also formulas representing *conditions* (or *correspondences*) in more arguments than just one: $y \notin x$ expresses the condition on y of not being an element of x.

In particular we said that Φ represented an *operation* or *functional correspondence* when $\forall x\, \exists! y\, \Phi$ expressed a valid assertion. In conformity with mathematical practice we have introduced special symbols for these operations, such as $\mathcal{P}$, $\cup$, etc. Though the ontological status of concepts, such as property, correspondence, operation etc. in the Zermelo-Fraenkel theory of sets is nil (*to be*, *to exist* in this theory means: to be a *set*; in other words: the quantifiers '$\exists x$' and '$\forall x$' are always read as: 'there is a *set* x' respectively,

'for all *sets* x'; in other words: a variable always denotes a *set*), it is by all means very convenient to allow oneself the use of these concepts.
This also is true of the concepts described below, which may easily trick the innocent reader, as one easily forgets their shadowy ontological status.
It has frequently occurred in connection with properties of some importance (for instance the property of being an ordinal), that we have substantiated them by forming the *collection* of all sets with this property (such as OR, the collection of all ordinals), even in cases (like OR) where we knew the collection not to be a set. We want to justify (or excuse) this behaviour by sketching the linguistic conventions on which it is based; doing so it will become clear that we shall not walk into the old naive-cantorian trap of the familiar paradoxes.
To begin with, the use of collections goes hand in hand with an improper use of the epsilon-symbol: '$\alpha \in$ OR' just means 'α is an ordinal'. The use of $\in$ was, however, specifically reserved for sets, so here is an obvious (informal) extension of the language. Synonymous with 'collection' are the terms *class*, *multitude*, *totality* and *virtual class*. The adjective of the last name stresses the ontological status of the concept. The use of the term *class* is in the Zermelo-Fraenkel-context slightly undesirable since it has a clear, non-virtual status in the closely related theory of von Neumann-Bernays-Gödel.
Substantiation of properties finds its linguistic expression in the use of the abstraction-notation with which *class-terms* or *abstraction-terms* of the form $\{t(x) \mid \Phi(x)\}$ can be formed, denoting the collection of all sets $t(x)$ where x is such that $\Phi(x)$. Here t represents an operation and Φ a property. The related improper use of the epsilon-symbol is introduced as follows:
'$y \in \{t(x) \mid \Phi(x)\}$' is a suggestive notation for '$\exists x(\Phi(x) \wedge y = t(x))$'.
Equality between collections was introduced by borrowing 'the extensionality principle from the domain of sets': '$K = L$' stands for '$\forall y(y \in K \leftrightarrow y \in L)$', where K and L are any class-terms. And in case a set a happened to be co-extensional with a class K (i.e., $\forall y(y \in a \leftrightarrow y \in K)$) we simply identified K with a. In this way it is clear that talking about classes is only a way of speech, not presupposing the *existence* of these things in any concrete way .
Hoping that this discussion has sufficiently clarified the status of the concepts mentioned, we finally explain some difficulties in connection with the concept of a *property*, bearing on the foundations of the subject.

When Zermelo formulated his axioms in 1904 he stated the separation-axiom in the following way: to every property E and every set a there exists a set b, the elements of which are exactly those of a having the property E. The set-theoretic paradoxes still fresh in his memory, he declared that the admissible properties should be "definite", i.e., for any set x the question, whether or not such a property applies to x, should have a clear meaning and a definite answer (even though the correct answer might be (still) unknown in specific cases). The property S described below probably would not count as definite.
Of course this was rather vague and hence unsatisfactory.

In 1922 Skolem proposed therefore to interprete Zermelo's formulation in the way we have adopted in this book: "property" is: "property expressible by a formula of the set-theoretic formalism". The replacement-axiom (absent in Zermelo's system) also could be made precise in this way, using the notion of a functional correspondence. Though this interpretation is sufficient for everyday mathematical practice, nevertheless, there are reasons to believe that this explanation in terms of a (at least to a certain extent arbitrarily chosen) linguistic apparatus is too restrictive. That is, if one is of the opinion that a satisfactory set theory not only should codify mathematical practice (since then we can be satisfied with the theory in its present form), but also should describe "Cantor's absolutum" as completely as possible. (Of course this requires some degree of belief, or commitment, with respect to the *existence*, of this big world, *the* set-theoretic universe.)
The following two phenomena illustrate the obvious shortcomings of the formalism of Zermelo-Fraenkel (or any first order theory of sets, for that matter) when it comes to describing the set theoretic universe.
In 1922 Skolem remarked that (on the basis of the so-called Skolem-Löwenheim theorem) the theory of Zermelo-Fraenkel (in the present formulation) has a countable model if it has a model at all. (The main reason is that there is only a countable number of formulas. Use has been made of this fact in more or less the same way in the proof of theorem 11.11, where **ZF**-models are constructed with a set of ordinals of countable cofinality. This can be compared to Skolem's remark.) Though this remark, known as *Skolem's paradox* (for one can *prove* the existence of uncountably many sets, cf. I.15.7) was already baffling, things became really problematic when Gödel proved his

completeness theorem in 1930, thereby showing that the hypothesis of having some model could be replaced by mere *consistency* (freedom of contradiction). Faith in a theory begins with consistency; according to Gödel consistency amounts to having a model; and this in turn amounts to having a countable model. So Cantor's absolutum, provably uncountable as it is, is being described rather defectively by the Zermelo-Fraenkel theory. An escape from this evil appears to be impossible as an alternative linguistic framework -as long as it is first-order - will show the same defect.

Tarski has shown us a direct way to realize that the formula-interpretation of "property" is too restrictive. We notice that (at least in principle) it is not difficult to associate a constant with every formula of the **ZF**-formalism. (One does not even need arbitrary constants, as constants denoting natural numbers suffice - a so-called Gödel-numbering.) Suppose for instance that $\ulcorner\Phi\urcorner$ is the constant associated with Φ . Now take S to be the condition applying to the sets x and y whenever y is denoted by a constant $\ulcorner\Phi\urcorner$ associated with a formula Φ in the sense mentioned, which expresses a property applying to x (i.e., $\Phi(x)$ holds).

S is the so-called *satisfaction-relation.*

Bearing the meaning of S in mind we see that, whenever Φ expresses a property of x, $\forall x\ (S(x, \ulcorner\Phi\urcorner) \leftrightarrow \Phi)$ is valid.

On this basis we shall prove that S cannot possibly be expressed by a formula. For suppose this to be the case, then for this formula, say, Ψ, we also should have that $\forall x\ (\Psi(x, \ulcorner\Phi\urcorner) \leftrightarrow \Phi)$ where Φ has the form described. One such Φ is the formula $\neg\Psi(x,x)$, therefore we get in particular

$\forall x(\Psi(x, \ulcorner\neg\Psi(x,x)\urcorner) \leftrightarrow \neg\Psi(x,x))$.

As this holds for all x we obtain in the case $x = \ulcorner\neg\Psi(x,x)\urcorner$ the instance $\Psi(\ulcorner\neg\Psi(x,x)\urcorner, \ulcorner\neg\Psi(x,x)\urcorner) \leftrightarrow \neg\Psi(\ulcorner\neg\Psi(x,x)\urcorner, \ulcorner\neg\Psi(x,x)\urcorner)$, which is contradictory. This means that such a Ψ cannot be found unless of course the theory is inconsistent.

So either we are left, as Zermelo, with vague properties - or with clear ones *à la* Skolem, but with an unsatisfactory theory.

From the first alternative there is a way out, at least in principle. One of the reasons to axiomatize the concept of a set is that it is vague. Why not axiomatize the concept of a property? In fact this has been done in several versions of the set theory, where properties have become *classes*, as in the theories of von Neumann-Bernays-Gödel, Kelley-Morse-Mostowski, Ackermann and others.

Objections, as mentioned above, relating to the fact that theories necessarily have to be formulated in languages with innate limitations, however remain inescapable. But other objections pertaining more to mathematical insufficiencies, sometimes, can be met in this way.

SYMBOLS

LITERATURE

Bachmann, H. *Transfinite Zahlen*. Berlin 1955.

Barwise, J. *The Handbook of Mathematical Logic*. North-Holland, Amsterdam 1977.

Baumgartner, J.E. and Prikry, K.: Singular cardinals and the Generalized Continuum Hypothesis. *American Mathematical Monthly* 84 (1977) pp. 108 - 113.

Benacerraf, P. & Putnam , H. (editors) *Philosophy of Mathematics* (selected readings). New York 1964.

Bourbaki, N. Utilisation des nombres réels en topologie générale. In: *Eléments de Mathématique I. Les Structures Fondamentales de l'Analyse, Livre III. Topologie Générale*, CH. 9. Paris 1958.

Cantor, G. Beiträge zur Begründung der transfiniten Mengenlehre I. *Math. Ann. 46*: 481, 1895.

Cantor, G. *Gesammelte Abhandlungen*. Berlin 1932. Herdruk Hildesheim 1966.

Chang, C.C. & Keisler, H.J. *Model Theory*. Amsterdam 1973.

Cohen, P.J. *Set Theory and the Continuum Hypothesis*. New York, Amsterdam 1966.

Comfort, W.W. and Negrepontis, S. *The theory of ultrafilters*. Berlin 1974.

Dalen, D. van & Monna, A.F. *Sets and Integration. An Outline of the Development*. Groningen 1972.

Devlin, K.J. *The Axiom of Constructability. A Guide for the Mathematician*. Springer, Berlin 1977.

Drake, F.R. *Set Theory*. Amsterdam 1974.

Enderton, H.B. *A mathematical introduction to logic*. New York 1972.

Euclides *Elements* (± 300 B.C.).

Fenstad, J.E. The axiom of determinateness. In: *Proceedings of the Second Scandinavian Logic Symposium*. Amsterdam 1971.

Fraenkel, A.A., Bar-Hillel, Y., Levy, A. & Dalen, D. van. *Foundations of Set Theory*. Amsterdam 1973.

Gödel, K. *The Consistency of the Axiom of Choice and of the Generalized Continuum-Hypothesis with the Axioms of Set Theory*. Princeton 1940.

Halmos, P.R. *Lectures on Boolean Algebras*. Princeton 1963.

Hausdorff, F. *Mengenlehre*. Berlin 1935. New York 1957.

Heijenoort, J. van *From Frege to Gödel. A Source-Book in Mathematical Logic*. Cambridge 1967.

Hilbert, D. *Grundlagen der Geometrie*. Stuttgart 1968.

Jaegermann, M. The axiom of choice and two definitions of continuity. *Bull. Acad. Polon. Sci., 13-699*, 1965.

Jech, T.J. *The Axiom of Choice*. Amsterdam 1973.

Kleene, S.C. *Mathematical Logic*. New York 1967.

Kreisel, G. and Troelstra, A.S. Formal systems for some branches of intuitionistic analysis *Ann. Math. Logic, 1-229*, 1970.

Krivine, J.L. *Théorie Axiomatique des Ensembles*. Paris 1970.

Kuratowski, C. *Topologie I, II*. Warszawa 1952.

Kuratowski, C. & Mostowski, A. *Set Theory*. Amsterdam 1968.

Lyapunow, A.A., Stschegolkow, E.A. & Arsenin, W.J. *Arbeiten zur descriptiven Mengenlehre*. Berlin 1955.

Manin, Y.I. *A Course in Mathematical Logic*. Springer, Berlin 1977.

Martin, D.A. Borel determinacy. *Annals of Math. Logic* 102, 1975, pp. 363-371.

Monna, A.F. The concept of function in the 19th and 20th century, in particular with regard to the discussion between Baire, Borel and Lebesgue. *Archive for History of Exact Sciences, 9-57*, 1972.

Morse, A.P. *A Theory of Sets*. New York 1965.

Moschovakis, Y.N. *Elementary Induction on Abstract Structures*. Amsterdam 1974.

Mostowski, A. *Constructible Sets with Applications*. Amsterdam 1969.

Natanson, I.P. *Theorie der Funktionen einer reellen Veränderlichen*. Berlin 1961.

Reinhardt, W.N. Remarks on reflection principles, large cardinals and elementary embeddings. In *Axiomatic Set Theory. Proc. of symp. in pure math.* Vol XIII, part II. ed. T.J. Jech. AMS. Providence 1974.

Rubin, H. & Rubin, J. *Equivalents of the Axiom of Choice*. Amsterdam 1963.

Shoenfield, J. *Mathematical Logic*. London 1967.

Sierpinski, W. *General Topology*. Toronto 1952.

Sikorski, R. *Boolean Algebras*. Berlin 1960.

Solovay, R.M., Reinhardt, W.N. & Kanamo, A. Strong axioms of infinity and elementary embeddings. *Annals Math. Logic* 13, 1978, pp. 73-116.

Tarski, A. *Logic, Semantics, Metamathematics*. London 1956.

INDEX

OTHER TITLES IN THE SERIES IN PURE AND APPLIED MATHEMATICS

Vol. 1. WALLACE—*An Introduction to Algebraic Topology*
Vol. 2. PEDOE—*Circles*
Vol. 3. SPAIN—*Analytical Conics*
Vol. 4. MIKHLIN—*Integral Equations*
Vol. 5. EGGLESTON—*Problems in Euclidean Space: Application of Convexity*
Vol. 6. WALLACE—*Homology Theory on Algebraic Varieties*
Vol. 7. NOBLE—*Methods Based on the Wiener-Hopf Technique for the Solution of Partial Differential Equations*
Vol. 8. MIKUSINSKI—*Operational Calculus*
Vol. 9. HEINE—*Group Theory in Quantum Mechanics*
Vol.10. BLAND—*The Theory of Linear Viscoelasticity*
Vol.11. KURTH—*Axiomatics of Classical Statistical Mechanics*
Vol.12. FUCHS—*Abelian Groups*
Vol.13. KURATOWSKI—*Introduction to Set Theory and Topology*
Vol.14. SPAIN—*Analytical Quadrics*
Vol.15. HARTMAN and MIKUSINSKI—*Theory of Lebesgue Measure and Integration*
Vol.16. KULCZYCKI—*Non-Euclidean Geometry*
Vol.17. KURATOWSKI—*Introduction to Calculus*
Vol.18. GERONIMUS—*Polynomials Orthogonal on a Circle and Interval*
Vol.19. ELSGOLC—*Calculus of Variations*
Vol.20. ALEXITS—*Convergence Problems of Orthogonal Series*
Vol.21. FUCHS and LEVIN—*Functions of a Complex Variable, Volume II*
Vol.22. GOODSTEIN—*Fundamental Concepts of Mathematics*
Vol.23. KEENE—*Abstract Sets and Finite Ordinals*
Vol.24. DITKIN and PRUDNIKOV—*Operational Calculus in Two Variables and its Applications*
Vol.25. VEKUA—*Generalized Analytic Functions*
Vol.26. FASS and AMIR-MOÉZ—*Elements of Linear Spaces*
Vol.27. GRADSHTEIN—*Direct and Converse Theorems*
Vol.28. FUCHS—*Partially Ordered Algebraic Systems*
Vol.29. POSTNIKOV—*Foundations of Galois Theory*
Vol.30. BERMANT—*A Course of Mathematical Analysis, Part II*
Vol.31. LUKASIEWICZ—*Elements of Mathematical Logic*
Vol.32. VULIKH—*Introduction to Functional Analysis for Scientists and Technologists*
Vol.33. PEDOE—*An Introduction to Projective Geometry*
Vol.34. TIMAN—*Theory of Approximation of Functions of a Real Variable*
Vol.35. CSASZAR—*Foundations of General Topology*
Vol.36. BRONSHTEIN and SEMENDYAYEV—*A Guide-Book to Mathematics for Technologists and Engineers*
Vol.37. MOSTOWSKI and STARK—*Introduction to Higher Algebra*
Vol.38. GODDARD—*Mathematical Techniques of Operational Research*
Vol.39. TIKHONOV and SAMARSKII—*Equations of Mathematical Physics*
Vol.40. MCLEOD—*Introduction to Fluid Dynamics*
Vol.41. MOISIL—*The Algebraic Theory of Switching Circuits*
Vol.42. OTTO—*Nomography*
Vol.43. RANKIN—*An Introduction to Mathematical Analysis*
Vol.44. BERMANT—*A Course of Mathematical Analysis, Part I*
Vol.45. KRASNOSEL'SKII—*Topological Methods in the Theory of Nonlinear Integral Equations*
Vol.46. KANTOROVICH and AKHILOV—*Functional Analysis in Normed Spaces*
Vol.47. JONES—*The Theory of Electromagnetism*
Vol.48. FEJES TÓTH—*Regular Figures*
Vol.49. YANO—*Differential Geometry on Complex and Almost Complex Spaces*
Vol.50. MIKHLIN—*Variational Methods in Mathematical Physics*
Vol.51. FUCHS and SHABAT—*Functions of a Complex Variable and Some of their Applications, Volume I*
Vol.52. BUDAK, SAMARSKII and TIKHONOV—*A Collection of Problems on Mathematical Physics*
Vol.53. GILES—*Mathematical Foundations of Thermodynamics*
Vol.54. SAUL'YEV—*Integration of Equations of Parabolic Type by the Method of Nets*
Vol.55. PONTRYAGIN *et al.*—*The Mathematical Theory of Optimal Processes*
Vol.56. SOBOLEV—*Partial Differential Equations of Mathematical Physics*
Vol.57. SMIRNOV—*A Course of Higher Mathematics, Volume I*
Vol.58. SMIRNOV—*A Course of Higher Mathematics, Volume II*